A Mathematical Orchard
Problems and Solutions

© 2012 by
The Mathematical Association of America (Incorporated)

Library of Congress Catalog Card Number 2012938987

Print edition ISBN 978-0-88385-833-2

Electronic edition ISBN 978-1-61444-403-9

Printed in the United States of America

Current Printing (last digit):
10 9 8 7 6 5 4 3 2 1

A Mathematical Orchard

Problems and Solutions

Mark I. Krusemeyer
George T. Gilbert
Loren C. Larson

Published and Distributed by
Mathematical Association of America

Council on Publications and Communications
Frank Farris, *Chair*

Committee on Books
Gerald Bryce, *Chair*

Problem Books Editorial Board

Richard A. Gillman, *Editor*

Zuming Feng
Elgin H. Johnston
Roger Nelsen
Tatiana Shubin
Richard A. Stong
Paul A. Zeitz

MAA PROBLEM BOOKS SERIES

Problem Books is a series of the Mathematical Association of America consisting of collections of problems and solutions from annual mathematical competitions; compilations of problems (including unsolved problems) specific to particular branches of mathematics; books on the art and practice of problem solving, etc.

Aha! Solutions, Martin Erickson

The Alberta High School Math Competitions 1957–2006: A Canadian Problem Book, compiled and edited by Andy Liu

The Contest Problem Book VII: American Mathematics Competitions, 1995–2000 Contests, compiled and augmented by Harold B. Reiter

The Contest Problem Book VIII: American Mathematics Competitions (AMC 10), 2000–2007, compiled and edited by J. Douglas Faires & David Wells

The Contest Problem Book IX: American Mathematics Competitions (AMC 12), 2000–2007, compiled and edited by David Wells & J. Douglas Faires

First Steps for Math Olympians: Using the American Mathematics Competitions, by J. Douglas Faires

A Friendly Mathematics Competition: 35 Years of Teamwork in Indiana, edited by Rick Gillman

Hungarian Problem Book IV, translated and edited by Robert Barrington Leigh and Andy Liu

The Inquisitive Problem Solver, Paul Vaderlind, Richard K. Guy, and Loren C. Larson

International Mathematical Olympiads 1986–1999, Marcin E. Kuczma

Mathematical Olympiads 1998–1999: Problems and Solutions From Around the World, edited by Titu Andreescu and Zuming Feng

Mathematical Olympiads 1999–2000: Problems and Solutions From Around the World, edited by Titu Andreescu and Zuming Feng

Mathematical Olympiads 2000–2001: Problems and Solutions From Around the World, edited by Titu Andreescu, Zuming Feng, and George Lee, Jr.

A Mathematical Orchard: Problems and Solutions, by Mark I. Krusemeyer, George T. Gilbert, and Loren C. Larson

Problems from Murray Klamkin: The Canadian Collection, edited by Andy Liu and Bruce Shawyer

The William Lowell Putnam Mathematical Competition Problems and Solutions: 1938–1964, A. M. Gleason, R. E. Greenwood, L. M. Kelly

The William Lowell Putnam Mathematical Competition Problems and Solutions: 1965–1984, Gerald L. Alexanderson, Leonard F. Klosinski, and Loren C. Larson

The William Lowell Putnam Mathematical Competition 1985–2000: Problems, Solutions, and Commentary, Kiran S. Kedlaya, Bjorn Poonen, Ravi Vakil

USA and International Mathematical Olympiads 2000, edited by Titu Andreescu and Zuming Feng

USA and International Mathematical Olympiads 2001, edited by Titu Andreescu and Zuming Feng

USA and International Mathematical Olympiads 2002, edited by Titu Andreescu and Zuming Feng

USA and International Mathematical Olympiads 2003, edited by Titu Andreescu and Zuming Feng

USA and International Mathematical Olympiads 2004, edited by Titu Andreescu, Zuming Feng, and Po-Shen Loh

MAA Service Center
P. O. Box 91112
Washington, DC 20090-1112
1-800-331-1622   fax: 1-301-206-9789

PREFACE

Inside this book is an older book. In 1993, the MAA published "The Wohascum County Problem Book", and a few years ago we were asked to consider reissuing that book with a less rustic and more descriptive title. Meanwhile, we had many more problems to contribute, and so the original list of 130 has grown to 208. The new problems are, if anything, more likely to involve pattern finding and experimentation, although technology is generally not needed or even particularly helpful. In difficulty the new problems tend to be in the middle range of the original book, so anyone familiar with that book who looks only at the very beginning or the very end of the problem list may not notice much difference. From a geographical perspective, we haven't tried to move the problems that were originally set in Wohascum County, and we still can't tell you where to look for that setting on a map.

We have been asked, and in any case it is appropriate in a preface, to say something about the purpose of this particular collection. There are actually multiple purposes, and different users will no doubt have their own priorities. One purpose is entertainment; we think these problems are attractive and will provide mathematical pleasure to those who spend time with them. This has been confirmed over the years by undergraduates at Carleton and St. Olaf Colleges, where many of the problems were first posed as weekly challenges, by high-school age (but unusually talented and enthusiastic) participants at Canada/USA Mathcamp, and by a variety of others. Were you once, or are you now, a mathematics major? Do you teach mathematics at any level? Do you simply find pleasure in encountering mathematical problems that are accessible but not routine? We like to think that in all those cases, you will enjoy browsing through this book, whether in the usual way or by starting in the index and looking up problems that have unexpected words or phrases in them. (The numbers in the index refer to problems, not to pages.) In an ideal world, maybe no one would look at a solution before trying seriously to solve the problem, but if you're feeling curious and are pressed for time, you can still appreciate the problem by reading the solution and the underlying ideas behind it. Should you be looking for weekly challenges, you might consider that if you ration yourself to one problem a week, there are enough problems here for four calendar years — spanning your undergraduate career, perhaps?

Another purpose is instruction. There are books (and one of us has written such a book) that provide systematic introductions to problem solving, presenting various techniques together with problems that allow students to test their skill and ingenuity in using those newly acquired techniques. This is not such a book. In fact, it can be argued that one of the most important things for a problem solver to learn is the ability to recognize which of the many possible tools might be applicable. Success may come only after trying several approaches. The process of finding the key idea is often what makes the problem interesting and ultimately valuable. Thus the order in which the problems first appear does not provide any clues as to their method of solution. However, the appendices in the back of the book do provide information about topics and prerequisites, and we hope this will be helpful to people teaching problem solving classes and to teams or individuals preparing for contests such as the Putnam. Specifically, the first appendix lists the prerequisites for each problem, while the second has the problems arranged by general topic.

So, one might ask, in what sense is this book providing instruction? Not surprisingly, most of the book is taken up by solutions rather than by problems; we hope that most of these solutions are instructive and/or elegant as well as being clear. For many problems, multiple solutions are presented. Sometimes the solutions are preceded by "Ideas", which can serve as hints, or followed by "Comments", which often provide a broader perspective. We have tried to provide enough details and motivation so that even the messier and more involved solutions, which may run to several pages, can be read and enjoyed by inexperienced problem solvers. If the ways of thinking suggested by these solutions turn out to be helpful and appealing to the readers of the book, we will be delighted.

Some of our problems have a third purpose, namely to generate more problems, that is, to suggest variations or extensions for further investigation. In particular, the comments to Problems 55, 106, 144, 158, 164, and 176 indicate related questions which we have so far been unable to solve.

All problems presented here are due to the authors. While the majority are taken from the Wohascum County Problem Book, as far as we know almost none of the others have appeared in print. Some of them can be found on the Web (but without solutions) in old issues of our respective department newsletters (the Carleton "Goodsell Gazette", the St. Olaf "Math Mess", and the TCU "Math Newsletter"). For a few of the problems, some knowledge of linear or abstract algebra is needed, but most require nothing beyond calculus, and many should be accessible to high school students. However, there is a wide range of difficulty, with some problems requiring considerable "mathematical maturity". For most students, few, if any, of the problems should be routine. We have tried to put easier problems before harder ones, while keeping the list as varied as possible, and we expect that nearly everyone will find Problem 150 more difficult than Problem 75. Nevertheless, a particular solver might find 150 easier than 140 or even 125.

As for the title, it has been a struggle to find an appropriate balance between the fanciful and the matter-of-fact. An earlier candidate, "Intuition and Proof", was perhaps a bit stark, and did not make clear reference to problems (or even mathematics). We hope the collective noun "Orchard" will evoke images of such good things as vigorous growth, thoughtful care, and delectable fruit; if so, the metaphor will have taken us as far as intended.

It is a pleasure to thank Don Albers, whose initiative led to this book. Our heartfelt thanks go, also, to the many people—our parents, spouses, colleagues and friends—whose vital encouragement and support we have been privileged to receive. The following reviewers each read substantial portions of the manuscript, and their thoughtful suggestions led to many improvements: Professors Irl Bivens, Joe Buhler, David Callan, Paul Campbell, Barry Cipra, the late William Firey, Paul Fjelstad, the late Steven Galovich, Gerald Heuer, the late Abraham Hillman, the late Meyer Jerison, Elgin Johnston, Eugene Luks, the late Murray Klamkin, Bruce Reznick, the late Ian Richards, John Schue, Allen Schwenk, Steven Tschanz, and William Waterhouse. Beverly Ruedi of the MAA brought unfailing good humor and patience to the nitty-gritty of manuscript production. Finally, our thanks to all past, present, and future students who respond to these and other challenge problems. Enjoy!

Mark Krusemeyer
George Gilbert
Loren Larson

Northfield, Minnesota
Fort Worth, Texas
October 15, 2011

CONTENTS

Preface vii
The Problems* 1
The Solutions 45
Appendix 1: Prerequisites by Problem Number 379
Appendix 2: Problem Numbers by Subject 389
Index 391
About the Authors 397

---

* The page number of the solution appears at the end of each problem.

THE PROBLEMS

1. Find all solutions in integers of $x^3 + 2y^3 = 4z^3$. (p. 45)

2. The Wohascum County Board of Commissioners, which has 20 members, recently had to elect a President. There were three candidates (A, B, and C); on each ballot the three candidates were to be listed in order of preference, with no abstentions. It was found that 11 members, a majority, preferred A over B (thus the other 9 preferred B over A). Similarly, it was found that 12 members preferred C over A. Given these results, it was suggested that B should withdraw, to enable a runoff election between A and C. However, B protested, and it was then found that 14 members preferred B over C! The Board has not yet recovered from the resulting confusion. Given that every possible order of A, B, C appeared on at least one ballot, how many board members voted for B as their first choice? (p. 46)

3. If $A = (0, -10)$ and $B = (2, 0)$, find the point(s) C on the parabola $y = x^2$ which minimizes the area of triangle ABC. (p. 47)

4. Does there exist a continuous function $y = f(x)$, defined for all real x, whose graph intersects every non-vertical line in infinitely many points? (Note that because f is a function, its graph will intersect every vertical line in exactly one point.) (p. 48)

5. A child on a pogo stick jumps 1 foot on the first jump, 2 feet on the second jump, 4 feet on the third jump, ..., 2^{n-1} feet on the nth jump. Can the child get back to the starting point by a judicious choice of directions? (p. 49)

6. Let $S_n = \{1, n, n^2, n^3, \ldots\}$, where n is an integer greater than 1. Find the smallest number $k = k(n)$ such that there is a number which may be expressed as a sum of k (possibly repeated) elements of S_n in more than one way (rearrangements are considered the same). (p. 49)

7. Find all integers a for which $x^3 - x + a$ has three integer roots. (p. 50)

8. At an outdoor concert held on a huge lawn in Wohascum Municipal Park, three speakers were set up in an equilateral triangle; the idea was that the audience would be between the speakers, and anyone at the exact center of the triangle would hear each speaker at an equal "volume" (sound level). Unfortunately, an electronic malfunction caused one of the speakers to play four times as loudly as the other two. As a result, the audience tended to move away from this speaker (with some people going beyond the original triangle). This helped considerably, because the sound level from a speaker is inversely proportional to the square of the distance to that speaker (we are assuming that the sound levels depend only on the distance to the speakers). This raised a question: Where should one sit so that each speaker could be heard at the same sound level? (p. 51)

9. Ten (not necessarily all different) integers have the property that if all but one of them are added, the possible results (depending on which one is omitted) are: 82, 83, 84, 85, 87, 89, 90, 91, 92. (This is not a misprint; there are only nine possible results.) What are the ten integers? (p. 53)

10. For what positive values of x does the infinite series $\sum_{n=1}^{\infty} x^{\log_{2012} n}$ converge? (p. 53)

11. Let A be a 4×4 matrix such that each entry of A is either 2 or -1. Let $d = \det(A)$; clearly, d is an integer. Show that d is divisible by 27. (p. 53)

12. Consider the $n \times n$ array whose entry in the ith row, jth column is $i + j - 1$. What is the smallest product of n numbers from this array, with one coming from each row and one from each column? (p. 54)

13. Let A and B be sets with the property that there are exactly 144 sets which are subsets of at least one of A or B. How many elements does the union of A and B have? (p. 54)

14. Find $\int (x^6 + x^3) \sqrt[3]{x^3 + 2} \, dx$. (p. 55)

15. Hidden among the fields of Wohascum County are some missile silos, and it recently came to light that they had attracted the attention of an invidious, but somewhat inept, foreign agent living nearby in the guise of a solid citizen. Soon the agent had two microfilms to hide, and he decided to hide them in two dark squares of a chessboard—the squares being opposite each other, and adjacent to opposite corners, as shown in the diagram. The chessboard was of a cheap collapsible variety which folded along both center lines. The agent's four-year-old daughter explored this possibility so often that the board came apart at the folds, and then she played with the four resulting pieces so they ended up in a random configuration. Her mother, unaware of the microfilms, then glued the pieces back together to form a chessboard which appeared just like the original (if somewhat the worse for wear). Considerable complications resulted when the agent could not find the films. How likely would it have been that the agent could have found them, that is, that the two squares in which they were hidden would have ended up adjacent to opposite corners? (p. 57)

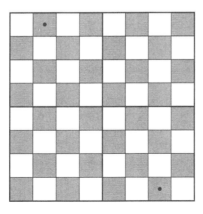

16. For what values of n is it possible to split up the (entire) set $\{1, 2, \ldots, n\}$ into three (disjoint) subsets so that the sum of the integers in each of the subsets is the same? (p. 58)

17. Let $f(x)$ be a positive, continuously differentiable function, defined for all real numbers, whose derivative is always negative. For any real number x_0, will the sequence (x_n) obtained by Newton's method (that is, given by $x_{n+1} = x_n - f(x_n)/f'(x_n)$) always have limit ∞? (p. 60)

18. Find all solutions in nonnegative integers to the system of equations

$$3x^2 - 2y^2 - 4z^2 + 54 = 0, \qquad 5x^2 - 3y^2 - 7z^2 + 74 = 0. \qquad \text{(p. 61)}$$

19. If n is a positive integer, how many real solutions are there, as a function of n, to $e^x = x^n$? (p. 61)

20. The swimming coach at Wohascum High has a bit of a problem. The swimming pool, like much of the building, is not in good repair, and in fact only three of the lanes are really usable. The school swimming championship is coming up, and while the coach expects only a small turnout for this event it is not at all sure that the number of participants will be divisible by 3, let alone a power of 3 (which would make it easy to arrange a "single-elimination" format). Instead, in order to choose three participants to compete in the final, the coach intends to have all participants swim in an equal number of preliminary races (of course, that number should be positive). Also, she wants each preliminary race to have exactly three swimmers in it. Finally, she does not want any two particular swimmers to compete against each other in more than one of these preliminary races. Can all this be arranged

a. if there are five participants;

b. if there are ten participants? (p. 62)

21. Does there exist a positive integer whose prime factors include at most the primes 2, 3, 5, and 7 and which ends in the digits 11? If so, find the smallest such positive integer; if not, show why none exists. (p. 65)

22. The Wohascum County Fish and Game Department issues four types of licenses, for deer, grouse, fish, and wild turkey; anyone can purchase any combination of licenses. In a recent year, (exactly) half the people who bought a grouse license also bought a turkey license. Half the people who bought a turkey license also bought a deer license. Half the people who bought a fish license also bought a grouse license, and one more than half the people who bought a fish license also bought a deer license. One third of the people who bought a deer license also bought both grouse and fish licenses. Of the people who bought deer licenses, the same number bought a grouse license as bought a fish license; a similar statement was true of buyers of turkey licenses. Anyone who bought both a grouse and a fish license also bought either a deer or a turkey license, and of these people the same number bought a deer license as bought a turkey license. Anyone who bought both a deer and a turkey license either bought both a grouse and a fish license or neither. The number of people buying a turkey license was equal to the number of people who bought some license but not a fish license. The number of people buying a grouse license was equal to the number of people buying some license but not a turkey license. The number of deer licenses sold was one more than the number of grouse licenses sold. Twelve people bought either a grouse or a deer license (or both). How many people in all bought licenses? How many licenses in all were sold? (p. 65)

23. Given three lines in the plane which form a triangle (that is, every pair of the lines intersects, and the three intersection points are distinct), what is the set of points for which the sum of the distances to the three lines is as small as possible? (Be careful not to overlook special cases.) (p. 67)

24. For the purposes of this problem, we'll say that two positive integers a and b are "one step apart" if $ab + 1$ is a perfect square. For example, 2 and 24 are one step apart because $2 \cdot 24 + 1 = 49$ is a perfect square. Also, 24 and 7 are one step apart, so it seems natural to say that 2 and 7 are at most two steps apart. In fact, since 15 is not a square, 2 and 7 are separated by exactly two steps; that is, they are two steps apart.

a. Show that any two (distinct) positive integers are separated by a finite number of steps.

b. How many steps separate two consecutive positive integers m and $m+1$?

c. How many steps separate 1 and 4? (p. 68)

25. Suppose the plane $x + 2y + 3z = 0$ is a perfectly reflecting mirror. Suppose a ray of light shines down the positive x-axis and reflects off the mirror. Find the direction of the reflected ray. (Assume the law of optics which asserts that the angle of incidence equals the angle of reflection.)

(p. 69)

26. Starting at $(a, 0)$, Jessie runs along the x-axis toward the origin at a constant speed s. Starting at $(0, 1)$ at the same time Jessie starts, Riley runs at a constant speed 1 counterclockwise around the unit circle until Jessie is far down the negative x-axis. For what integers $a > 1$ is there a speed s so that Jessie and Riley reach $(1, 0)$ at the same time and $(-1, 0)$ at the same time? (p. 70)

27. Find the set of all solutions to $x^{y/z} = y^{z/x} = z^{x/y}$, with x, y, and z positive real numbers. (p. 71)

28. Find all perfect squares whose base 9 representation consists only of ones. (p. 72)

29. Show that any polygon can be tiled by convex pentagons. (p. 73)

30. The following is an excerpt from a recent article in the *Wohascum Times*. (Names have been replaced by letters.) "Because of the recent thaw, the trail for the annual Wohascum Snowmobile Race was in extremely poor condition, and it was impossible for more than two competitors to be abreast each other anywhere on the trail. Nevertheless, there was frequent passing. ... After a few miles A pulled ahead of the pack, closely followed by B and C in that order, and thereafter these three did not relinquish the top three positions in the field. However, the lead subsequently changed hands nine times among these three; meanwhile, on eight different occasions the vehicles that were running second and third at the times changed places. ... At the end of the race, C complained that B had driven recklessly just before the finish line to keep C, who was immediately behind B at the finish, from passing. ..." Can this article be accurate? If so, can you deduce who won the race? If the article cannot be accurate, why not? (p. 75)

31. At a recent trade fair in Wohascum Center, an inventor showed a device called a "trisector," with which any straight line segment can be divided into three equal parts. The following dialogue ensued. Customer: "But I need to find the midpoint of a segment, not the points 1/3 and 2/3 of the way from one end of the segment to the other!" Inventor: "Sorry, I hadn't realized there was a market for that. I'll guess that you'll have to get some compasses and use the usual construction." Show that the inventor was wrong, that is, show how to construct the midpoint of any given segment using only a straightedge (but no compasses) and the "trisector." (p. 75)

32. Find a positive integer n such that $2011n+1$ and $2012n+1$ are both perfect squares, or show that no such positive integer n exists. (p. 77)

33. Consider a 12×12 chessboard (consisting of 144 1×1 squares). If one removes 3 corners, can the remainder be covered by 47 1×3 tiles? (p. 78)

34. Is there a function f, differentiable for all real x, such that

$$|f(x)| < 2 \quad \text{and} \quad f(x)f'(x) \geq \sin x?$$

(p. 79)

35. Let N be the largest possible number that can be obtained by combining the digits 1, 2, 3, and 4 using the operations addition, multiplication, and exponentiation, if the digits can be used only once. Operations can be used repeatedly, parentheses can be used, and digits can be juxtaposed (put next to each other). For instance, 12^{34}, $1+(2 \times 3 \times 4)$, and $2^{31 \times 4}$ are all candidates, but none of these numbers is actually as large as possible. Find N. (All numbers are to be construed in base ten.) (p. 79)

36. The *Wohascum Times* has a filing system with 366 slots, each of which corresponds to one date of the year and should contain the paper that appeared most recently on that date, except that the February 29 paper is taken out, leaving an empty slot, one year after it is filed in a given leap year. It was discovered some time in the summer of a non-leap year that some prankster had scrambled the papers; for example, the March 10 paper was in the April 18 slot, but there was still exactly one paper in each slot, with the exception of the February 29 slot, which was vacant, as it should be. A junior employee was given the task of putting the papers back in order. He decided to try to do this in a series of moves, each move consisting of transferring some paper from the slot it was in to the slot that was vacant (so that the slot the paper was moved from becomes the new vacant slot).

a. Is it possible, using this method, to unscramble the papers, and if so, what is the maximum number of moves that might be needed, assuming that the papers were moved as efficiently as possible?

b. What would the maximum number of moves be if there were n slots instead of 366 (that is, there is one slot for a "leap date" which is empty at the beginning and the end, and $n-1$ papers are scrambled among the other $n-1$ slots)? (p. 80)

37. Babe Ruth's batting performance in the 1921 baseball season is often considered the best in the history of the game. In home games, his batting average was .404; in away games it was .354. Furthermore, his slugging percentage at home was a whopping .929, while in away games it was .772. This was based on a season total of 44 doubles, 16 triples, and 59 home runs. He had 30 more at bats in away games than in home games. What were his overall batting average and his slugging percentage for the year? (Batting average is defined to be the number of hits divided by the number of at bats. One way of defining slugging percentage is the number of hits plus the number of doubles plus twice the number of triples plus three times the number of home runs, all divided by the number of at bats. Both of these percentages are rounded to three decimal places.) (p. 82)

38. Let C be a circle with center O, and Q a point inside C different from O. Where should a point P be located on the circumference of C to maximize $\angle OPQ$? (p. 83)

39. With technology you can find that $\int_0^1 \frac{\sqrt{x^4+2x^2+2}}{x^2+1} dx \approx 1.27798$ and $\frac{1}{4}\sqrt{16+\pi^2} \approx 1.27155$. Clearly, then, $\int_0^1 \frac{\sqrt{x^4+2x^2+2}}{x^2+1} dx > \frac{1}{4}\sqrt{16+\pi^2}$. Prove that this is indeed correct, without using any technology. (p. 84)

40. Let $f_1(x) = x^2 + 4x + 2$, and for $n \geq 2$, let $f_n(x)$ be the n-fold composition of the polynomial $f_1(x)$ with itself. For example,

$$f_2(x) = f_1(f_1(x)) = x^4 + 8x^3 + 24x^2 + 32x + 14.$$

Let s_n be the sum of the coefficients of the terms of even degree in $f_n(x)$. For example, $s_2 = 1 + 24 + 14 = 39$. Find s_{2012}.

(p. 85)

41. Find all real solutions of the equation $\sin(\cos x) = \cos(\sin x)$. (p. 86)

42. For what real numbers α does the series

$$\frac{1}{2^\alpha} + \frac{1}{(2+\frac{1}{2^\alpha})^\alpha} + \frac{1}{\left(2+\frac{1}{2^\alpha}+\frac{1}{(2+\frac{1}{2^\alpha})^\alpha}\right)^\alpha} + \cdots$$

converge? (p. 86)

43. For a natural number $n \geq 2$, let $0 < x_1 \leq x_2 \leq \cdots \leq x_n$ be real numbers whose sum is 1. If $x_n \leq 2/3$, prove that there is some k, $1 \leq k \leq n$, for which $1/3 \leq \sum_{j=1}^{k} x_j < 2/3$. (p. 87)

44. Is there a cubic curve $y = ax^3 + bx^2 + cx + d$, $a \neq 0$, for which the tangent lines at two distinct points coincide? (p. 88)

45. Show that for any positive integer n,
$$\sum_{i=0}^{n-1} \arcsin\left(\frac{i(i+1) + n^2}{\sqrt{i^2 + n^2}\sqrt{(i+1)^2 + n^2}}\right) = \frac{(2n-1)\pi}{4}.$$
(p. 88)

46. Digital watches have become the norm even in Wohascum County. Recently three friends there were comparing their watches and found them reasonably well synchronized. In fact, all three watches were perfectly accurate (those amazing silicon chips!) in the sense that the length of a second was the same according to each watch. The time indicated shifted from one watch to the next, but in such a way that any two watches would show the same time in minutes for part of each minute. (A different pair of watches might show the same time in minutes for a different part of the minute.)
  a. Show that there was at least one pair of watches that showed the same time in minutes for more than half of each minute.
  b. Suppose there were n watches, rather than three, such that once again any two watches would show the same time in minutes for part of each minute. Find the largest number x such that at least one pair of watches necessarily showed the same time in minutes for more than the fraction x of each minute. (p. 90)

47. Let C be a circle with center O, and Q a point inside C different from O. Show that the area enclosed by the locus of the centroid of triangle OPQ as P moves about the circumference of C is independent of Q. (p. 91)

48. Suppose you have an unlimited supply of identical barrels. To begin with, one of the barrels contains n ounces of liquid, where n is a positive integer, and all the others are empty. You are allowed to redistribute the liquid between the barrels in a series of steps, as follows. If a barrel contains k ounces of liquid and k is even, you may pour exactly half that amount into an empty barrel (leaving the other half in the original barrel). If k is odd, you may pour the *largest* integer that is less than half that amount into an empty barrel. No other operations are allowed. Your object is to "isolate" a total of m ounces of liquid, where m is a positive integer less than n; that is, you need to get a situation in which the sum of the amounts in certain barrels (which can then be set aside) is exactly m.

a. What is the least number of steps (as a function of n) in which this can be done for $m = n - 1$?

b. What is the smallest number of steps in which it can be done regardless of m, as long as m is known in advance and is some positive integer less than n? (p. 92)

49. Describe the set of points (x, y) in the plane for which

$$\sin(x + y) = \sin x + \sin y.$$ (p. 94)

50. For n a positive integer, find the smallest positive integer $d = d(n)$ for which there exists a polynomial of degree d whose graph passes through the points $(1, 2), (2, 3), \ldots, (n, n+1)$, and $(n+1, 1)$ in the plane. (p. 95)

51. It is shown early on in most linear algebra courses that every invertible matrix can be written as a product of elementary matrices (or, equivalently, that every invertible matrix can be reduced to the identity matrix by a finite number of row reduction steps). Show that every 2×2 matrix of determinant 1 is the product of *three* elementary matrices. (2×2 elementary matrices are matrices of types

$$\begin{pmatrix} 1 & x \\ 0 & 1 \end{pmatrix}, \quad \begin{pmatrix} 1 & 0 \\ x & 1 \end{pmatrix}, \quad \begin{pmatrix} 0 & 1 \\ 1 & 0 \end{pmatrix}, \quad \begin{pmatrix} y & 0 \\ 0 & 1 \end{pmatrix}, \quad \begin{pmatrix} 1 & 0 \\ 0 & y \end{pmatrix},$$

where $y \neq 0$ and x are arbitrary. The standard row reduction of $\begin{pmatrix} a & b \\ c & d \end{pmatrix}$ would ordinarily use *four* row reduction steps.) (p. 96)

52. Let $ABCD$ be a convex quadrilateral (a four-sided figure with angles less than $180°$). Find a necessary and sufficient condition for a point P to exist inside $ABCD$ such that the four triangles ABP, BCP, CDP, DAP all have the same area. (p. 97)

53. Let k be a positive integer. Find the largest power of 3 which divides $10^k - 1$. (p. 99)

54. Consider the parabola $y = x^2$. For different points P in the plane, there may be different numbers of normal lines to the parabola that pass through P.

a. Show that there is always at least one normal line, and that there are at most three normal lines, to $y = x^2$ that pass through any given point P.
b. Show that there are exactly two normal lines to $y = x^2$ that pass through $Q = (4, 7/2)$, and find and sketch the set of all points Q with this property.
c. Are there any such points Q for which *both* coordinates of Q are integers?
(p. 99)

55. Every week, the Wohascum Folk Dancers meet in the high school auditorium. Attendance varies, but since the dancers come in couples, there is always an even number n of dancers. In one of the dances, the dancers are in a circle; they start with the two dancers in each couple directly opposite each other. Then two dancers who are next to each other change places while all others stay in the same place; this is repeated with different pairs of adjacent dancers until, in the ending position, the two dancers in each couple are once again opposite each other, but in the opposite of the starting position (that is, every dancer is halfway around the circle from her/his original position). What is the least number of interchanges (of two adjacent dancers) necessary to do this? (p. 102)

56. If n points in the plane are such that all the distances between them are equal, it's easy to see that n can be at most 3 (which occurs for an equilateral triangle). Now suppose that n points in the plane are such that there are just two different distances between them, that is, there are two numbers a and b such that whenever we choose any two of the n points, their distance to each other will be either a or b. What is the largest possible value of n? (p. 102)

57. Let L be a line in the plane; let A and B be points on L which are a distance 2 apart. If C is any point in the plane, there may or may not (depending on C) be a point X on the line L for which the distance from X to C is equal to the average of the distances from X to A and B. Give a precise description of the set of all points C in the plane for which there is no such point X on the line. (p. 104)

58. Show that there exists a positive number λ such that

$$\int_0^\pi x^\lambda \sin x \, dx = 3.$$

(p. 107)

59. There is no analog of the quadratic formula that solves polynomial equations of degree 5 and higher, such as $x^5 - 5x^4 + 8x^3 - 6x^2 + 3x + 3 = 0$. However, this particular polynomial has two roots that sum to 2. Using this information, find all solutions. (p. 109)

60. One regular n-gon is inscribed in another regular n-gon and the area of the large n-gon is twice the area of the small one.

a. What are the possibilities for n?
b. What are the possibilities for the angles that the sides of the large n-gon make with the sides of the small one? (p. 110)

61. Consider a rectangular array of numbers, extending infinitely to the left and right, top and bottom. Start with all the numbers equal to 0 except for a single 1. Then go through a series of steps, where at each step each number gets replaced by the sum of its four neighbors. For example, after one step the array will look like

$$\begin{array}{ccc} & 1 & \\ 1 & 0 & 1 \\ & 1 & \end{array}$$

surrounded by an infinite "sea" of zeros, and after two steps we will have

$$\begin{array}{ccccc} & & 1 & & \\ & 2 & 0 & 2 & \\ 1 & 0 & 4 & 0 & 1 \\ & 2 & 0 & 2 & \\ & & 1 & & \end{array}$$

a. After n steps, what will be the sum of all the numbers in the array, and why?
b. After n steps, what will be the number in the center of the array (at the position of the original 1)?
c. Can you describe the various nonzero numbers that will occur in the array after n steps? (p. 112)

62. What is the fifth digit from the end (the ten thousands digit) of the number $5^{5^{5^{5^{5}}}}$? (p. 114)

63. One of last week's deals at the Wohascum Bridge Club will long be remembered. [As you may know, bridge is a partnership game; two players sitting in the North and South positions around a square table join forces against East and West. A standard deck of 52 cards is dealt, so each player starts with 13 cards. After an episode called "bidding" which is irrelevant to this problem, a deal is played out in 13 rounds or "tricks" of 4 cards each, with each player contributing one card to each trick. The first card to be played in each trick determines the suit (spades, hearts, diamonds, or clubs) of that trick; the other three players must contribute a card of the same suit ("follow suit") if they are able to do so, else they may choose any card. The highest card *in the suit of a trick* wins that trick (at least in "no trump", a circumstance that happened on this deal); "highest" is according to the usual ranking: ace (high), king, queen, jack, $10, 9, \ldots, 2$ (low). Also, the player winning a trick thereby becomes the first player to play to the next trick.]

On this particular deal, West had to play the very first card (on the first trick). One of her options was to play the king of hearts, and it turned out that if West did so, the East-West partnership would take all thirteen tricks, *no matter what* (assuming legal, but not necessarily intelligent, play). On the other hand, if West started the play by contributing any of her other twelve cards, the North-South partnership would take all thirteen tricks, *no matter what!* Given that North had the two of spades and the jack of clubs, who had the five of diamonds? (p. 116)

64. Describe the set of all points P in the plane such that exactly two tangent lines to the curve $y = x^3$ pass through P. (p. 118)

65. The sets $\{1, 8, 12\}$ and $\{2, 3, 16\}$ have the property that all their elements are distinct but the two sets have the same sum and the same product. Show that there exist such sets of size n for any $n \geq 3$. That is, show there exist $2n$ distinct positive integers $a_1, a_2, \ldots, a_n, b_1, b_2, \ldots, b_n$ such that

$$\sum_{i=1}^n a_i = \sum_{i=1}^n b_i \quad \text{and} \quad \prod_{i=1}^n a_i = \prod_{i=1}^n b_i.$$ (p. 119)

66. Let f and g be odd functions (that is, $f(-x) = -f(x)$ and $g(-x) = -g(x)$ for all x) that are infinitely differentiable at $x = 0$, and assume that $f'(0) = g'(0) = 1$. Consider the compositions $F = f \circ g$ and $G = g \circ f$.

a. Show that $F'(0) = G'(0)$, $F^{(3)}(0) = G^{(3)}(0)$, and $F^{(5)}(0) = G^{(5)}(0)$.

b. Show that for all even n, $F^{(n)}(0) = G^{(n)}(0)$.

c. Is it always true that for all odd n, $F^{(n)}(0) = G^{(n)}(0)$? If so, prove it; if not, give a counterexample. (p. 121)

67. In how many ways can the integers $1, 2, \ldots, n$, $n \geq 2$, be listed (once each) so that as you go through the list, there is exactly one integer which is immediately followed by a smaller integer? (p. 123)

68. A two-player game is played as follows. The players take turns changing a positive integer to a smaller one and then passing that smaller integer back to their opponent. If the integer is even, the two legal moves are (i) subtracting 1 from the integer and (ii) halving the integer. If the integer is odd, the two legal moves are (i) subtracting 1 from the integer and (ii) subtracting 1 and then halving the result. The game ends when the integer reaches 0, and the player making the last move wins. For example, if the starting integer is 15, the first player might move to 7, the second player to 6, the first player to 3, the second player to 2, the first player to 1 and now the second player moves to 0 and wins. However, in this sample game the first player could have played better!

a. Given best play, if the starting integer is 1000, should the first or second player win? How about if the starting integer is N?

b. If we take a starting integer at random, say from all integers from 1 to n inclusive, we can consider the probability that the second player should win. This probability will fluctuate as n increases, but what is its limit as $n \to \infty$? (p. 126)

69. Find a solution to the system of simultaneous equations

$$\begin{cases} x^4 - 6x^2y^2 + y^4 = 1 \\ 4x^3y - 4xy^3 = 1, \end{cases}$$

where x and y are real numbers. (p. 127)

70. Show that if $p(x)$ is a polynomial of odd degree greater than 1, then through any point P in the plane, there will be at least one tangent line to the curve $y = p(x)$. Is this still true if $p(x)$ is of even degree? (p. 129)

71. For each positive integer n, let

$$N(n) = \left\lceil \frac{n}{2} \right\rceil + \left\lceil \frac{n}{4} \right\rceil + \left\lceil \frac{n}{8} \right\rceil + \cdots + \left\lceil \frac{n}{2^k} \right\rceil,$$

where k is the unique integer such that $2^{k-1} \leq n < 2^k$, and $\lceil x \rceil$ denotes the smallest integer greater than or equal to x. For which numbers n is $N(n) = n$? (p. 129)

72. Call a convex pentagon (five-sided figure with angles less than 180°) "parallel" if each diagonal is parallel to the side with which it does not have a vertex in common. That is, $ABCDE$ is parallel if the diagonal AC is parallel to the side DE and similarly for the other four diagonals. It is easy to see that a regular pentagon is parallel, but is a parallel pentagon necessarily regular? (p. 130)

73. Given a permutation of the integers $1, 2, \ldots, n$, define the *total fluctuation* of that permutation to be the sum of all the differences between successive numbers along the permutation, where all differences are counted positively regardless of which of the two successive numbers is larger. For example, for the permutation $5, 3, 1, 2, 6, 4$ the differences would be $2, 2, 1, 4, 2$ and the total fluctuation would be $2 + 2 + 1 + 4 + 2 = 11$. What is the greatest possible total fluctuation, as a function of n, for permutations of $1, 2, \ldots, n$? (p. 132)

74. Find the sum of the infinite series

$$\sum_{n=1}^{\infty} \frac{1}{2n^2 - n} = 1 + \frac{1}{6} + \frac{1}{15} + \frac{1}{28} + \cdots.$$ (p. 136)

75. It is not hard to show that any integer either is a square, or one can get it by adding and/or subtracting *distinct* squares of integers. For one thing, any odd positive integer is a difference of consecutive squares (for example, $13 = 49 - 36$), and any even positive integer can be found from the odd one just before it by adding the square 1 (for example, $14 = 49 - 36 + 1$). This leaves two awkward cases in which we have used the square 1 twice ($2 = 1 - 0 + 1$ and $4 = 4 - 1 + 1$), but we also have $2 = 16 - 9 - 4 - 1$, and 4 is itself a square. Now for the problem: Can one get any integer that is not itself a cube by adding and/or subtracting distinct cubes of integers? (p. 138)

76. For any vector $\mathbf{v} = (x_1, \ldots, x_n)$ in $\mathbb{R}^n$ and any permutation σ of $1, 2, \ldots, n$, define $\sigma(\mathbf{v}) = (x_{\sigma(1)}, \ldots, x_{\sigma(n)})$. Now fix $\mathbf{v}$ and let V be the span of $\{\sigma(\mathbf{v}) \mid \sigma \text{ is a permutation of } 1, 2, \ldots, n\}$. What are the possibilities for the dimension of V? (p. 140)

77. Note that the set $\{1, 2, 3, 4\}$ can be split into two sets $S = \{1, 3\}$ and $T = \{2, 4\}$ with the property that the average of the elements of S is an element of T while the average of the elements of T is an element of S. For this problem, say a positive integer n is *suitable* if there is a way to write the set $\{1, 2, \ldots, n\}$ as the union of sets S and T which have no elements in common and such that the average of all the elements of S is an element of T and vice versa. (It is not necessary that S and T have the same number of elements.) Show that, with one exception, composite numbers are suitable and prime numbers are not suitable. (p. 141)

78. Suppose three circles, each of radius 1, go through the same point in the plane. Let A be the set of points which lie inside at least two of the circles. What is the smallest area A can have? (p. 144)

79. Do there exist two different positive integers (written as usual in base 10) with an equal number of digits so that the square of each of the integers starts with the other? For example, we could try 21 as one of the integers; the square is 441, which starts with 44, but alas, the square of 44, which is 1936, starts with 19 rather than 21, so 21 and 44 won't do. If two such integers exist, give an example; if not, show why not. (p. 147)

80. How many real solutions does the equation

$$\sqrt[7]{x} - \sqrt[5]{x} = \sqrt[3]{x} - \sqrt{x}$$

have? (p. 148)

81. Suppose you draw n parabolas in the plane. What is the largest number of (connected) regions that the plane may be divided into by those parabolas? (The parabolas can be positioned in any way; in particular, their axes need not be parallel to either the x- or the y-axis.) (p. 150)

82. A particle starts somewhere in the plane and moves 1 unit in a straight line. Then it makes a "shallow right turn," abruptly changing direction by an acute angle α, and moves 1 unit in a straight line in the new direction.

Then it again changes direction by α (to the right) and moves 1 unit, and so forth. In all, the particle takes 9 steps of 1 unit each, with each direction at an angle α to the previous direction.

a. For which value(s) of α does the particle end up exactly at its starting point?

b. For how many values of the acute angle α does the particle end up at a point whose (straight-line) distance to the starting point is exactly 1 unit? (p. 152)

83. Let $A \neq 0$ and B_1, B_2, B_3, B_4 be 2×2 matrices (with real entries) such that

$$\det(A + B_i) = \det A + \det B_i \qquad \text{for } i = 1, 2, 3, 4.$$

Show that there exist real numbers k_1, k_2, k_3, k_4, not all zero, such that

$$k_1 B_1 + k_2 B_2 + k_3 B_3 + k_4 B_4 = 0.$$

(0 is the zero matrix, all of whose entries are 0.) (p. 155)

84. Let g be a continuous function defined on the positive real numbers. Define a sequence (f_n) of functions as follows. Let $f_0(x) = 1$, and for $n \geq 0$ and $x > 0$, let

$$f_{n+1}(x) = \int_1^x f_n(t) g(t)\, dt.$$

Suppose that for all $x > 0$, $\sum_{n=0}^{\infty} f_n(x) = x$. Find the function g. (p. 155)

85. Let r and s be specific positive integers. Let F be a function from the set of all positive integers to itself with the following properties:

(i) F is one-to-one and onto;
(ii) For every positive integer n, either $F(n) = n + r$ or $F(n) = n - s$.

a. Show that there exists a positive integer k such that the k-fold composition of F with itself is the identity function.

b. Find the smallest such positive integer k. (The answer will depend on r and s.) (p. 156)

86. As you might expect, ice fishing is a popular "outdoor" pastime during the long Wohascum County winters. Recently two ice fishermen arrived at Round Lake, which is perfectly circular, and set up their ice houses in exactly opposite directions from the center, two-thirds of the way from the center to the lakeshore. The point of this symmetrical arrangement was that any fish that could be lured would (perhaps) swim toward the closest lure, so that both fishermen would have equal expectations of their catch. Some time later, a third fisherman showed up, and since the first two adamantly refused to move their ice houses, the following problem arose. Could a third ice house be put on the lake in such a way that all three fishermen would have equal expectations at least to the extent that the three regions, each consisting of all points on the lake for which one of the three ice houses was closest, would all have the same area? (p. 158)

87. a. Define sequences (a_n) and (b_n) as follows: a_n is the result of writing down the first n odd integers in order (for example, $a_7 = 135791113$), while b_n is the result of writing down the first n even integers in order. Evaluate $\lim_{n\to\infty} \frac{a_n}{b_n}$.

b. Now suppose we do the same thing, but we write all the odd and even integers in base B (and we interpret the fractions a_n/b_n in base B). For example, if $B = 9$ we will now have $a_2 = 13$, $a_7 = 1357101214$. Show that for any base $B \geq 2$, $\lim_{n\to\infty} \frac{a_n}{b_n}$ exists. For what values of B will the limit be the same as for $B = 10$? (p. 159)

88. Note that the integers $2, -3,$ and 5 have the property that the difference of any two of them is an integer times the third:

$$2 - (-3) = 1 \times 5, \qquad (-3) - 5 = (-4) \times 2, \qquad 5 - 2 = (-1) \times (-3).$$

Suppose three distinct integers a, b, c have this property.

a. Show that a, b, c cannot all be positive.

b. Now suppose that a, b, c, in addition to having the above property, have no common factors (except $1, -1$). (For example, $20, -30, 50$ would not qualify, because although they have the above property, they have the common factor 10.) Is it true that one of the three integers has to be either $1, 2, -1,$ or -2? (p. 161)

89. Start with four numbers arranged in a circle. Form the average of each pair of adjacent numbers, and put these averages in the circle between the original numbers; then delete the original numbers so that once again there are four numbers in the circle. Repeat. Suppose that after twenty steps, you find the numbers $1, 2, 3, 4$ in *some* order. What numbers did you have after one step? Can you recover the original numbers from this information?
(p. 162)

90. Suppose all the integers have been colored with the three colors red, green and blue such that each integer has exactly one of those colors. Also suppose that the sum of any two (unequal or equal) green integers is blue, the sum of any two blue integers is green, the opposite of any green integer is blue, and the opposite of any blue integer is green. Finally, suppose that 1492 is red and that 2011 is green. Describe precisely which integers are red, which integers are green, and which integers are blue. (p. 165)

91. Let A be an $m \times n$ matrix with every entry either 0 or 1. How many such matrices A are there for which the number of 1's in each row and each column is even? (p. 167)

92. For $0 \leq x \leq 1$, let $T(x) = \begin{cases} x & \text{if } x \leq 1/2 \\ 1 - x & \text{if } x \geq 1/2 \end{cases}$. (You can think of $T(x)$ as the distance from x to the nearest integer.) Define $f(x) = \sum_{n=1}^{\infty} T(x^n)$.

a. Evaluate $f\left(\dfrac{1}{\sqrt[3]{2}}\right)$.

b. Find all x ($0 \leq x \leq 1$) for which $f(x) = 2012$. (p. 169)

93. For three points P, Q, and R in $\mathbb{R}^3$ (or, more generally, in $\mathbb{R}^n$) we say that R is *between* P and Q if R is on the line segment connecting P and Q ($R = P$ and $R = Q$ are allowed). A subset A of $\mathbb{R}^3$ is called *convex* if for any two points P and Q in A, every point R which is between P and Q is also in A. For instance, an ellipsoid is convex, a banana is not. Now for the problem: Suppose A and B are convex subsets of $\mathbb{R}^3$. Let C be the set of all points R for which there are points P in A and Q in B such that R lies between P and Q. Does C have to be convex? (p. 170)

94. Start with a circle and inscribe a regular n-gon in it, then inscribe a circle in that regular n-gon, then inscribe a regular n-gon in the new circle, then a third circle in the second n-gon, and so forth. Continuing in this way, the region (disk) inside the original circle will be divided into infinitely many smaller regions, some of which are bounded by a circle on the outside and one side of a regular n-gon on the inside (call these "type I" regions) while others are bounded by two sides of a regular n-gon on the outside and a circle on the inside ("type II" regions).

Let $f(n)$ be the fraction of the area of the original disk that is occupied by type I regions. What is the limit of $f(n)$ as n tends to infinity? (p. 171)

95. Suppose we have a configuration (set) of finitely many points in the plane which are not all on the same line. We call a point in the plane a *center* for the configuration if for every line through that point, there is an equal number of points of the configuration on either side of the line.

a. Give a necessary and sufficient condition for a configuration of four points to have a center.

b. Is it possible for a finite configuration of points (not all on the same line) to have more than one center? (p. 173)

96. Define a sequence of matrices by $A_1 = \begin{pmatrix} 0 & 1/2 \\ -1/2 & 1 \end{pmatrix}$, $A_2 = \begin{pmatrix} 3 & -2 \\ 1 & 0 \end{pmatrix}$, and for $n \geq 1$, $A_{n+2} = A_{n+1} A_n A_{n+1}^{-1}$. Find approximations to the matrices A_{2010} and A_{2011}, with each entry correct to within 10^{-300}. (p. 174)

97. Find all real solutions x of the equation
$$x^{10} - x^8 + 8x^6 - 24x^4 + 32x^2 - 48 = 0.$$
(p. 176)

98. The proprietor of the Wohascum Puzzle, Game and Computer Den, a small and struggling but interesting enterprise in Wohascum Center, recently was trying to design a novel set of dice. An ordinary die, of course, is cubical, with each face showing one of the numbers 1, 2, 3, 4, 5, 6. Since each face borders on four other faces, each number is "surrounded" by four of the other numbers. The proprietor's plan was to have each die in the shape of a regular dodecahedron (with twelve pentagonal faces). Each of the numbers 1, 2, 3, 4, 5, 6 would occur on two different faces and be "surrounded" both times by all five other numbers. Is this possible? If so, in how many essentially different ways can it be done? (Two ways are considered essentially the same if one can be obtained from the other by rotating the dodecahedron.) (p. 177)

99. Arrange the positive integers in an array with three columns, as follows. The first row is [1, 2, 3]; for $n > 1$, row n is $[a, b, a+b]$, where a and b, with $a < b$, are the two smallest positive integers that have not yet appeared as entries in rows $1, 2, \ldots, n-1$. The first four rows of the array are

	Column 1	Column 2	Column 3
Row 1	1	2	3
Row 2	4	5	9
Row 3	6	7	13
Row 4	8	10	18

Note that after row 3 is in place, 8 and 10 are the two smallest positive integers that have not been placed yet, so they appear in columns 1 and 2 of row 4. For each non-zero digit d and each positive integer m, in which column will the m-digit number $\underbrace{ddd\ldots d}_{m}$ end up, and why? (p. 179)

100. For what positive integers n is $17^n - 1$ divisible by 2^n? (p. 182)

101. Let f be a continuous function on the real numbers. Define a sequence of functions $f_0 = f, f_1, f_2, \ldots$ by repeated integration, as follows:

$$f_0(x) = f(x) \quad \text{and} \quad f_{i+1}(x) = \int_0^x f_i(t)\,dt, \quad \text{for } i = 0, 1, 2, \ldots .$$

Show that for *any* continuous function f and any real number x,

$$\lim_{n \to \infty} f_n(x) = 0. \qquad \text{(p. 183)}$$

102. Consider an arbitrary circle of radius 2 in the coordinate plane. Let n be the number of lattice points (points whose coordinates are both integers) inside, but not on, the circle.
a. What is the smallest possible value for n?
b. What is the largest possible value for n? (p. 184)

103. Let a and b be nonzero real numbers and (x_n) and (y_n) be sequences of real numbers. Given that
$$\lim_{n\to\infty} \frac{ax_n + by_n}{\sqrt{x_n^2 + y_n^2}} = 0$$
and that x_n is never 0, show that
$$\lim_{n\to\infty} \frac{y_n}{x_n}$$
exists and find its value. (p. 185)

104. Note that if we tile the plane with black and white squares in a regular "checkerboard" pattern, then every square has an equal number of black and of white neighbors (four each), where two squares are considered neighbors if they are not the same but they have at least one common point. If we try the analogous pattern of cubes in 3-space, it no longer works this way: every white cube has 14 black neighbors and only 12 white neighbors, and vice versa.

a. Show that there is a different color pattern of black and white "grid" cubes in 3-space for which every cube does have exactly 13 neighbors of each color.

b. What happens in n-space for $n > 3$? Is it still possible to find a color pattern for a regular grid of "hypercubes" so that every hypercube, whether black or white, has an equal number of black and white neighbors? If so, show why; if not, give an example of a specific n for which it is impossible.
(p. 187)

105. The MAA Student Chapter at Wohascum College is about to organize an icosahedron-building party. Each participant will be provided twenty congruent equilateral triangles cut from old ceiling tiles. The edges of the triangles are to be beveled so they will fit together at the correct angle to form a regular icosahedron. What is this angle (between adjacent faces of the icosahedron)? (p. 189)

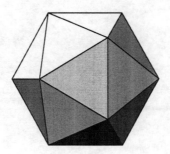

106. Consider the following procedure for unscrambling any permutation of the integers from 1 through n into increasing order: Pick any number that's out of place, and wedge it into its "proper" position, shifting others over to make room for it. Repeat this procedure as long as there are numbers that are out of place. For example, for $n = 5$, take the permutation 3 5 1 2 4 and choose the underlined out-of-place number at each step.

$$\begin{array}{ccccc} 3 & 5 & 1 & 2 & \underline{4} \\ 3 & 5 & 1 & 4 & \underline{2} \\ \underline{3} & 2 & 5 & 1 & 4 \\ 2 & \underline{5} & 3 & 1 & 4 \\ 2 & \underline{3} & 1 & 4 & 5 \\ \underline{2} & 1 & 3 & 4 & 5 \\ 1 & 2 & 3 & 4 & 5 \end{array}$$

In this case, it's taken 6 steps to sort the numbers into their proper order. We could have done it in no more than 5 steps, by putting $1, 2, 3, 4, 5$ in their proper places in that order. On the other hand, some choice of numbers might have taken more than 6 steps, or this procedure might not terminate at all for some choice of out-of-place numbers.

Find a permutation of $1, 2, 3, 4, 5$ that may take as many as 15 steps to sort using this procedure. (p. 190)

107. Given a constant C, find all functions f such that

$$f(x) + C f(2 - x) = (x - 1)^3 \qquad \text{for all } x. \qquad \text{(p. 192)}$$

108. The Wohascum Center branch of Wohascum National Bank recently installed a digital time/temperature display which flashes back and forth between time, temperature in degrees Fahrenheit, and temperature in degrees Centigrade (Celsius). Recently one of the local college mathematics professors became concerned when she walked by the bank and saw readings of 21°C and 71°F, especially since she had just taught her precocious five-year-old that same day to convert from degrees C to degrees F by multiplying by 9/5 and adding 32 (which yields 21°C = 69.8°F, which should be rounded to 70°F). However, a bank officer explained that both readings were correct; the apparent error was due to the fact that the display device converts before rounding either Fahrenheit or Centigrade temperature to a whole number. (Thus, for example, 21.4°C = 70.52°F.) Suppose that over the course of a week in summer, the temperatures measured are between 15°C and 25°C and that they are randomly and uniformly distributed over that interval. What is the probability that at any given time the display will appear to be in error for the reason above, that is, that the rounded value in degrees F of the converted temperature is not the same as the value obtained by first rounding the temperature in degrees C, then converting to degrees F and rounding once more? (p. 193)

109. Consider the sequence 4, 1/3, 4/3, 4/9, 16/27, 64/81, ... , in which each term (after the first two) is the product of the two previous ones. Note that for this particular sequence, the first and third terms are greater than 1 while the second and fourth terms are less than 1. However, after that the "alternating" pattern fails: the fifth and all subsequent terms are less than 1. Do there exist sequences of positive real numbers in which each term is the product of the two previous terms and for which *all* odd-numbered terms are greater than 1, while all even-numbered terms are less than 1? If so, find all such sequences. If not, prove that no such sequence is possible. (p. 196)

110. Sketch the set of points (x, y) in the plane which satisfy

$$(x^2 - y^2)^{2/3} + (2xy)^{2/3} = (x^2 + y^2)^{1/3}.$$ (p. 198)

111. Let $f_1, f_2, \ldots, f_{2012}$ be functions such that the derivative of each of them is the sum of the others. Let $F = f_1 f_2 \cdots f_{2012}$ be the *product* of all these functions. Find all possible values of r, given that $\lim_{x \to -\infty} F(x)e^{rx}$ is a finite nonzero number. (p. 199)

112. On a table in a dark room there are n hats, each numbered clearly with a different number from the set $\{1, 2, \ldots, n\}$. A group of k intelligent students, with $k < n$, comes into the room, and each student takes a hat at random and puts it on his or her head. The students go back outside, where they can see the numbers on each other's hats (but, naturally, no one can see her/his own hat number). Each student now looks carefully at all the other students and announces

(i) the largest hat number that (s)he can see, as well as
(ii) the smallest hat number that (s)he can see.

After all these announcements are made, what is the probability that *all* students should be able to deduce their own hat numbers:
a. if $n = 6, k = 5$;
b. in general, as a function of n and k? (p. 201)

113. Find all integer solutions to $x^2 + 615 = 2^n$. (p. 204)

114. Sum the infinite series
$$\sum_{n=1}^{\infty} \sin \frac{2\alpha}{3^n} \sin \frac{\alpha}{3^n}.$$
(p. 204)

115. A fair coin is flipped repeatedly. Starting from $x = 0$, each time the coin comes up "heads," 1 is added to x, and each time the coin comes up "tails," 1 is subtracted from x. Let a_n be the expected value of $|x|$ after n flips of the coin. Does $a_n \to \infty$ as $n \to \infty$? (p. 205)

116. a. Show that there is no cubic polynomial whose graph passes through the points $(0,0)$, $(1,1)$, and $(2,16)$ and which is increasing for *all* x.
b. Show that there is a polynomial whose graph passes through $(0,0)$, $(1,1)$, and $(2,16)$ and which is increasing for all x.
c. Show that if $x_1 < x_2 < x_3$ and $y_1 < y_2 < y_3$, there is always an increasing polynomial (for all x) whose graph passes through (x_1, y_1), (x_2, y_2), and (x_3, y_3). (p. 206)

117. Do there exist five rays emanating from the origin in $\mathbb{R}^3$ such that the angle between any two of these rays is obtuse (greater than a right angle)? (p. 208)

118. Note that the integers $a = 1, b = 5, c = 7$ have the property that the square of b (namely, 25) is the average of the square of a and the square of c (1 and 49). Of course, from this one example we can get infinitely many examples by multiplying all three integers by the same factor. But if we don't allow this, will there still be infinitely many examples? That is, are there infinitely many triples (a, b, c) such that the integers a, b, c have no common factors and the square of b is the average of the squares of a and c? (p. 208)

119. Find all twice continuously differentiable functions f for which there exists a constant c such that, for all real numbers a and b,
$$\left| \int_a^b f(x)\,dx - \frac{b-a}{2}\Big(f(b) + f(a)\Big) \right| \leq c(b-a)^4. \qquad \text{(p. 211)}$$

120. Let A be a set of n real numbers. Because A has 2^n subsets, we can get 2^n sums by choosing a subset B of A and taking the sum of the numbers in B. (By convention, if B is the empty set, that sum is 0.) What is the least number of *different* sums we must get (as a function of n) by taking the 2^n possible sums of subsets of a set with n numbers? (p. 212)

121. The proprietor of the Wohascum Puzzle, Game, and Computer Den has invented a new two-person game, in which players take turns coloring edges of a cube. Three colors (red, green, and yellow) are available. The cube starts off with all edges uncolored; once an edge is colored, it cannot be colored again. Two edges with a common vertex are not allowed to have the same color. The last player to be able to color an edge wins the game.

a. Given best play on both sides, should the first or the second player win? What is the winning strategy?

b. There are twelve edges in all, so a game can last at most twelve turns (whether or not the players use optimal strategies); it is not hard to see that twelve turns are possible. How many twelve-turn end positions are essentially different? (Two positions are considered essentially the same if one can be obtained from the other by rotating the cube.) (p. 214)

122. As is well known, the unit circle has the property that distances along the curve are numerically equal to the difference of the corresponding angles (in radians) at the origin; in fact, this is how angles are often defined. (For example, a quarter of the unit circle has length $\pi/2$ and corresponds to an angle $\pi/2$ at the origin.) Are there other differentiable curves in the plane with this property? If so, what do they look like? (p. 216)

123. Fifty-two is the sum of two squares;
And three less is a square! So who cares?
You may think it's curious,
Perhaps it is spurious,
Are there other such numbers somewheres?

Are there other solutions in integers? If so, how many? (p. 218)

THE PROBLEMS

124. Let E be an ellipse in the plane. Describe the set S of all points P outside the ellipse such that the two tangent lines to the ellipse that pass through P form a right angle. (p. 219)

125. Starting with a positive number $x_0 = a$, let $(x_n)_{n \geq 0}$ be the sequence of numbers such that
$$x_{n+1} = \begin{cases} x_n^2 + 1 & \text{if } n \text{ is even,} \\ \sqrt{x_n} - 1 & \text{if } n \text{ is odd.} \end{cases}$$
For what positive numbers a will there be terms of the sequence arbitrarily close to 0? (p. 222)

126. Suppose you pick the one million entries of a 1000×1000 matrix independently and at random from the set of digits. Is the determinant of the resulting matrix more likely to be even or odd? (p. 223)

127. a. Find all positive numbers T for which
$$\int_0^T x^{-\ln x} dx = \int_T^\infty x^{-\ln x} dx.$$
b. Evaluate the above integrals for all such T, given that
$$\int_0^\infty e^{-x^2} dx = \frac{\sqrt{\pi}}{2}.$$
(p. 225)

128. Let a_n be the number of different strings of length n that can be formed from the symbols X and O with the restriction that a string may not consist of identical smaller strings. For example, XXXX and XOXO are not allowed. The possible strings of length 4 are

XXXO, XXOX, XXOO, XOXX, XOOX, XOOO,
OXXX, OXXO, OXOO, OOXX, OOXO, OOOX,

so $a_4 = 12$. Here is a table showing a_n and the ratio $\dfrac{a_{n+1}}{a_n}$ for $n = 1, 2, \ldots, 13$.

n	1	2	3	4	5	6	7	8	9	10	11	12	13
a_n	2	2	6	12	30	54	126	240	504	990	2046	4020	8190
$\dfrac{a_{n+1}}{a_n}$	1	3	2	2.5	1.8	2.33	1.90	2.1	1.96	2.07	1.96	2.03	1.98

The table suggests two conjectures:

a. For any $n > 2$, a_n is divisible by 6.

b. $\lim\limits_{n \to \infty} \dfrac{a_{n+1}}{a_n} = 2$.

Prove or disprove each of these conjectures. (p. 225)

129. Let $f(x,y) = x^2 + y^2$ and $g(x,y) = x^2 - y^2$. Are there differentiable functions $F(z), G(z)$, and $z = h(x,y)$ such that

$$f(x,y) = F(z) \quad \text{and} \quad g(x,y) = G(z)?$$

(p. 227)

130. For $x \geq 0$, let $y = f(x)$ be continuously differentiable, with positive, increasing derivative. Consider the ratio between the distance from $(0, f(0))$ to $(x, f(x))$ along the curve $y = f(x)$ (the arc length from 0 to x) and the straight-line distance from $(0, f(0))$ to $(x, f(x))$. Must this ratio have a limit as $x \to \infty$? If so, what is the limit? (p. 228)

131. Note that the three positive integers $1, 24, 120$ have the property that the sum of any two of them is a different perfect square. Do there exist four positive integers such that the sum of any two of them is a perfect square and such that the six squares found in this way are all different? If so, exhibit four such positive integers; if not, show why this cannot be done. (p. 230)

132. A point $P = (a, b)$ in the plane is *rational* if both a and b are rational numbers. Find all rational points P such that the distance between P and every rational point on the line $y = 13x$ is a rational number. (p. 231)

133. For what positive numbers x does the series

$$\sum_{n=1}^{\infty}(1 - \sqrt[n]{x}) = (1 - x) + (1 - \sqrt{x}) + (1 - \sqrt[3]{x}) + \cdots$$

converge? (p. 234)

134. Note that a triangle ABC is isosceles, with equal angles at A and B, if and only if the median from C and the angle bisector at C are the same. This suggests a measure of "scalenity": For each vertex of a triangle ABC, measure the distance along the opposite side from the midpoint to the "end" of the angle bisector, as a fraction of the total length of that opposite side.

This yields a number between 0 and 1/2; take the least of the three numbers found in this way. The triangle is isosceles if and only if this least number is zero. What are the possible values of this least number if ABC can be any triangle in the plane? (p. 236)

135. Let R be a commutative ring with at least one, but only finitely many, (nonzero) zero divisors. Prove that R is finite. (p. 238)

136. Consider the transformation of the plane (except for the coordinate axes) defined by sending the point (x, y) to the point $(y + 1/x, x + 1/y)$. Suppose we apply this transformation repeatedly, starting with some specific point (x_0, y_0), to get a sequence of points (x_n, y_n).

a. Show that if (x_0, y_0) is in the first or the third quadrant, the sequence of points will tend to infinity.

b. Show that if (x_0, y_0) is in the second or the fourth quadrant, either the sequence will terminate because it lands at the origin, or the sequence will be eventually periodic with period 1 or 2, or there will be infinitely many n for which (x_n, y_n) is further from the origin than (x_{n-1}, y_{n-1}). (p. 239)

137. Let $(a_n)_{n \geq 0}$ be a sequence of positive integers given recursively by $a_{n+1} = 2a_n + 1$. Is there an a_0 such that the sequence consists only of prime numbers? (p. 241)

138. Suppose $c > 0$ and $0 < x_1 < x_0 < 1/c$. Suppose also that

$$x_{n+1} = cx_n x_{n-1} \qquad \text{for } n = 1, 2, \ldots.$$

a. Prove that

$$\lim_{n \to \infty} x_n = 0.$$

b. Let $\phi = (1 + \sqrt{5})/2$. Prove that

$$\lim_{n \to \infty} \frac{x_{n+1}}{x_n^\phi}$$

exists, and find it. (p. 241)

139. Consider the number of solutions to the equation

$$\sin x + \cos x = \alpha \tan x$$

for $0 \leq x \leq 2\pi$, where α is an unspecified real number. What are the possibilities for this number of solutions, as α is allowed to vary?

(p. 242)

140. Given 64 points in the plane which are positioned so that 2001, but no more, distinct lines can be drawn through pairs of points, prove that at least four of the points are collinear. (p. 245)

141. Let $f_1, f_2, \ldots, f_n$ be linearly independent, differentiable functions. Prove that some $n-1$ of their derivatives $f'_1, f'_2, \ldots, f'_n$ are linearly independent. (p. 246)

142. Find all real numbers A and B such that

$$\left| \int_1^x \frac{1}{1+t^2} dt - A - \frac{B}{x} \right| < \frac{1}{3x^3}$$

for all $x > 1$. (p. 247)

143. The following figure shows a closed knight's tour of the chessboard which is symmetric under a 180° rotation of the board. (A closed knight's tour is a sequence of consecutive knight moves that visits each square of the chessboard exactly once and returns to the starting point.)
a. Prove that there is no closed knight's tour of the chessboard which is symmetric under a reflection in one of the main diagonals of the board.
b. Prove that there is no closed knight's tour of the chessboard which is symmetric under a reflection in the horizontal axis through the center of the board. (p. 249)

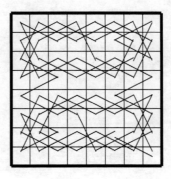

144. A function f on the rational numbers is defined as follows. Given a rational number $x = \frac{m}{n}$, where m and n are relatively prime integers and $n > 0$, set $f(x) = \frac{3m-1}{2n+1}$. Now, starting with a rational number x_0, apply f repeatedly to get a sequence $x_1 = f(x_0)$, $x_2 = f(x_1)$, ..., $x_{n+1} = f(x_n)$, ... Find all rational numbers x_0 for which that infinite sequence is periodic. (p. 250)

145. a. Find a sequence (a_n), $a_n > 0$, such that

$$\sum_{n=1}^{\infty} \frac{a_n}{n^3} \quad \text{and} \quad \sum_{n=1}^{\infty} \frac{1}{a_n}$$

both converge.

b. Prove that there is no sequence (a_n), $a_n > 0$, such that

$$\sum_{n=1}^{\infty} \frac{a_n}{n^2} \quad \text{and} \quad \sum_{n=1}^{\infty} \frac{1}{a_n}$$

both converge. (p. 252)

146. It is a standard result that the limit of the indeterminate form x^x, as x approaches zero from above, is 1. What is the limit of the repeated power $x^{x^{\cdot^{\cdot^{\cdot^{x}}}}}$ with n occurrences of x, as x approaches zero from above? (p. 253)

147. Show that there exist an integer N and a rational number r such that $\sum_{n=2012}^{\infty} \frac{(-1)^n}{\binom{n}{2012}} = N \ln 2 + r$, and find the integer N. (p. 255)

148. A new subdivision is being laid out on the outskirts of Wohascum Center. There are ten north–south streets and six east–west streets, forming blocks which are exactly square. The Town Council has ordered that fire hydrants be installed at some of the intersections, in such a way that no intersection will be more than two "blocks" (really sides of blocks) away from an intersection with a hydrant. (Thus, no house will be more than $2\frac{1}{2}$ blocks from a hydrant. The blocks need not be in the same direction.) What is the smallest number of hydrants that could be used? (p. 257)

149. Consider a continuous function $f : \mathbb{R}^+ \longrightarrow \mathbb{R}^+$ with the following properties:
  (i) $f(2) = 3$,
  (ii) For all $x, y > 0$, $f(xy) = f(x)f(y) - f\left(\frac{x}{y}\right)$.

a. Show that if such a function f exists, it is unique.
b. Find an explicit formula for such a function. (p. 259)

150. A can is in the shape of a right circular cylinder of radius r and height h. An intelligent ant is at a point on the edge of the top of the can (that is, on the circumference of the circular top) and wants to crawl to the point on the edge of the bottom of the can that is diametrically opposite to its starting point. As a function of r and h, what is the minimum distance the ant must crawl? (p. 261)

151. Consider a triangle ABC whose angles α, β, and γ (at A, B, C respectively) satisfy $\alpha \leq \beta \leq \gamma$. Under what conditions on α, β, and γ can a beam of light placed at C be aimed at the segment AB, reflect to the segment BC, and then reflect to the vertex A? (Assume that the angle of incidence of a beam of light equals the angle of reflection.) (p. 263)

152. It's not hard to see that in the plane, the largest number of nonzero vectors that can be chosen so that any two of the vectors make the same nonzero angle with each other is 3 (and the only possible nonzero angle for three such vectors to make is $2\pi/3$). Now suppose we have vectors in n-dimensional space. What is the largest possible number of nonzero vectors in n-space so that the angle between any two of the vectors is the same (and not zero)? In that situation, what are the possible values for the angle? (p. 265)

153. Let S be a set of numbers which includes the elements 0 and 1. Suppose S has the property that for any nonempty finite subset T of S, the average of all the numbers in T is an element of S. Prove or disprove: S must contain all the rational numbers between 0 and 1. (p. 267)

154. Find
$$\lim_{n \to \infty} \left(\sum_{k=1}^{n} \frac{1}{\binom{n}{k}} \right)^n,$$
or show that this limit does not exist. (p. 270)

155. A person starts at the origin and makes a sequence of moves along the real line, with the kth move being a change of $\pm k$.

a. Prove that the person can reach any integer in this way.

b. If $m(n)$ is the least number of moves required to reach a positive integer n, prove that
$$\lim_{n \to \infty} \frac{m(n)}{\sqrt{n}}$$
exists and evaluate this limit. (p. 271)

156. Let $S(n)$ be the number of solutions of the equation $e^{\sin x} = \sin(e^x)$ on the interval $[0, 2n\pi]$. Find $\lim_{n \to \infty} \frac{S(n)}{e^{2n\pi}}$, or show that the limit does not exist. (p. 272)

157. Consider the functions $\varepsilon \colon \mathbb{Z} \to \{1, -1\}$ having period N, where $N > 1$ is a positive integer. For which periods N does there exist an infinite series $\sum_{n=1}^{\infty} a_n$ with the following properties: $\sum_{n=1}^{\infty} a_n$ diverges, whereas $\sum_{n=1}^{\infty} \varepsilon(n) a_n$ converges for all nonconstant ε (of period N)? (p. 275)

158. Suppose you form a sequence of quadrilaterals as follows. The first quadrilateral is the unit square. To get from each quadrilateral to the next, pick a vertex of your quadrilateral and a side that is not adjacent to that vertex, and then connect the midpoint of that side to that vertex. This will divide the quadrilateral into a triangle and a new quadrilateral; discard the triangle, and repeat the process with the new quadrilateral. (The second quadrilateral will be a rectangular trapezoid, for instance with vertices $(0,0), (1,0), (1,1/2), (0,1)$. The size and shape of the third quadrilateral will depend on what vertex and what side of the trapezoid you choose.) Show that there are at most $2^{n-1} - n + 1$ possibilities for the area of the nth quadrilateral, and state explicitly what the $2^{n-1} - n + 1$ candidates are. (p. 275)

159. Suppose f is a continuous, increasing, bounded, real-valued function, defined on $[0, \infty)$, such that $f(0) = 0$ and $f'(0)$ exists. Show that there exists $b > 0$ for which the volume obtained by rotating the area under f from 0 to b about the x-axis is half that of the cylinder obtained by rotating $y = f(b)$, $0 \le x \le b$, about the x-axis. (p. 278)

160. In general, composition of functions is not commutative. For example, for the functions f and g given by $f(x) = x+1$, $g(x) = 2x$, we have $f(g(x)) = 2x+1$ and $g(f(x)) = 2x+2$. Now suppose that we have three functions f, g, h. Then there are six possible compositions of the three, given by $f(g(h(x))), g(h(f(x))), \ldots$. Give an example of three continuous functions that are defined for all real x and for which exactly five of the six compositions are the same. (Reprinted with the permission of the Canadian Mathematical Society, this problem was originally published in the Mathematical Mayhem section of *Crux Mathematicorum with Mathematical Mayhem*, vol. 25, 1999, p. 293, problem C87.) (p. 279)

161. Does the Maclaurin series (Taylor series at 0) for e^{x-x^3} have any zero coefficients? (p. 280)

162. Let a and d be relatively prime positive integers, and consider the sequence $a, a+d, a+4d, a+9d, \ldots, a+n^2 d, \ldots$. Given a positive integer b, can one always find an integer in the sequence which is relatively prime to b? (p. 282)

163. The other day, in the honors calculus class at Wohascum College, the instructor asked the students to compute $\int_0^\infty \frac{e^{-x} - e^{-2x}}{x} \, dx$. One student split up the integral and made the substitution $u = 2x$ in the second part, concluding that

$$\int_0^\infty \frac{e^{-x} - e^{-2x}}{x} \, dx = \int_0^\infty \frac{e^{-x}}{x} \, dx - \int_0^\infty \frac{e^{-u}}{u} \, du = 0.$$

The instructor was not too impressed by this, pointing out that for all positive values of x, $\frac{e^{-x} - e^{-2x}}{x}$ is positive, so how could the integral be zero?

a. Resolve this paradox.

b. Eventually a student gave up and asked *Mathematica* to compute the integral, and an exact answer appeared on the screen: Log[2] (which is *Mathematica's* notation for $\ln 2$). Is *this* answer correct? (p. 282)

164. Find the smallest possible n for which there exist integers $x_1, x_2, \ldots, x_n$ such that each integer between 1000 and 2000 (inclusive) can be written as the sum, without repetition, of one or more of the integers $x_1, x_2, \ldots, x_n$. (It is not required that all such sums lie between 1000 and 2000, just that any integer between 1000 and 2000 be such a sum.) (p. 284)

165. Two d-digit integers (with first digit $\neq 0$) are chosen randomly and independently, then multiplied together. Let P_d be the probability that the first digit of the product is 9. Find $\lim_{d \to \infty} P_d$. (p. 285)

166. Define $(x_n)_{n \geq 1}$ by $x_1 = 1$, $x_{n+1} = \dfrac{1}{\sqrt{2}}\sqrt{1 - \sqrt{1 - x_n^2}}$.

a. Show that $\lim_{n \to \infty} x_n$ exists and find this limit.

b. Show that there is a unique number A for which $L = \lim_{n \to \infty} \dfrac{x_n}{A^n}$ exists as a finite nonzero number. Evaluate L for this value of A. (p. 286)

167. Consider the line segments in the xy-plane formed by connecting points on the positive x-axis with x an integer to points on the positive y-axis with y an integer. We call a point in the first quadrant an *I-point* if it is the intersection of two such line segments. We call a point an *L-point* if there is a sequence of distinct I-points whose limit is the given point. Prove or disprove: If (x, y) is an L-point, then either x or y (or both) is an integer. (p. 287)

168. a. Find all lines which are tangent to both of the parabolas

$$y = x^2 \quad \text{and} \quad y = -x^2 + 4x - 4.$$

b. Now suppose $f(x)$ and $g(x)$ are any two quadratic polynomials. Find geometric criteria that determine the number of lines tangent to both of the parabolas $y = f(x)$ and $y = g(x)$. (p. 289)

169. Suppose we are given an m-gon (polygon with m sides, and including the interior for our purposes) and an n-gon in the plane. Consider their intersection; assume this intersection is itself a polygon (other possibilities would include the intersection being empty or consisting of a line segment).

a. If the m-gon and the n-gon are convex, what is the maximal number of sides their intersection can have?

b. Is the result from (a) still correct if only one of the polygons is assumed to be convex?

(Note: A subset of the plane is *convex* if for every two points of the subset, every point of the line segment between them is also in the subset. In particular, a polygon is convex if each of its interior angles is less than $180°$.) (p. 292)

170. Suppose we start with a Pythagorean triple (a, b, c) of positive integers, that is, positive integers a, b, c such that $a^2 + b^2 = c^2$ and which can therefore be used as the side lengths of a right triangle. Show that it is not possible to have another Pythagorean triple (b, c, d) with the same integers b and c; that is, show that $b^2 + c^2$ can never be the square of an integer.

(p. 294)

171. Every year, the first warm days of summer tempt Lake Wohascum's citizens to venture out into the local parks; in fact, one day last May, the MAA Student Chapter held an impromptu picnic. A few insects were out as well, and at one point an insect dropped from a tree onto a paper plate (fortunately an empty one) and crawled off. Although this did not rank with Newton's apple as a source of inspiration, it did lead the club to wonder: If an insect starts at a random point inside a circle of radius R and crawls in a straight line in a random direction until it reaches the edge of the circle, what will be the average distance it travels to the perimeter of the circle? ("Random point" means that given two equal areas within the circle, the insect is equally likely to start in one as in the other; "random direction" means that given two equal angles with vertex at the point, the insect is equally likely to crawl off inside one as the other.) (p. 296)

172. Let $ABCD$ be a parallelogram in the plane. Describe and sketch the set of all points P in the plane for which there is an ellipse with the property that the points A, B, C, D, and P all lie on the ellipse. (p. 298)

173. Find $\lim\limits_{n \to \infty} \int_0^\infty \dfrac{n \cos\left(\sqrt[4]{x/n^2}\right)}{1 + n^2 x^2} \, dx.$ (p. 301)

174. Let x_0 be a rational number, and let $(x_n)_{n \geq 0}$ be the sequence defined recursively by

$$x_{n+1} = \left| \frac{2x_n^3}{3x_n^2 - 4} \right|.$$

Prove that this sequence converges, and find its limit as a function of x_0.

(p. 302)

175. Let f be a continuous function on $[0, 1]$, which is bounded below by 1, but is not identically 1. Let R be the region in the plane given by $0 \leq x \leq 1$, $1 \leq y \leq f(x)$. Let

$$R_1 = \{(x, y) \in R \mid y \leq \bar{y}\} \quad \text{and} \quad R_2 = \{(x, y) \in R \mid y \geq \bar{y}\},$$

where $\bar{y}$ is the y-coordinate of the centroid of R. Can the volume obtained by rotating R_1 about the x-axis equal that obtained by rotating R_2 about the x-axis? (p. 304)

176. Let $n \geq 3$ be a positive integer. Begin with a circle with n marks about it. Starting at a given point on the circle, move clockwise, skipping over the next two marks and placing a new mark; the circle now has $n + 1$ marks. Repeat the procedure beginning at the new mark. Must a mark eventually appear between each pair of the original marks? (p. 306)

177. Let

$$c = \sum_{n=1}^{\infty} \frac{1}{n(2^n - 1)} = 1 + \frac{1}{6} + \frac{1}{21} + \frac{1}{60} + \cdots.$$

Show that

$$e^c = \frac{2}{1} \cdot \frac{4}{3} \cdot \frac{8}{7} \cdot \frac{16}{15} \cdots. \qquad \text{(p. 307)}$$

178. Let $q(x) = x^2 + ax + b$ be a quadratic polynomial with real roots. Must all roots of $p(x) = x^3 + ax^2 + (b - 3)x - a$ be real? (p. 309)

179. a. For what real numbers α is $\int_0^\infty \left(\frac{\pi}{2} - \arctan(x^\alpha)\right) dx$ convergent?

b. Evaluate $\lim_{\alpha \to \infty} \int_0^\infty \left(\frac{\pi}{2} - \arctan(x^\alpha)\right) dx$. (p. 310)

180. Let $p(x) = x^3 + a_1 x^2 + a_2 x + a_3$ have rational coefficients and have roots r_1, r_2, r_3. If $r_1 - r_2$ is rational, must r_1, r_2, and r_3 be rational?

(p. 312)

181. Let $f(x) = x^3 - 3x + 3$. Prove that for any positive integer P, there is a "seed" value x_0 such that the sequence $x_0, x_1, x_2, \ldots$ obtained from Newton's method, given by
$$x_{n+1} = x_n - \frac{f(x_n)}{f'(x_n)},$$
has period P. (p. 313)

182. Show that $\displaystyle\sum_{k=0}^{n} \frac{(-1)^k}{2n+2k+1} \binom{n}{k} = \frac{\left(2^n (2n)!\right)^2}{(4n+1)!}$. (p. 315)

183. Suppose a and b are distinct real numbers such that
$$a - b, \ a^2 - b^2, \ \ldots, \ a^k - b^k, \ \ldots$$
are all integers.
a. Must a and b be rational?
b. Must a and b be integers? (p. 317)

184. The mayor of Wohascum Center has ten pairs of dress socks, ranging through ten shades of color from medium gray (1) to black (10). When he has worn all ten pairs, the socks are washed and dried together. Unfortunately, the light in the laundry room is very poor and all the socks look black there; thus, the socks get paired at random after they are removed from the drier. A pair of socks is unacceptable for wearing if the colors of the two socks differ by more than one shade.

What is the probability that the socks will be paired in such a way that all ten pairs are acceptable? (p. 318)

185. Let $p(x, y)$ be a real polynomial.
a. If $p(x, y) = 0$ for infinitely many (x, y) on the unit circle $x^2 + y^2 = 1$, must $p(x, y) = 0$ on the unit circle?
b. If $p(x, y) = 0$ on the unit circle, is $p(x, y)$ necessarily divisible by $x^2 + y^2 - 1$? (p. 320)

186. For a real number $x > 1$, we repeatedly replace x by $x - \sqrt[2011]{x}$ until the result is at most 1. Let $N(x)$ be the number of replacement steps that is needed. Determine, with proof, whether the improper integral $\displaystyle\int_1^\infty \frac{N(x)}{x^2} dx$ converges. (p. 322)

187. Find all real polynomials $p(x)$, whose roots are real, for which
$$p(x^2 - 1) = p(x)p(-x). \quad \text{(p. 324)}$$

188. Consider sequences of points in the plane that are obtained as follows: The first point of each sequence is the origin. The second point is reached from the first by moving one unit in any of the four "axis" directions (east, north, west, south). The third point is reached from the second by moving $1/2$ unit in any of the four axis directions (but not necessarily in the same direction), and so on. Thus, each point is reached from the previous point by moving in any of the four axis directions, and each move is half the size of the previous move. We call a point *approachable* if it is the limit of some sequence of the above type.

Describe the set of all approachable points in the plane. That is, find a necessary and sufficient condition for (x, y) to be approachable. (p. 325)

189. A gambling game is played as follows: D dollar bills are distributed in some manner among N indistinguishable envelopes, which are then mixed up in a large bag. The player buys random envelopes, one at a time, for one dollar and examines their contents as they are purchased. If the player can buy as many or as few envelopes as desired, and, furthermore, knows the initial distribution of the money, then for what distribution(s) will the player's expected net return be maximized? (p. 327)

190. Let $\alpha = 0.d_1 d_2 d_3 \ldots$ be a decimal representation of a real number between 0 and 1. Let r be a real number with $|r| < 1$.
a. If α and r are rational, must $\sum_{i=1}^{\infty} d_i r^i$ be rational?
b. If α and r are rational, must $\sum_{i=1}^{\infty} i d_i r^i$ be rational?
c. If r and $\sum_{i=1}^{\infty} d_i r^i$ are rational, must α be rational? (p. 329)

191. Let $\mathcal{L}_1$ and $\mathcal{L}_2$ be skew lines in space (that is, straight lines which do not lie in the same plane). How many straight lines $\mathcal{L}$ have the property that every point on $\mathcal{L}$ has the same distance to $\mathcal{L}_1$ as to $\mathcal{L}_2$? (p. 332)

192. We call a sequence $(x_n)_{n \geq 1}$ a *superinteger* if (i) each x_n is a nonnegative integer less than 10^n and (ii) the last n digits of x_{n+1} form x_n. One example of such a sequence is $1, 21, 021, 1021, 21021, 021021, \ldots$, which we abbreviate by $\ldots 21021$. Note that the digit 0 is allowed (as in the example) and that (unlike in the example) there may not be a pattern to the digits. The ordinary positive integers are just those superintegers with only finitely many nonzero digits. We can do arithmetic with superintegers; for instance, if x is the superinteger above, then the product xy of x with the superinteger $y = \ldots 66666$ is found as follows:

$1 \times 6 = 6$: the last digit of xy is 6.
$21 \times 66 = 1386$: the last two digits of xy are 86.
$021 \times 666 = 13986$: the last three digits of xy are 986.
$1021 \times 6666 = 6805986$: the last four digits of xy are 5986, etc.

Is it possible for two nonzero superintegers to have product $0 = \ldots 00000$?

(p. 334)

193. If $\sum a_n$ converges, does there have to exist a periodic function $\varepsilon : \mathbb{Z} \to \{1, -1\}$ such that $\sum \varepsilon(n)|a_n|$ converges? (p. 335)

194. Let $f(x) = x - 1/x$. For any real number x_0, consider the sequence defined by $x_0, x_1 = f(x_0), \ldots, x_{n+1} = f(x_n), \ldots$, provided $x_n \neq 0$. Define x_0 to be a *T-number* if the sequence terminates, that is, if $x_n = 0$ for some n. (For example, -1 is a T-number because $f(-1) = 0$, but $\sqrt{2}$ is not, because the sequence

$$\sqrt{2}, \ 1/\sqrt{2} = f(\sqrt{2}), \ -1/\sqrt{2} = f(1/\sqrt{2}), \ 1/\sqrt{2} = f(-1/\sqrt{2}), \ \ldots$$

does not terminate.)

a. Show that the set of all T-numbers is countably infinite (denumerable).
b. Does every open interval contain a T-number? (p. 336)

195. For n a positive integer, show that the number of integral solutions (x, y) of $x^2 + xy + y^2 = n$ is finite and a multiple of 6. (p. 338)

196. For what real numbers x can one say the following?
a. For each positive integer n, there exists an integer m such that

$$\left| x - \frac{m}{n} \right| < \frac{1}{3n}.$$

b. For each positive integer n, there exists an integer m such that

$$\left| x - \frac{m}{n} \right| \leq \frac{1}{3n}.$$

(p. 341)

197. Starting with an empty $1 \times n$ board (a row of n squares), we successively place 1×2 dominoes to cover two adjacent squares. At each stage, the placement of the new domino is chosen at random, with all available pairs of adjacent empty squares being equally likely. The process continues until no further dominoes can be placed. Find the limit, as $n \to \infty$, of the expected fraction of the board that is covered when the process ends. (p. 342)

198. Let $\mathbb{Z}/n\mathbb{Z}$ be the set $\{0, 1, \ldots, n-1\}$ with addition modulo n. Consider subsets S_n of $\mathbb{Z}/n\mathbb{Z}$ such that $(S_n + k) \cap S_n$ is nonempty for every k in $\mathbb{Z}/n\mathbb{Z}$. Let $f(n)$ denote the minimal number of elements in such a subset. Find
$$\lim_{n \to \infty} \frac{\ln f(n)}{\ln n},$$
or show that this limit does not exist. (p. 344)

199. a. If a rational function (a quotient of two real polynomials) takes on rational values for infinitely many rational numbers, prove that it may be expressed as the quotient of two polynomials with rational coefficients.

b. If a rational function takes on integral values for infinitely many integers, prove that it must be a polynomial with rational coefficients. (p. 345)

200. Can there be a multiplicative $n \times n$ magic square ($n > 1$) with entries $1, 2, \ldots, n^2$? That is, does there exist an integer $n > 1$ for which the numbers $1, 2, \ldots, n^2$ can be placed in a square so that the product of all the numbers in any row or column is always the same? (p. 348)

201. Define a function f by $f(x) = x^{1/x^{x^{1/x^{x^{1/x^{\cdots}}}}}}$ ($x > 0$). That is to say, for a fixed x, let

$$a_1 = x, \quad a_2 = x^{1/x}, \quad a_3 = x^{1/x^x} = x^{1/x^{a_1}}, \quad a_4 = x^{1/x^{x^{1/x}}} = x^{1/x^{a_2}}, \quad \ldots$$

and, in general, $a_{n+2} = x^{1/x^{a_n}}$, and take $f(x) = \lim_{n \to \infty} a_n$.

a. Assuming that this limit exists, let M be the maximum value of f as x ranges over all positive real numbers. Evaluate M^M.

b. Prove that $f(x)$ is well defined; that is, that the limit exists. (p. 350)

202. Note that if the edges of a regular octahedron have length 1, then the distance between any two of its vertices is either 1 or $\sqrt{2}$. Are there other configurations of six points in $\mathbb{R}^3$ for which the distance between any two of the points is either 1 or $\sqrt{2}$? If so, find them. (p. 353)

203. Let a and b be positive real numbers, and define a sequence (x_n) by

$$x_0 = a, \; x_1 = b, \; x_{n+1} = \frac{1}{2}\left(\frac{1}{x_n} + x_{n-1}\right).$$

a. For what values of a and b will this sequence be periodic?

b. Show that given a, there exists a unique b for which the sequence converges. (p. 357)

204. Consider the equation $x^2 + \cos^2 x = \alpha \cos x$, where α is some positive real number.

a. For what value or values of α does the equation have a unique solution?

b. For how many values of α does the equation have precisely four solutions? (p. 361)

205. Fast Eddie needs to double his money; he can only do so by playing a certain win-lose game, in which the probability of winning is p. However, he can play this game as many or as few times as he wishes, and in a particular game he can bet any desired fraction of his bankroll. The game pays even money (the odds are one-to-one). Assuming he follows an optimal strategy if one is available, what is the probability, as a function of p, that Fast Eddie will succeed in doubling his money? (p. 364)

206. Define a *die* to be a convex polyhedron. For what n is there a fair die with n faces? By fair, we mean that, given any two faces, there exists a symmetry of the polyhedron which takes the first face to the second. (p. 366)

207. Prove that

$$\det \begin{pmatrix} 1 & 4 & 9 & \cdots & n^2 \\ n^2 & 1 & 4 & \cdots & (n-1)^2 \\ \vdots & \vdots & \vdots & \vdots & \vdots \\ 4 & 9 & 16 & \cdots & 1 \end{pmatrix}$$

$$= (-1)^{n-1} \frac{n^{n-2}(n+1)(2n+1)\big((n+2)^n - n^n\big)}{12}.$$

(p. 368)

208. Let $a_0 = 0$ and let a_n be equal to the smallest positive integer that cannot be written as a sum of n numbers (with repetitions allowed) from among $a_0, a_1, \ldots, a_{n-1}$. The sequence (a_n) starts off with

$$a_0 = 0, \quad a_1 = 1, \quad a_2 = 3, \quad a_3 = 8, \quad a_4 = 21.$$

Find a formula for a_n. (p. 371)

THE SOLUTIONS

Problem 1

Find all solutions in integers of $x^3 + 2y^3 = 4z^3$.

Answer. The only solution is $(x, y, z) = (0, 0, 0)$.

Solution 1. First note that if (x, y, z) is a solution in integers, then x must be even, say $x = 2w$. Substituting this, dividing by 2, and subtracting y^3, we see that $(-y)^3 + 2z^3 = 4w^3$, so $(-y, z, w) = (-y, z, x/2)$ is another solution. Now repeat this process to get (x, y, z), $(-y, z, x/2)$, $(-z, x/2, -y/2)$, $(-x/2, -y/2, -z/2)$ as successive integer solutions. Conclusion: If (x, y, z) is a solution, then so is $(-x/2, -y/2, -z/2)$, and in particular, x, y, and z are all even. But if x, y, and z were not all zero, we could keep replacing (x, y, z) by $(-x/2, -y/2, -z/2)$ and eventually arrive at a solution containing an odd integer, a contradiction.

Solution 2. Suppose (x, y, z) is a nonzero solution for which $|x|^3 + 2|y|^3 + 4|z|^3$ is minimized. Clearly x is even, say $x = 2w$. We have $8w^3 + 2y^3 = 4z^3$ or $(-y)^3 + 2z^3 = 4w^3$. Then $(-y, z, w)$ is a nonzero solution and

$$|-y|^3 + 2|z|^3 + 4|w|^3 = \frac{|x|^3 + 2|y|^3 + 4|z|^3}{2},$$

a contradiction. Therefore $(0, 0, 0)$ is the only solution.

45

Comment. This solution method, in which it is shown that any solution gives rise to a "smaller" solution, is known as the *method of infinite descent*. It was introduced by Pierre de Fermat (1601–1665), who used intricate cases of the method to solve diophantine equations that were beyond the reach of any of his contemporaries.

Problem 2

Each of twenty commissioners ranked three candidates (A, B, and C) in order of preference, with no abstentions. It was found that 11 commissioners preferred A to B, 12 preferred C to A, but 14 preferred B to C. Given that every possible order of A, B, and C appeared on at least one ballot, how many commissioners voted for B as their first choice?

Solution. Eight commissioners voted for B. To see this, we will use the given information to study how many voters chose each order of A, B, C.

The six orders of preference are $ABC, ACB, BAC, BCA, CAB, CBA$; assume they receive a, b, c, d, e, f votes respectively. We know that

$$a + b + e = 11 \quad (A \text{ over } B) \tag{1}$$
$$d + e + f = 12 \quad (C \text{ over } A) \tag{2}$$
$$a + c + d = 14 \quad (B \text{ over } C). \tag{3}$$

Because 20 votes were cast, we also know that

$$c + d + f = 9 \quad (B \text{ over } A) \tag{4}$$
$$a + b + c = 8 \quad (A \text{ over } C) \tag{5}$$
$$b + e + f = 6 \quad (C \text{ over } B). \tag{6}$$

There are many ways one might proceed. For example, equations (3) and (4) imply that $a = 5 + f$, and equations (1) and (5) imply that $e = c + 3$. Substituting these into (1) yields $f + b + c = 3$, and therefore $b = c = f = 1$. It follows that $a = 6$, $e = 4$, and $d = 7$.

The number of commissioners voting for B as their first choice is therefore $c + d = 1 + 7 = 8$ (A has 7 first place votes, and C has 5).

Comments. The answer to this question would have been the same had we known only that *at least* 14 commissioners preferred B over C.

The seemingly paradoxical nature of the commissioners' preferences (A preferred to B, B preferred to C, C preferred to A), an example of "Condorcet's paradox," is not uncommon when individual choices are pooled.

Problem 3

If $A = (0, -10)$ and $B = (2, 0)$, find the point(s) C on the parabola $y = x^2$ which minimizes the area of triangle ABC.

Answer. The area of triangle ABC is minimized when $C = (5/2, 25/4)$.

Solution 1. The area of triangle ABC is half the product of the length of AB, which is fixed, and the length of the altitude from C to AB, which varies with C. The length of this altitude is the distance between C and the line AB. Note that the line AB, given by $y = 5x - 10$, does not intersect the parabola $y = x^2$. Thus, for $C = (x, x^2)$ on the parabola to minimize the area, the tangent line at C must be parallel to AB. This occurs when $2x = 5$, or $C = (5/2, 25/4)$.

Solution 2. (Gerald A. Heuer, Concordia College) For a straightforward solution based on the properties of the cross product, recall that

$$\text{Area } ABC = \frac{1}{2}\left|\overrightarrow{AB} \times \overrightarrow{AC}\right|.$$

Now

$$\overrightarrow{AB} \times \overrightarrow{AC} = \det\begin{pmatrix} \mathbf{i} & \mathbf{j} & \mathbf{k} \\ 2 & 10 & 0 \\ x & x^2+10 & 0 \end{pmatrix} = 2(x^2 - 5x + 10)\mathbf{k}.$$

Thus, Area $ABC = x^2 - 5x + 10 = (x - 5/2)^2 + 15/4$, and this is minimized when $x = 5/2$.

Solution 3. (Les Nelson, St. Olaf College) The transformation (a shear followed by a vertical translation)

$$\begin{pmatrix} X \\ Y \end{pmatrix} = \begin{pmatrix} 1 & 0 \\ -5 & 1 \end{pmatrix}\begin{pmatrix} x \\ y \end{pmatrix} + \begin{pmatrix} 0 \\ 10 \end{pmatrix}$$

is area-preserving because

$$\det\begin{pmatrix} 1 & 0 \\ -5 & 1 \end{pmatrix} = 1,$$

and it takes $A = (0, -10)$ to $A' = (0, 0)$, $B = (2, 0)$ to $B' = (2, 0)$, and $C = (x, x^2)$ to $C' = (x, x^2 - 5x + 10)$. Therefore,

$$\text{Area } ABC = \text{Area } A'B'C' = x^2 - 5x + 10 = (x - 5/2)^2 + 15/4,$$

which is minimized when $x = 5/2$.

Problem 4

Does there exist a continuous function $y = f(x)$, defined for all real x, whose graph intersects every non-vertical line in infinitely many points?

Solution. Yes, there is such a function; an example is $f(x) = x^2 \sin x$. The graph of this function oscillates between the graph of $y = x^2$ (which it intersects when $\sin x = 1$, that is, when $x = \pi/2 + 2k\pi$) and the graph of $y = -x^2$ (which it intersects at $x = 3\pi/2 + 2k\pi$).

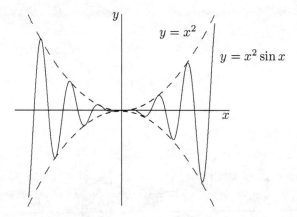

Suppose $y = mx + b$ is the equation of a non-vertical line. Because

$$\lim_{x \to \infty} \frac{mx + b}{x^2} = 0,$$

we know that for x large enough

$$\left| \frac{mx + b}{x^2} \right| < 1,$$

or equivalently $-x^2 < mx + b < x^2$. Therefore, the line $y = mx + b$ will intersect the graph $y = x^2 \sin x$ in each interval $(\pi/2 + 2k\pi, 3\pi/2 + 2k\pi)$ for sufficiently large integers k.

Problem 5

A child on a pogo stick jumps 1 foot on the first jump, 2 feet on the second jump, 4 feet on the third jump, ..., 2^{n-1} feet on the nth jump. Can the child get back to the starting point by a judicious choice of directions?

Solution. No. The child will always "overshoot" the starting point, because after n jumps the distance from the starting point to the child's location is at most $1 + 2 + 4 + \ldots + 2^{n-1} = 2^n - 1$ feet. Since the $(n+1)$st jump is 2^n feet, it can never return the child exactly to the starting point.

Problem 6

Let $S_n = \{1, n, n^2, n^3, \ldots\}$, where n is an integer greater than 1. Find the smallest number $k = k(n)$ such that there is a number which may be expressed as a sum of k (possibly repeated) elements of S_n in more than one way (rearrangements are considered the same).

Answer. We will show that $k(n) = n + 1$.

Solution 1. Suppose we have $a_1, a_2, \ldots, a_k$ and $b_1, b_2, \ldots, b_k$ in S_n such that

$$a_1 + a_2 + \cdots + a_k = b_1 + b_2 + \cdots + b_k,$$

where $a_i \leq a_{i+1}$, $b_i \leq b_{i+1}$, and, for some i, $a_i \neq b_i$. If k is the smallest integer for which such a sum exists, then clearly $a_i \neq b_j$ for all i, j, since otherwise we could just drop the equal terms from the sums. By symmetry, we may now assume $a_1 < b_1$. Upon dividing both sides of the equation by a_1, we may assume $a_1 = 1$. Since, at this point, every b_j is divisible by n, there must be at least n 1's on the left side of the equation. If we had $k = n$, then the right side would be strictly greater than the left. Therefore, $k \geq n + 1$. The expression

$$\underbrace{1 + 1 + \cdots + 1}_{n \text{ times}} + n^2 = \underbrace{n + n + \cdots + n}_{n+1 \text{ times}}$$

shows that $k = n + 1$.

Solution 2. First, $k(n) \leq n+1$ because

$$\underbrace{1+1+\cdots+1}_{n \text{ times}}+n^2 = \underbrace{n+n+\cdots+n}_{n+1 \text{ times}}.$$

Next, $k(n) \geq n$. To see this, recall that every positive integer N has a *unique* base n representation; that is, there are *unique* "digits" $d_0, d_1, \ldots, d_s$, $0 \leq d_i < n$, such that

$$N = d_0 + d_1 n + d_2 n^2 + \cdots + d_s n^s$$
$$= \underbrace{1+\cdots+1}_{d_0}+\underbrace{n+\cdots+n}_{d_1}+\underbrace{n^2+\cdots+n^2}_{d_2}+\cdots+\underbrace{n^s+\cdots+n^s}_{d_s}.$$

This means that if N has a second representation as a sum of elements from S_n, at least one of the elements of S_n must occur at least n times. Thus $k(n) \geq n$.

Finally, $k(n) \neq n$. For suppose $k(n) = n$ and that M is an integer with two different representations,

$$M = s_1 + \cdots + s_n = t_1 + \cdots + t_n, \qquad s_i, t_i \in S_n.$$

By the uniqueness of base n representation, at least one of these representations, say the left side, is *not* the base n representation. As argued in the preceding paragraph, all of the s_i must be equal, say to n^s. Thus, $M = n^{s+1}$. But this means that the right side is not the base n representation of M either, so again, all of the t_i are equal, say to n^t. It follows that $M = n^{s+1} = n^{t+1}$ and therefore $s = t$, $s_1 = \cdots = s_n = t_1 = \cdots = t_n$. This contradicts our assumption that the representations are different.

Thus, $n < k(n) \leq n+1$, so $k(n) = n+1$.

Problem 7

Find all integers a for which $x^3 - x + a$ has three integer roots.

Answer. The only integer a for which $x^3 - x + a$ has three integer roots is $a = 0$. The roots are then $-1, 0, 1$.

Solution 1. The factorization $x^3 - x = (x-1)x(x+1)$ clearly implies $x^3 - x$ is strictly increasing for $x \geq 1$ and also for $x \leq -1$. (One could prove this using calculus, as well.) Thus, the polynomial $x^3 - x + a$ can have

THE SOLUTIONS 51

at most one positive integral root and one negative integral root. A third integral root must then be 0, hence $a = 0$.

Solution 2. (Suggested by students at the 1990 United States Mathematical Olympiad training program) Let r_1, r_2, r_3 be the integral roots of $x^3 - x + a$. Writing the coefficients of the polynomial as symmetric functions of its roots, we have

$$r_1 + r_2 + r_3 = 0 \quad \text{and} \quad r_1 r_2 + r_1 r_3 + r_2 r_3 = -1.$$

Combining these yields

$$r_1^2 + r_2^2 + r_3^2 = (r_1 + r_2 + r_3)^2 - 2(r_1 r_2 + r_1 r_3 + r_2 r_3) = 2.$$

We conclude that one of the roots is 0, forcing $a = 0$, and the other two roots are -1 and 1.

Solution 3. Let r_1, r_2, r_3 be the roots of $x^3 - x + a$. The discriminant of this cubic is

$$\Big((r_1 - r_2)(r_1 - r_3)(r_2 - r_3)\Big)^2 = -4(-1)^3 - 27a^2 = 4 - 27a^2.$$

If all three roots are integral, then the above discriminant is a perfect square, say s^2. Then $4 = 27a^2 + s^2$, which implies $a = 0$.

Problem 8

Three loudspeakers are placed so as to form an equilateral triangle. One of the speakers plays four times as loudly as the other two. Assuming that the sound level from a speaker is inversely proportional to the square of the distance to that speaker (and that the sound levels depend only on the distance to the speakers), where should one sit so that each speaker will be heard at the same sound level?

Solution 1. The listener should sit at the reflection of the center of the triangle in the side opposite the louder speaker.

Because the sound level for a speaker is inversely proportional to the square of the distance from that speaker, we seek a point twice as far from the louder speaker as from the other two speakers. Recall that the intersection of the medians of a triangle occurs at the point two-thirds of the way from a vertex of the triangle to the midpoint of the opposite side. Furthermore, in an equilateral triangle, each median is perpendicular to its corresponding side. Thus, the reflection of the center of the triangle in the side opposite the louder speaker is twice as far from that speaker as is the center of the triangle. On the other hand, noting that the reflection of a given point in a line has the same distance to any point on the line as the given point, we see that the reflected point and the center of the triangle are equidistant from the two other speakers. We have found a point where the sound levels from the three speakers are equal.

Solution 2. Let the louder speaker be at point A, and the other two speakers be at B and C. Then the desired point is the intersection of the perpendicular to AB at B and the perpendicular bisector of BC.

We again begin with the observation that we want a point P for which $PB = PC = \frac{1}{2}PA$. The set of points equidistant from B and C is the perpendicular bisector of the line segment BC, which passes through A since $AB = AC$. If P is the intersection of the perpendicular to AB at B and the perpendicular bisector of BC, clearly $PB = PC$. In addition, the triangle APB is a 30°–60°–90° triangle, hence $PB = \frac{1}{2}PA$, as required.

It is now easy to go on to show that P is the only such point. Suppose Q is on the line PA. If A is between P and Q, then $QA < QB = QC$. If Q is on the same side of A as P, then $QB \geq QA \sin 30° = \frac{1}{2}QA$, with equality if and only if the lines QB and AB are perpendicular.

Solution 3. If the identical speakers are at the points $(-1, 0)$ and $(1, 0)$ in the plane, and the louder speaker is at $(0, \sqrt{3})$, then the unique point we seek is at $(0, -1/\sqrt{3})$.

Since the points $(-1, 0)$, $(1, 0)$, and $(0, \sqrt{3})$ form the vertices of an equilateral triangle, there is no loss of generality in choosing coordinates so that the speakers are at these points. By looking at the squares of the distances from a point (x, y) to these points, we arrive at the condition

$$(x+1)^2 + y^2 = (x-1)^2 + y^2 = \tfrac{1}{4}\left(x^2 + (y-\sqrt{3})^2\right).$$

The first equality implies $x = 0$, which we substitute into the second equality, obtaining

$$1 + y^2 = \tfrac{1}{4}\left(y^2 - 2\sqrt{3}y + 3\right).$$

After combining terms, we find that $y = -1/\sqrt{3}$ is the only solution to this last equation. Thus, the only point where the sound levels from the three speakers are equal is $(0, -1/\sqrt{3})$.

Problem 9

Ten (not necessarily all different) integers have the property that if all but one of them are added, the possible results (depending on which one is omitted) are: 82, 83, 84, 85, 87, 89, 90, 91, 92. What are the ten integers?

Solution. The integers are 5, 6, 7, 7, 8, 10, 12, 13, 14, 15.

Let $n_1 \leq n_2 \leq \cdots \leq n_{10}$ be the ten integers and $S = n_1 + n_2 + \cdots + n_{10}$ be their sum. If all but n_1 are added, the result is $S - n_1$; similarly, if all but n_2 are added, the result is $S - n_2$; and so forth. If the ten results are added, we get $(S - n_1) + (S - n_2) + \cdots + (S - n_{10}) = 10S - S = 9S$. Now we are given that there are only nine possible different results: 82, 83, 84, 85, 87, 89, 90, 91, 92. Let x denote the sum that occurs twice. Then the sum of the ten results is $82 + 83 + 84 + 85 + 87 + 89 + 90 + 91 + 92 + x$; by the above, this sum is also $9S$. Therefore, $783 + x = 9S$; equivalently, $x = 9(S - 87)$. This shows that x must be divisible by 9. However, of the given results 82, 83, ..., 92, only 90 is divisible by 9, so $x = 90$, and $S = 97$. Subtracting 82, 83, ..., 92, with 90 occurring twice, from 97, we find the ten integers listed above.

Problem 10

For what positive values of x does the infinite series $\sum_{n=1}^{\infty} x^{\log_{2012} n}$ converge?

Answer. For $0 < x < \frac{1}{2012}$.

Solution. Note that
$$x^{\log_{2012} n} = x^{\ln n / \ln 2012} = e^{(\ln x)(\ln n)/\ln 2012} = n^{\ln x / \ln 2012}.$$
Therefore, the series is a "p-series" $\sum_{n=1}^{\infty} \frac{1}{n^p}$ with $p = -\ln x / \ln 2012$, so it converges if and only if $-\ln x / \ln 2012 > 1$. This condition is equivalent to $\ln x < -\ln 2012 = \ln \frac{1}{2012}$, or $0 < x < \frac{1}{2012}$.

Problem 11

Let A be a 4×4 matrix such that each entry of A is either 2 or -1. Let $d = \det(A)$; clearly, d is an integer. Show that d is divisible by 27.

Solution. Let B be the matrix obtained from A by subtracting row one of A from each of the other three rows. Then $\det A = \det B$. Each entry in the last three rows of B is -3, 0, or 3, and therefore is divisible by 3. Now let C be the matrix obtained from B by dividing all these entries (in the last three rows) by 3. Then all entries of C are integers, so $\det C$ is an integer; moreover, $\det A = \det B = 3^3 \det C$, so $\det A$ is divisible by 27.

Problem 12

Consider the $n \times n$ array whose entry in the ith row, jth column is $i + j - 1$. What is the smallest product of n numbers from this array, with one coming from each row and one from each column?

Solution. The smallest product is $1 \cdot 3 \cdot 5 \cdots (2n-1)$. This value of the product occurs when the n numbers are on the main diagonal of the array (positions $(1,1), (2,2), \ldots, (n,n)$).

Because there are only finitely many possible products, there is a smallest product. We show this could not occur for any other choice of numbers.

Assume $n \geq 2$, and suppose we choose the n numbers from row i, column i', $i = 1, 2, \ldots, n$. Unless $i' = i$ for all i, there will be some i and j, $i < j$, for which $i' > j'$. If we choose entries (i, j') and (j, i') instead of (i, j) and (i', j'), keeping the other $n - 2$ entries the same, the product of the two entries will be $(i + j' - 1)(j + i' - 1)$ instead of $(i + i' - 1)(j + j' - 1)$. Since

$$(i + j' - 1)(j + i' - 1) - (i + i' - 1)(j + j' - 1) = (j - i)(j' - i') < 0,$$

the new product will be smaller, so we are done.

Problem 13

Let A and B be sets with the property that there are exactly 144 sets which are subsets of at least one of A or B. How many elements does the union of A and B have?

Answer. There are exactly 8 elements in the union of A and B.

Solution 1. Let a denote the number of elements in A but not in B, b the number of elements in B but not in A, and c the number of elements in both A and B.

THE SOLUTIONS

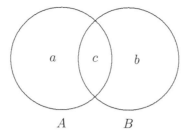

The number of subsets of A is 2^{a+c}, of B is 2^{b+c}, and of both A and B is 2^c. The collection of all subsets of A together with the collection of all subsets of B will contain duplications, namely, those subsets that are subsets of both A and B. Taking this into account, we have

$$2^{a+c} + 2^{b+c} - 2^c = 144$$
$$2^c(2^a + 2^b) = 2^c + 2^4 3^2$$
$$2^a + 2^b = 1 + 2^{4-c} 3^2. \qquad (*)$$

If $a = 0$ then $2^b = 2^{4-c} 3^2$, or equivalently, $2^{b+c-4} = 3^2$ which is impossible, regardless of the value of b or c. Similarly, we get a contradiction if $b = 0$. Therefore, $a > 0$ and $b > 0$, so the left side of $(*)$ is even. Hence $c = 4$ and $2^a + 2^b = 10$. The only solutions to this are $a = 1, b = 3$ or $a = 3, b = 1$. In either case the number of elements in the union is $a+b+c = 1+3+4 = 8$.

Solution 2. We may assume that $|A| \geq |B|$. The number of subsets of a set with c elements is 2^c. Thus,

$$2^{|A|} \leq 144 < 2^{|A|} + 2^{|B|} \leq 2^{|A|+1}.$$

Hence, $|A| = 7$. The 16 missing subsets must be subsets of B but not of A. Thus,

$$16 = 2^{|B|} - 2^{|A \cap B|} = 2^{|A \cap B|}(2^{|B|-|A \cap B|} - 1).$$

The only possibility is $|A \cap B| = 4$ and $|B| - |A \cap B| = 1$, so

$$|A \cup B| = |A| + |B| - |A \cap B| = 7 + 1 = 8.$$

Problem 14

Find $\int (x^6 + x^3) \sqrt[3]{x^3 + 2}\, dx$.

Answer.

$$\int (x^6 + x^3) \sqrt[3]{x^3 + 2}\, dx = \frac{1}{8}(x^6 + 2x^3)^{4/3} + C = \frac{1}{8}(x^7 + 2x^4)\sqrt[3]{x^3 + 2} + C.$$

Solution 1. We start by bringing a factor x inside the cube root:

$$\int (x^6 + x^3)\sqrt[3]{x^3 + 2}\, dx = \int (x^5 + x^2)\sqrt[3]{x^6 + 2x^3}\, dx.$$

The substitution $u = x^6 + 2x^3$ now yields the answer.

Solution 2. Setting $u = x^3 + 2$, we find

$$du = 3x^2\, dx, \qquad dx = \frac{du}{3(u-2)^{2/3}}.$$

Thus,

$$\int (x^6 + x^3)\sqrt[3]{x^3 + 2}\, dx = \int \frac{(u-2)(u-1)u^{1/3}}{3(u-2)^{2/3}}\, du$$

$$= \frac{1}{3}\int (u^2 - 2u)^{1/3}(u - 1)\, du$$

$$= \frac{1}{6}\left(\frac{(u^2 - 2u)^{4/3}}{4/3}\right) + C$$

$$= \frac{1}{8}(x^6 + 2x^3)^{4/3} + C.$$

Solution 3. (Eugene Luks, University of Oregon) Starting with the term of highest degree and integrating by parts, we get

$$\int x^6 \sqrt[3]{x^3 + 2}\, dx = \int x^4 \left(x^2 \sqrt[3]{x^3 + 2}\right) dx$$

$$= \frac{1}{4} x^4 (x^3 + 2)^{4/3} - \int x^3 (x^3 + 2)^{4/3}\, dx$$

$$= \frac{1}{4} x^4 (x^3 + 2)^{4/3} - \int (x^6 + 2x^3)\sqrt[3]{x^3 + 2}\, dx.$$

Combining the integrals then yields

$$\int (x^6 + x^3)\sqrt[3]{x^3 + 2}\, dx = \frac{1}{8} x^4 (x^3 + 2)^{4/3} + C.$$

Solution 4. (George Andrews, Pennsylvania State University) Given that we can find an antiderivative, we expect it to have the form $p(x)\sqrt[3]{x^3+2}$ for some polynomial $p(x)$. Thus we look for a polynomial

$$p(x) = a_0 + a_1 x + a_2 x^2 + a_3 x^3 + \cdots$$

for which

$$\frac{d}{dx}\left(p(x)\sqrt[3]{x^3+2}\right) = (x^6+x^3)\sqrt[3]{x^3+2}.$$

This leads to the differential equation

$$(x^3+2)p'(x) + x^2 p(x) = (x^6+x^3)(x^3+2),$$

that is,

$$(x^3+2)(a_1+2a_2x+3a_3x^2+\cdots)+(a_0x^2+a_1x^3+a_2x^4+\cdots) = x^9+3x^6+2x^3.$$

Comparing coefficients of x^n, $n \geq 0$, and using the convention that $a_j = 0$ for $j < 0$, we find that

$$(2n+2)a_{n+1} + (n-1)a_{n-2} = \begin{cases} 2 & \text{if } n = 3, \\ 3 & \text{if } n = 6, \\ 1 & \text{if } n = 9, \\ 0 & \text{otherwise.} \end{cases}$$

This is equivalent to

$$a_{n+3} = -\frac{n+1}{2n+6}a_n + \frac{\delta_n}{2n+6}, \quad \text{where } \delta_n = \begin{cases} 2 & \text{if } n = 1, \\ 3 & \text{if } n = 4, \\ 1 & \text{if } n = 7, \\ 0 & \text{otherwise.} \end{cases}$$

In particular, for $p(x)$ to be a polynomial, we must have $a_0 = a_3 = a_6 = \cdots = 0$ and $a_2 = a_5 = a_8 = \cdots = 0$, while from $a_1 = 0$ we get $a_4 = 1/4$, $a_7 = 1/8$, $a_{10} = a_{13} = a_{16} = \cdots = 0$. Thus,

$$\int (x^6+x^3)\sqrt[3]{x^3+2}\,dx = \left(\frac{1}{8}x^7 + \frac{1}{4}x^4\right)\sqrt[3]{x^3+2} + C.$$

Comment. Solution 4 illustrates the *Risch algorithm*, which finds, when possible, antiderivatives of elementary functions.

Problem 15

Two microfilms are hidden in two dark squares of a chessboard; the squares are adjacent to opposite corners and symmetric to each other about the center of the board. The chessboard is a collapsible variety which folds along both center lines. Suppose the four quarters of the chessboard come apart and then are reassembled in a random manner (but so as to maintain the checkerboard pattern of light and dark squares). What is the probability that the two squares which hide the microfilms are again adjacent to opposite corners?

Solution. The probability is 1/24.

When the board is reassembled so there is a light square in the lower right-hand corner, there are eight possible positions that can have a film (indicated by black dots in the figure). Of these eight positions, two are adjacent to a corner, so there is a probability of $2/8 = 1/4$ that a given film will end up adjacent to a corner. If this happens, there are six possibilities remaining for the second film since the two films cannot end up in the same "quadrant" of the board. Of these six possibilities, only one is adjacent to a corner. Thus the probability of both films ending up in acceptable positions is $(1/4)(1/6) = 1/24$.

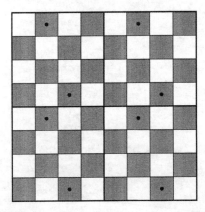

Problem 16

For what values of n is it possible to split up the (entire) set $\{1, 2, \ldots, n\}$ into three (disjoint) subsets so that the sum of the integers in each of the subsets is the same?

Answer. It can be done if and only if $n \geq 5$ and $n \equiv 0$ or 2 modulo 3.

Solution 1. A necessary condition is that the sum of the elements in $\{1, 2, \ldots, n\}$ be divisible by 3. In other words, $n(n + 1)/2$ must be divisible by 3, which is true if and only if $n \equiv 0$ or 2 modulo 3. Thus, the set cannot be split up as desired if $n \equiv 1$ modulo 3. Also, it's easy to see that the sets $\{1, 2\}$ and $\{1, 2, 3\}$ cannot be split up as desired.

To see that it can be done for $n \geq 5$ if $n \equiv 0$ or 2 modulo 3, check the first few cases:

$$n = 5: \quad \{1, 2, 3, 4, 5\} = \{1, 4\} \cup \{2, 3\} \cup \{5\};$$
$$n = 6: \quad \{1, 2, 3, 4, 5, 6\} = \{1, 6\} \cup \{2, 5\} \cup \{3, 4\};$$
$$n = 8: \quad \{1, 2, 3, 4, 5, 6, 7, 8\} = \{1, 2, 3, 6\} \cup \{4, 8\} \cup \{5, 7\};$$
$$n = 9: \quad \{1, 2, 3, 4, 5, 6, 7, 8, 9\} = \{1, 2, 3, 4, 5\} \cup \{6, 9\} \cup \{7, 8\}.$$

Now we can use an inductive argument. Suppose we can do it for n. Then to do it for $n+6$, just add $n+1, n+6$ to the first subset for n, add $n+2, n+5$ to the second subset for n, and add $n+3, n+4$ to the third subset for n. Since to each subset we're adding numbers that sum to $(n+1) + (n+6) = (n+2) + (n+5) = (n+3) + (n+4) = 2n+7$, the three new subsets will all add to the same number, and between them they will account for all integers in the new set $\{1, 2, \ldots, n, n+1, n+2, n+3, n+4, n+5, n+6\}$. Because it can be done for $n = 5, 6, 8$ and 9, it can also be done for $n = 11, 17, 23, \ldots$, for $n = 12, 18, 24, \ldots$, for $n = 14, 20, 26, \ldots$, and for $n = 15, 21, 27, \ldots$, and our proof is complete.

Solution 2. As we've seen in the first solution, we may assume that $n \geq 5$ and that $n \equiv 0$ or 2 modulo 3. We will partition $S = \{1, 2, \ldots, n\}$ by applying a "greedy" strategy, as illustrated here for $n = 21$. In this case, each subset must sum to $\frac{1}{3} \cdot \left(\frac{1}{2} \cdot (21 \cdot 22) \right) = 77$. Begin by choosing the largest possible numbers from S: start with 21, then adjoin 20, then 19 (our total is now 60) and the largest number we can choose is 17, so let $A = \{21, 20, 19, 17\}$. Now construct B by, once again, choosing the largest possible among the remaining elements; this yields $B = \{18, 16, 15, 14, 13, 1\}$. The remaining elements, which will automatically add to 77, compose the third set: $C = \{12, 11, 10, 9, 8, 7, 6, 5, 4, 3, 2\}$.

For a general n, let $s = n(n+1)/6$ denote the common sum of the three subsets. The choices for the first set begin with n, then $n{-}1$, then $n{-}2$, and so forth, until we get to m, where $a = m+(m{+}1)+\cdots+(n{-}1)+n$ is such that $a \leq s < a+(m{-}1)$. If $a = s$, then our first set is $A = \{n, n{-}1, \ldots, m\}$; if $a < s$, then $A = \{n, n{-}1, \ldots, m, r\}$, where $r = s{-}a$. Continue with the construction for $B : \{m-1,\ldots,\hat{r},\ldots,k\}$, where we use the notation $\hat{r}$ to mean that r is not chosen if it is between k and $m-1$ (it has already been put into set A), and where k is such that $b = k + (k+1) + \cdots + \hat{r} + \cdots + (m-1)$ satisfies $b \leq s < b + (k-1)$. If $b = s$ then our second set is $B = \{m-1,\ldots,\hat{r},\ldots,k\}$. If $s > b$ then set $t = s - b$; note that $t < \left(b + (k-1)\right) - b = k-1$. There are two cases.

Case 1. $t \neq r$: In this case, t has not yet been chosen as an element of either A or B, so we can take $B = \{m-1,\ldots,\hat{r},\ldots,k,t\}$.

Case 2. $t = r$: Since $r = t < k-1$, we know that $t-1$ has not yet been chosen for A or B, so if $r \geq 3$ we can take $B = \{m-1,\ldots k, t-1, 1\}$. If $r = 2$, then in A, replace m with $m-1$ and 2 with 3, so that $A = \{n, n-1, \ldots, m+1, m-1, 3\}$, and in B (from Case 1), replace $m-1$ with m and 2 with 1, so that $B = \{m, m-2, \ldots, k, 1\}$. Subsets A and B are disjoint and both sum to s. Finally, if $r = 1$, do a similar thing: change A to $\{n, n-1, \ldots, m+1, m-1, 2\}$, and B to $\{m, m-2, \ldots, k\}$. Once again, A and B are disjoint and both sum to s.

Finally, we take C to consist of the remaining elements in S, and we are done.

Problem 17

Let $f(x)$ be a positive, continuously differentiable function, defined for all real numbers, whose derivative is always negative. For any real number x_0, will the sequence (x_n) obtained by Newton's method ($x_{n+1} = x_n - f(x_n)/f'(x_n)$) always have limit ∞?

Solution. Yes. To see why, note that since $f'(x)$ is always negative while $f(x)$ is always positive, the sequence (x_n) is increasing. Therefore, if the sequence did not have limit ∞, it would be bounded and have a finite limit L. But then taking the limit of both sides of the equation

$$x_{n+1} = x_n - \frac{f(x_n)}{f'(x_n)}$$

(and using the fact that f is continuously differentiable) would yield

$$L = L - \frac{f(L)}{f'(L)}$$

and thus $f(L) = 0$. This is a contradiction, because f is supposed to be a positive function.

Problem 18

Find all solutions in nonnegative integers to the system of equations
$$3x^2 - 2y^2 - 4z^2 + 54 = 0, \qquad 5x^2 - 3y^2 - 7z^2 + 74 = 0.$$

Solution. The two solutions are $(x, y, z) = (4, 7, 1)$ and $(16, 13, 11)$.

We eliminate x by subtracting 3 times the second equation from 5 times the first, obtaining $-y^2 + z^2 + 48 = 0$, or
$$48 = y^2 - z^2 = (y+z)(y-z).$$
Since $(y+z) - (y-z) = 2z$ is even, both $y+z$ and $y-z$ must be even. The possibilities are $(y+z, y-z) = (24, 2)$, $(12, 4)$, and $(8, 6)$. Solving for y and z yields $(y, z) = (13, 11)$, $(8, 4)$, and $(7, 1)$. From either original equation, we find the corresponding x values to be 16, $\sqrt{46}$, and 4, respectively. Thus, the only nonnegative integral solutions to the system of equations are $(16, 13, 11)$ and $(4, 7, 1)$.

Comment. If we eliminate y instead, we get $x^2 - 2z^2 = 14$, a Pellian equation with infinitely many solutions, which is more difficult to solve than $y^2 - z^2 = 48$.

Problem 19

If n is a positive integer, how many real solutions are there, as a function of n, to $e^x = x^n$?

Solution. For $n = 1$, there are no solutions; for $n = 2$, there is one solution; for odd $n \geq 3$, there are two solutions; for even $n \geq 4$, there are three solutions.

Let us first consider negative solutions. For odd n, there are none, since $x^n < 0$ when $x < 0$. So suppose n is even. Then x^n decreases for $x < 0$. Because e^x increases, $x^n - e^x$ decreases for $x < 0$. (This can also be seen by examining the derivative.) Since $x^n - e^x$ approaches ∞ as x approaches $-\infty$ and since $0^n - e^0 = -1 < 0$, there must be exactly one negative x for which $x^n - e^x = 0$.

Clearly, $x = 0$ is never a solution, so we have only positive solutions left to consider. For these there is no need to distinguish between even and odd n; in fact, we will prove the following claim:

If $a > 0$ is any real number, then $e^x = x^a$ has no positive solutions if $a < e$, one positive solution ($x = e$) if $a = e$, and two positive solutions if $a > e$.

Because 1 and 2 are the only positive integers less than e, we will be done once we prove this.

To prove the claim, consider the ratio

$$f(x) = \frac{x^a}{e^x}.$$

We are interested in the number of solutions of $f(x) = 1$. Note that $f(0) = 0$ and $\lim_{x \to \infty} f(x) = 0$. Since

$$f'(x) = \frac{x^{a-1}(a-x)}{e^x},$$

f is increasing for $0 < x < a$ and decreasing for $x > a$. Therefore, $f(x) = 1$ has two solutions if $f(a) > 1$, one if $f(a) = 1$, and none if $f(a) < 1$. The claim now easily follows from $f(a) = (a/e)^a$.

Problem 20

Three final participants for a swimming championship are to be chosen from an initial set of contestants by having all contestants swim in an equal number of preliminary races. The race director wants exactly three swimmers in each preliminary race with no two particular swimmers competing against each other in more than one of them. Can such a schedule of preliminary races be arranged

a. if there are five participants;

b. if there are ten participants?

Answer. It cannot be done with five participants, but it can be done for ten.

Solution. a. Suppose such a schedule could be arranged for five participants A, B, C, D, E. The first race must feature three swimmers, say A, B, C. One of these, say A, must be in the second race. Since A has already competed with B and C, the second race must be between A, D, E. Then B must be in another race (because A has now competed in two races and all contestants participate in an equal number of races), and since B has already competed with A and C, the other entrants in this race must be D

and E. But this means that D and E will compete against each other a second time, contradiction.

b. If it can be done, each participant must swim in r preliminary races, for some $r \geq 1$. This makes for $10r$ "participant entries" (counting one entry for each time a participant is in a race). On the other hand, each race involves exactly 3 participants, so if there are k races in all, there are $3k$ participant entries. We now see that $10r = 3k$, so r must be divisible by 3, and thus $r \geq 3$. But $r < 6$ because a participant has two different opponents each time and this takes at least $2r$ different opponents. The only possibility is therefore $r = 3$, $k = 10$.

(Alternatively, and more generally, suppose there are n participants, each competing in r races. Then the number of races is $rn/3$, so either n or r is divisible by 3. In order to avoid competing against the same opponent twice, a participant's number of opponents must satisfy $2r \leq n - 1$, which means that $r \leq (n-1)/2$. When $n = 5$, we must have r divisible by 3 and $r \leq 2$, which can't happen. When $n = 10$, we must have r divisible by 3 and $r \leq 9/2$, which means that $r = 3$.)

To find such a schedule we apply a "greedy" strategy. Arrange the swimmers in order from 1 to 10, and construct a 10×10 matrix of 0s and 1s, where a 1 in row i and column j means that swimmer j participates in the ith race. Starting in the upper left corner of the matrix, assign swimmers to races $1, 2, \ldots$, in order, at each step taking the *next* available swimmer (in cyclic order $1, 2, \ldots$) who meets the conditions of the problem. This leads to the following schedule, which satisfies the conditions of the problem. [For each race, the first swimmer chosen is indicated by a boldface **1** and the other two are indicated by 1.]

Participants

	1	2	3	4	5	6	7	8	9	10
1	**1**	1	1	0	0	0	0	0	0	0
2	0	0	0	**1**	1	1	0	0	0	0
3	0	0	0	0	0	0	**1**	1	1	0
4	1	0	0	1	0	0	0	0	0	**1**
5	0	0	0	0	1	0	1	0	0	1
6	1	0	0	0	1	0	0	1	0	0
7	0	1	0	0	0	0	0	0	1	1
8	0	0	1	1	0	0	1	0	0	0
9	0	1	0	0	0	1	0	1	0	0
10	0	0	1	0	0	1	0	0	1	0

Races (label for rows)

Another approach is to construct a connected configuration of 10 points and 10 "lines", where each line contains 3 points, for example, as shown in the following figure (the circle in the center represents the line containing the points labeled 3, 6, and 7). Note that any two lines in this figure intersect in at most one point, and any two points are contained in at most one line, and, as such, the configuration is an example of a finite geometry.

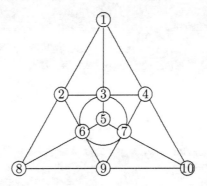

Once the points are labeled from 1 to 10, for example as shown, the points of the lines will identify the participants in the preliminary races. This labeling results in the following schedule.

		\multicolumn{10}{c}{Participants}
		1	2	3	4	5	6	7	8	9	10
Races	1	1	1	0	0	0	0	0	1	0	0
	2	1	0	1	0	1	0	0	0	0	0
	3	1	0	0	1	0	0	0	0	0	1
	4	0	1	1	1	0	0	0	0	0	0
	5	0	1	0	0	0	1	0	0	1	0
	6	0	0	1	0	0	1	1	0	0	0
	7	0	0	0	1	0	0	1	0	1	0
	8	0	0	0	0	1	1	0	1	0	0
	9	0	0	0	0	1	0	1	0	0	1
	10	0	0	0	0	0	0	0	1	1	1

Problem 21

Does there exist a positive integer whose prime factors include at most the primes 2, 3, 5, and 7 and which ends in the digits 11? If so, find the smallest such positive integer; if not, show why none exists.

Solution. There is no such positive integer.

The last digit of a multiple of 2 is even; the last digit of a multiple of 5 is 0 or 5. Therefore, 2 and 5 cannot divide a number ending in the digits 11. Thus, we consider numbers of the form $3^m 7^n$.

A check on small numbers of this form having 1 in the units position shows that $3^4 = 81$, $3 \cdot 7 = 21$, and $7^4 = 2401$. Each of these numbers has the correct units digit, but in each case, the tens digit is even instead of 1.

Now note that integers r and $r(20k+1)$ have the same remainder upon division by 20. Since 3^4, $3 \cdot 7$, and 7^4 have the form $20k + 1$, we conclude that if $3^m 7^n$ has the form $20k + 11$ (that is, has odd tens digit and final digit 1), then so do $3^{m-4} 7^n$, $3^{m-1} 7^{n-1}$, and $3^m 7^{n-4}$, provided the respective exponents are nonnegative.

By repeated application of this observation, we can say that if any integer $3^m 7^n$ has the form $20k+11$, then so will one of 1, 3, 3^2, 3^3, 7, 7^2, 7^3. However, among these, only 1 has the correct last digit. Thus, $3^m 7^n$ can never have remainder 11 upon division by 20, hence cannot end in the digits 11.

Comment. This proof can be streamlined by using congruences, as follows. If $3^m 7^n = 100k + 11$ for some nonnegative integers m, n, k, then, in particular, $3^m 7^n \equiv 11 \pmod{20}$. On the other hand, since $3^4 \equiv 3 \cdot 7 \equiv 7^4 \equiv 1 \pmod{20}$, we have $3^m 7^n \equiv 3^{m-4} 7^n \equiv 3^{m-1} 7^{n-1} \equiv 3^m 7^{n-4} \pmod{20}$, provided the respective exponents are nonnegative. By repeated application of these identities we find that $3^m 7^n$ is congruent modulo 20 to one of the numbers 1, 3, 3^2, 3^3, 7, 7^2, 7^3. Since none of these is 11 modulo 20, neither is $3^m 7^n$.

Problem 22

The Wohascum County Fish and Game Department issues four types of licenses, for deer, grouse, fish, and wild turkey; anyone can purchase any combination of licenses. In a recent year, (exactly) half the people who bought a grouse license also bought a turkey license. Half the people who bought a turkey license also bought a deer license. Half the people who bought a fish license also bought a grouse license, and one more than half the people who bought a fish license also bought a deer license. One third of the people who bought a deer license also bought both grouse and fish licenses. Of the people who bought deer licenses, the same number bought a grouse license as bought a fish license; a similar statement was true of buyers of turkey licenses. Anyone who bought both a grouse and a fish license also bought either a deer or a turkey license, and of these people the same number bought a deer license as bought a turkey license. Anyone who bought both a deer and a turkey license either bought both a grouse and a fish license or neither. The number of people buying a turkey license was equal to the number of people who bought some license but not a fish license. The number of people buying a grouse license was equal to the number of people buying some license but not a turkey license. The number of deer licenses sold was one more than the number of grouse licenses sold. Twelve people bought either a grouse or a deer license (or both). How many people in all bought licenses? How many licenses in all were sold?

Solution. Fourteen people bought a total of thirty-one licenses.

We let n stand for the number of people buying licenses. We let d, f, g, and t stand for the number of deer, fish, grouse, and wild turkey licenses sold, respectively. Since we will never multiply two variables, we can let dg be the number of people buying both deer and grouse licenses, $dfgt$ be the number of people buying all four licenses, etc. If we then translate the given information into equations, we get

$$\tfrac{1}{2}g = gt, \tag{1}$$

$$\tfrac{1}{2}t = dt, \tag{2}$$

$$\tfrac{1}{2}f = fg, \tag{3}$$

$$\tfrac{1}{2}f + 1 = df, \tag{4}$$

$$\tfrac{1}{3}d = dfg, \tag{5}$$

$$dg = df, \tag{6}$$

$$gt = ft, \tag{7}$$

$$fg = dfg + fgt - dfgt, \tag{8}$$

$$dfg = fgt, \tag{9}$$

$$dgt = dft = dfgt, \tag{10}$$

$$t = n - f, \tag{11}$$
$$g = n - t, \tag{12}$$
$$d = g + 1, \tag{13}$$
$$12 = g + d - dg. \tag{14}$$

We can also express n in terms of the other variables using the inclusion-exclusion principle. This yields

$$\begin{aligned}n = (d + g + f + t) - (dg + df + dt + fg + gt + ft) \\ + (dfg + dgt + dft + fgt) - dfgt.\end{aligned} \tag{15}$$

Equations (13), (12), and (11) imply that d, t, and f can be expressed in terms of n and g. Doing so and substituting the results into equations (4) and (14), we find

$$df = \tfrac{1}{2}g + 1 \tag{16}$$

and

$$dg = 2g - 11. \tag{17}$$

Equations (6), (16), and (17) combine to yield $g = 8$. Substituting this into our expressions for d, t, and f gives

$$d = 9, \quad t = n - 8, \quad f = 8.$$

Equations (4), (6), (2), (3), (1), and (7) yield

$$df = 5, \quad dg = 5, \quad dt = \tfrac{1}{2}n - 4, \quad fg = 4, \quad gt = 4, \quad ft = 4.$$

Equations (5), (9), (8), and (10) then yield

$$dfg = 3, \quad fgt = 3, \quad dfgt = 2, \quad dgt = 2, \quad dft = 2.$$

Finally, substitution into equation (15) implies

$$n = (17 + n) - (18 + \tfrac{1}{2}n) + 10 - 2 = 7 + \tfrac{1}{2}n.$$

We conclude that 14 people bought a total of $17 + 14 = 31$ licenses.

Problem 23

Given three lines in the plane which form a triangle, what is the set of points for which the sum of the distances to the three lines is as small as possible?

Solution. The set consists of the vertex opposite the longest side of the triangle, assuming that side is uniquely determined. If there are exactly two such sides (of equal length), the set is the remaining (shortest) side of the triangle (including endpoints); if the triangle is equilateral, the set consists of the entire triangle and its interior.

Let P be a point for which the sum of the three distances is minimized. If A, B, and C are the points of intersection of the three lines, then it is easy to see that P cannot be outside the triangle ABC. For if P were outside the triangle, say between the lines AB and AC, inclusive, then we could reduce the sum of the distances by moving P toward A.

Let a, b, and c be the lengths of BC, AC, and AB, respectively. There is no loss of generality in assuming $a \leq b \leq c$. If K is twice the area of triangle ABC, and x, y, z are the distances from a point P to BC, AC, and AB, respectively, then we wish to minimize $x + y + z$ subject to the constraint $ax + by + cz = K$, and conceivably to other geometric constraints.

Now (x, y, z) is a nonnegative triple satisfying the constraint if and only if $(0, 0, (a/c)x + (b/c)y + z)$ is such a triple. But

$$x + y + z \geq (a/c)x + (b/c)y + z,$$

with necessary and sufficient conditions for equality given by

$$a = c \quad \text{or} \quad x = 0,$$

and

$$b = c \quad \text{or} \quad y = 0.$$

We conclude that the set P consists of C if $a \leq b < c$, the line segment BC if $a < b = c$, and the triangle ABC and its interior if $a = b = c$.

Comment. As we saw, this problem translates to a linear programming problem for which the geometry of the situation makes the solution fairly apparent. The triangle and its interior correspond to the part of the plane $ax + by + cz = K$ with $x, y, z \geq 0$. The minimum of $x + y + z$ can occur at a point, on a side of the triangle, or on the entire triangle, depending on the relative orientations of the planes $ax + by + cz = K$ and $x + y + z = K/c$.

Problem 24

Define two positive integers a and b to be "one step apart" if and only if $ab + 1$ is a perfect square.

THE SOLUTIONS 69

a. Show that any two (distinct) positive integers are separated by a finite number of steps.

b. How many steps separate two consecutive positive integers m and $m+1$?

c. How many steps separate 1 and 4?

Solution. a. First of all, for any positive integer n, $(n+1)^2 = n(n+2)+1$ is a perfect square, so n and $n+2$ are always one step apart. So if the distinct integers a, b have the same parity, we can get from a to b in $|a-b|/2$ steps (by repeatedly adding or subtracting 2). If a and b have opposite parity, say a is even and b is odd, then we can get from a to 8 in this way, from 8 to 3 in one step, and then from 3 to b in the same way. So we can always get from a to b in a finite number of steps.

b. If m and $m+1$ were one step apart, $m(m+1)+1 = m^2+m+1$ would be a perfect square. However, this number is strictly between the consecutive perfect squares m^2 and $(m+1)^2 = m^2+2m+1$, so m^2+m+1 cannot be a perfect square. On the other hand, $m(16m+8)+1 = (4m+1)^2$ and $(m+1)(16m+8)+1 = 16m^2+24m+9 = (4m+3)^2$, so we can get from m to $m+1$ in two steps, so m to $m+1$ are separated by exactly two steps.

c. Because 1 and 2 are two steps apart and 2 and 4 are one step apart, 1 and 4 are at most three steps apart. They're not just one step apart because $1 \cdot 4 + 1 = 5$ is not a perfect square, so the only question is whether they could be two steps apart. If that were the case, there would be a positive integer c so that $1 \cdot c + 1$ and $4 \cdot c + 1$ are both perfect squares, say $c+1 = k^2$ and $4c+1 = \ell^2$. Then $c = k^2-1$ and $4(k^2-1) = \ell^2-1$, or $4k^2 - \ell^2 = 3$. But the only two squares that differ by 3 are 1 and 4, so we would have $k = \ell = 1$ and $c = 0$, contradiction. So no such positive integer c exists, and 1 and 4 are exactly three steps apart.

Problem 25

Suppose the plane $x+2y+3z = 0$ is a perfectly reflecting mirror. Suppose a ray of light shines down the positive x-axis and reflects off the mirror. Find the direction of the reflected ray.

Answer. The ray has the direction of the unit vector $\frac{1}{7}(-6, 2, 3)$.

Solution 1. Let **u** be the unit vector in the direction of the reflected ray. The vector $(1,2,3)$ is normal to the plane, so by vector addition and geometry, $(1,0,0) + \mathbf{u}$ is twice the projection of the vector $(1,0,0)$ onto $(1,2,3)$. Using the dot product formula for projection, we obtain

$$(1,0,0) + \mathbf{u} = 2\frac{(1,0,0)\cdot(1,2,3)}{(1,2,3)\cdot(1,2,3)}(1,2,3) = \frac{1}{7}(1,2,3),$$

or $\mathbf{u} = \frac{1}{7}(-6,2,3)$.

Solution 2. (The late Murray Klamkin, University of Alberta)
Let (u,v,w) denote the reflection (mirror image) of the point $(1,0,0)$ in the plane $x + 2y + 3z = 0$. Because the incoming ray is directed from $(1,0,0)$ to $(0,0,0)$, the reflected ray is in the direction from (u,v,w) to $(0,0,0)$, that is, in the direction of the vector $(-u,-v,-w)$.

To find (u,v,w), note that the vector from $(1,0,0)$ to (u,v,w) is normal to the plane, hence

$$(u-1, v, w) = \lambda(1,2,3) \qquad \text{for some } \lambda.$$

This yields $u = \lambda+1$, $v = 2\lambda$, $w = 3\lambda$. Also, (u,v,w) has the same distance to the origin as $(1,0,0)$, so $u^2+v^2+w^2 = 1$. Substituting for u, v, w and solving for λ, we find $\lambda = -\frac{1}{7}$ (since $\lambda \neq 0$). Therefore, $(u,v,w) = (\frac{6}{7}, -\frac{2}{7}, -\frac{3}{7})$, and the reflected ray has the direction of $\frac{1}{7}(-6,2,3)$.

Problem 26

Starting at $(a,0)$, Jessie runs along the x-axis toward the origin at a constant speed s. Starting at $(0,1)$ at the same time Jessie starts, Riley runs at a constant speed 1 counterclockwise around the unit circle until Jessie is far down the negative x-axis. For what integers $a > 1$ is there a speed s so that Jessie and Riley reach $(1,0)$ at the same time and $(-1,0)$ at the same time?

Answer. There is such a speed s if and only if a is even.

Solution. For Jessie and Riley to meet at $(1,0)$, their speeds must be in the same ratio as the distances they traveled to get to the meeting. That distance is $a-1$ for Jessie and $(3/2 + 2m)\pi$ for Riley, where $m \geq 0$ is the number of times Riley runs around the unit circle after arriving at $(1,0)$ the

THE SOLUTIONS

first time. Thus we must have

$$s = \frac{s}{1} = \frac{a-1}{(3/2 + 2m)\pi}$$

for some integer $m \geq 0$.

For Jessie and Riley to meet again at $(-1, 0)$, the additional distances they travel between their meetings must also be in the ratio above, so we must have

$$s = \frac{2}{(2n+1)\pi}$$

for some integer $n \geq 0$.

Thus the condition for a speed s to exist is that there are nonnegative integers m and n with

$$\frac{a-1}{(3/2+2m)\pi} = \frac{2}{(2n+1)\pi},$$

that is,

$$a = 1 + \frac{4m+3}{2n+1}.$$

Because $a > 1$ is an integer, $\frac{4m+3}{2n+1}$ must be a positive integer. Clearly, that integer is odd, so a must be even. Conversely, if a is even, we can always write a in the desired form $1 + \frac{4m+3}{2n+1}$. Specifically, if a is divisible by 4, say $a = 4k$, we can take $m = k-1, n = 0$, whereas if a has the form $4k+2$, we can take $m = 3k, n = 1$; we are done.

Problem 27

Find the set of all solutions to $x^{y/z} = y^{z/x} = z^{x/y}$, with x, y, and z positive real numbers.

Solution. The only solutions are $x = y = z$, with z equal to an arbitrary positive number. These are obviously solutions; we show there are no others.

For positive numbers a and b, $a^b < 1$ if and only if $a < 1$. Thus, if one of x, y, z is less than 1, all three are. If (x, y, z) is a solution to the system of equations, then it is easy to verify that $(1/x, 1/z, 1/y)$ is also a solution. Therefore, we may assume $x, y, z \geq 1$. The symmetry of the equations allows us to assume $x \leq z$ and $y \leq z$ (but then we may not assume any relationship

between x and y). We then have
$$x \geq x^{y/z} = y^{z/x} \geq y.$$
This in turn implies
$$x \geq x^{y/z} = z^{x/y} \geq z,$$
hence $x = z$. This makes the original equations
$$x^{y/x} = y = x^{x/y}.$$
If $x = 1$, then $y = 1$. If $x > 1$, then $y/x = x/y$, hence $x = y$. In either case, $x = y = z$, as asserted.

Problem 28

Find all perfect squares whose base 9 representation consists only of ones.

Solution. The only such square is 1.

We are asked to find nonnegative integers x and n such that
$$x^2 = 1 + 9 + 9^2 + \cdots + 9^n.$$
Summing the finite geometric series on the right and multiplying by 8, we arrive at the equation
$$8x^2 = 9^{n+1} - 1 = (3^{n+1} + 1)(3^{n+1} - 1).$$
Because the greatest common divisor of $3^{n+1} + 1$ and $3^{n+1} - 1$ is 2, there must exist positive integers c and d such that either

(i) $\qquad 3^{n+1} + 1 = 4c^2 \qquad$ and $\qquad 3^{n+1} - 1 = 2d^2, \qquad$ or

(ii) $\qquad 3^{n+1} + 1 = 2c^2 \qquad$ and $\qquad 3^{n+1} - 1 = 4d^2.$

In case (i), $3^{n+1} = 4c^2 - 1 = (2c+1)(2c-1)$. Since $2c+1$ and $2c-1$ are relatively prime, we must have $2c + 1 = 3$, $2c - 1 = 1$, and it follows that $n = 0$ and $x = 1$.

In case (ii), n must be odd, since $3^{n+1} - 1$ is divisible by 4. But then
$$1 = 3^{n+1} - 4d^2 = \left(3^{(n+1)/2} + 2d\right)\left(3^{(n+1)/2} - 2d\right),$$
which is impossible for $d > 0$.

Problem 29

Show that any polygon can be tiled by convex pentagons.

Solution 1. Because any polygon can be tiled by triangles, it's enough to show that any triangle can be tiled by convex pentagons. Now, given any two triangles in the plane, we can convert one into the other by an affine transformation, that is, a linear transformation followed by a shift (translation). Such a transformation sends convex pentagons into convex pentagons, so it's enough to show that there is *one* triangle that can be subdivided into convex pentagons. The figure shows one way to do this.

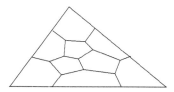

Solution 2. (Haggai Nuchi '08, Carleton College) As in the first solution, it's enough to prove that every triangle can be tiled by convex pentagons. So consider triangle ABC. Trisect AB, BC, CA at D, D', E, E', F, F'; then DE', EF', FD' are concurrent at G. Bisect DG at H, $D'G$ at H', EG at I, $E'G$ at I', FG at J and $F'G$ at J', and let K, L, M be the centroids of triangles $HH'G, II'G, JJ'G$, respectively. Using these points, we get the convex pentagonal tiling shown in the figure.

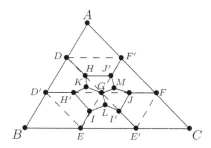

Solution 3. In fact, any polygon can be tiled with convex pentagons which have two right angles, by using the following construction for a triangle. In triangle ABC, let m_A, m_B, m_C denote the median lines from vertices A, B, C

with feet at D, E, F respectively. Let G (the centroid) be their common intersection and construct lines ℓ_A, ℓ_B, ℓ_C perpendicular to m_A, m_B, m_C at the midpoints of AG, BG, CG, respectively (see figure (i)).

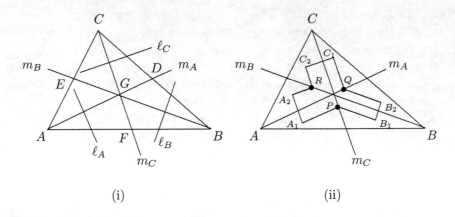

(i)　　　　　　　　　　　　(ii)

Now choose P on CF so that the line through P parallel to m_B intersects ℓ_B at B_1 within triangle BGF and the line through P parallel to m_A intersects ℓ_A at A_1 within triangle AGF. (This is possible by placing P appropriately close to G.) Similarly, choose point Q on AD so that the lines through Q parallel to m_B and m_C intersect ℓ_B and ℓ_C at B_2 and C_1 within triangles BGD and CGD respectively, and point R on BE so the lines through R parallel to m_C and m_A intersect ℓ_C and ℓ_A at C_2 and A_2 within triangles CGE and AGE respectively (see figure (ii)).

Finally, drop perpendiculars from A_1 and B_1 to AB, from B_2 and C_1 to BC, and from C_2 and A_2 to AC. The tiling obtained in this way will consist of convex pentagons that each have two right angles (see figure (iii)).

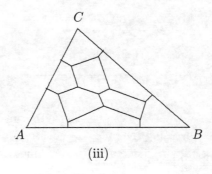

(iii)

Problem 30

In a recent snowmobile race, conditions were such that no more than two competitors could be abreast of each other anywhere along the trail. Nevertheless, there was frequent passing. The following is an excerpt from an article about the race. "After a few miles A pulled ahead of the pack, closely followed by B and C in that order, and thereafter these three did not relinquish the top three positions in the field. However, the lead subsequently changed hands nine times among these three; meanwhile, on eight different occasions the vehicles that were running second and third at the time changed places. ... At the end of the race, C complained that B had driven recklessly just before the finish line to keep C, who was immediately behind B at the finish, from passing. ..." Can this article be accurate? If so, can you deduce who won the race? If the article cannot be accurate, why not?

Solution. The article cannot be accurate.

At the point where A, B, and C pulled ahead of the pack, the order of the three was (A, B, C): A first, B second, and C third. At the end of the race, if the article is correct, the order must have been (A, B, C) or (B, C, A). Once the three pulled ahead of the pack, there were $9 + 8 = 17$ place changes among A, B, and C. If we let the set S consist of the three orders (A, B, C), (B, C, A), and (C, A, B), and T consist of the other three possible orders, (A, C, B), (B, A, C), and (C, B, A), then a single interchange of two of the snowmobiles will take an order in S to an order in T, and vice versa. Thus, to get from (A, B, C) to either itself or to (B, C, A) requires an even number of interchanges, rather than 17.

Comment. This solution proves a specific case of the following well-known theorem on permutations: A given permutation can never be obtained both by an even number of transpositions (interchanges of two elements) and by an odd number of transpositions.

Problem 31

Show how to construct the midpoint of any given segment using only a straightedge (but no compasses) and a "trisector," a device with which any straight line segment can be divided into three equal parts.

Solution 1. (See figure below.) Let AB denote the line segment we wish to bisect and let C be an arbitrary point off the line AB. Trisect AC at D and E, D being the closer point to C. Then E is the midpoint of AD, and BE is the median of the triangle ABD. Trisect the segment BE, letting F be the point two–thirds of the way to E. It is well known that F is the intersection of the medians of ABD. Hence, DF intersects AB at its midpoint.

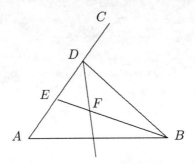

Solution 2. (See figure below.) Let AB be the line segment whose midpoint is to be constructed. Choose an arbitrary point C off the line AB; draw the line segments AC and BC. Trisect AC and BC; let P and Q be the trisection points closest to C on AC and BC, respectively. Now draw the line segments AQ and BP; let R be their point of intersection. The line CR will then intersect the segment AB at the desired midpoint M. This follows immediately from Ceva's theorem.

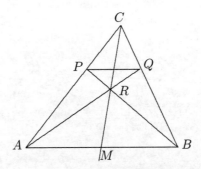

Comment. To prove the validity of the second construction without recourse to Ceva's theorem, one can use the following general result: Let A, B, and C be noncollinear. Let P lie on the line segment AC and Q lie on the line segment BC such that PQ is parallel to AB. Let R be the intersection of AQ and BP. Then R lies on the median from C to AB. To prove this, let

X on AC and Y on BC be on the line parallel to AB through R. Then the triangles PXR and PAB are similar, as are the triangles QRY and QAB. The ratios XR/AB and RY/AB are both equal to the ratio of the distance between the lines PQ and XY to the distance between the lines PQ and AB, hence are equal. Thus, $XR = RY$. If M is the intersection of AB and CR, the similarity of triangles CXR and CAM and of CRY and CMB imply that M is the midpoint of AB, proving the result.

Problem 32

Find a positive integer n such that $2011n+1$ and $2012n+1$ are both perfect squares, or show that no such positive integer n exists.

Answer. $2011n+1$ and $2012n+1$ are both squares for $n = 32184$.

Solution. Because we want $kn+1$ and $(k+1)n+1$ both to be perfect squares for $k = 2010$, we might start by trying to solve the analogous problem for smaller values of k. A little experimentation shows that:

$n+1$ and $2n+1$ are both squares for $n = 24$ (the squares are 5^2 and 7^2);

$2n+1$ and $3n+1$ are both squares for $n = 40$ (the squares are 9^2 and 11^2);

$3n+1$ and $4n+1$ are both squares for $n = 56$ (the squares are 13^2 and 15^2);

$4n+1$ and $5n+1$ are both squares for $n = 72$ (the squares are 17^2 and 19^2).

By now the pattern is clear: $kn+1$ and $(k+1)n+1$ are both squares for $n = 16k+8$; specifically, the squares are

$$k(16k+8)+1 = 16k^2 + 8k + 1 = (4k+1)^2,$$
$$(k+1)(16k+8)+1 = 16k^2 + 24k + 9 = (4k+3)^2.$$

In particular, for $n = 16(2011)+8 = 32184$, $2011n+1$ and $2012n+1$ are both perfect squares.

Comment. After we posed this problem we found that a stronger (and more difficult) version of this problem had been posed by Mihály Bencze as Problem 11508 in *The American Mathematical Monthly*, Vol. 117, No. 5, May 2010, p. 459. It states that for all positive integers k, there are *infinitely* many positive integers n such that $kn+1$ and $(k+1)n+1$ are both perfect squares.

Problem 33

Consider a 12 × 12 chessboard (consisting of 144 1 × 1 squares). If one removes 3 corners, can the remainder be covered by 47 1 × 3 tiles?

Idea. This is reminiscent of the old "chessnut": If opposite corners of an ordinary 8×8 chessboard are removed, the modified board cannot be covered with 31 1 × 2 tiles. The reason is that if the chessboard is colored in the usual alternating manner, opposite corners will have the same color. Thus, the modified board will consist of 30 squares of one color and 32 of the other. Since any tile in a tiling will cover one white square and one black square, no tiling is possible. We look for a similar coloring idea: What about three colors?

Solution. No. To see why, rotate the 12 × 12 chessboard so the non-deleted corner square is in the upper right corner, and color the board as shown below.

The board has 48 squares of each color, labeled here as 1, 2, and 3. We have deleted one square of color 1 and two of color 3, so 47 of the remaining squares are labeled 1, 48 are labeled 2, and 46 are labeled 3. Thus the board cannot be covered by 1 × 3 tiles, because for any such tiling there would be an equal number of squares of each color.

Comment. Our coloring scheme would not work if the only remaining corner were the lower right corner.

	1	2	3	1	2	3	1	2	3	1	2
2	3	1	2	3	1	2	3	1	2	3	1
1	2	3	1	2	3	1	2	3	1	2	3
3	1	2	3	1	2	3	1	2	3	1	2
2	3	1	2	3	1	2	3	1	2	3	1
1	2	3	1	2	3	1	2	3	1	2	3
1	1	2	3	1	2	3	1	2	3	1	2
2	3	1	2	3	1	2	3	1	2	3	1
1	2	3	1	2	3	1	2	3	1	2	3
1	1	2	3	1	2	3	1	2	3	1	2
2	3	1	2	3	1	2	3	1	2	3	1
	2	3	1	2	3	1	2	3	1	2	

Problem 34

Is there a function f, differentiable for all real x, such that

$$|f(x)| < 2 \quad \text{and} \quad f(x)f'(x) \geq \sin x \,?$$

Idea. Observe that $2f(x)f'(x)$ is the derivative of $(f(x))^2$, so we should perhaps be thinking about integration.

Solution. There is no such function. To see this, note that for $x > 0$, the second condition implies

$$\bigl(f(x)\bigr)^2 - \bigl(f(0)\bigr)^2 = \int_0^x 2f(t)f'(t)\,dt \geq \int_0^x 2\sin t\,dt = 2 - 2\cos x.$$

Thus, for $x > 0$,

$$\bigl(f(x)\bigr)^2 \geq 2 - 2\cos x + \bigl(f(0)\bigr)^2 \geq 2 - 2\cos x.$$

Therefore, in particular, $(f(\pi))^2 \geq 4$, so $|f(\pi)| \geq 2$, contradicting the first condition.

Comment. If the condition $|f(x)| < 2$ were replaced with $|f(x)| \leq 2$, then $f(x) = 2\sin(x/2)$ would satisfy both conditions.

Problem 35

Let N be the largest possible number that can be obtained by combining the digits 1, 2, 3 and 4 using the operations addition, multiplication, and exponentiation, if the digits can be used only once. Operations can be used repeatedly, parentheses can be used, and digits can be juxtaposed (put next to each other). For instance, 12^{34}, $1+(2\times 3\times 4)$, and $2^{31\times 4}$ are all candidates, but none of these numbers is actually as large as possible. Find N.

Solution. $N = 2^{(3^{41})}$; by a standard convention, the parentheses are actually unnecessary. We find that $N \approx 10^{1.1\times 10^{19}}$, which means that N has between 10^{19} and 10^{20} digits!

We first show that although addition and multiplication are allowed, these operations are not needed in forming N. In fact, for all $a, b \geq 2$, we have $a + b \leq ab \leq a^b$, so any addition or multiplication involving only (integer) terms or factors greater than 1 can be replaced by an exponentiation. Meanwhile, multiplication by 1 is useless, and the result of an addition of the form $a + 1$ will certainly be less than the result of juxtaposing 1 with one of the digits within the expression for a.

It follows that N can be formed by juxtaposition and exponentiation only. Clearly at least one exponentiation is needed (else $N \leq 4321$), and so N must be of the form a^b. Suppose the base, a, is formed by exponentiation, say $a = c^d$. Note that $b, d \geq 2$. Then $a^b = c^{db} \leq c^{(d^b)}$. Thus we may assume that a is not formed by exponentiation, but consists of one or more juxtaposed digits. If a has three digits, then b can only have one, and $a^b < 1000^4 = 10^{12}$; if a has two digits, then b is formed from the other two, so $b \leq 3^4 = 81$, and $a^b < 100^{81} = 10^{162}$. In both these cases, a^b is much less than $2^{3^{41}}$. Therefore, a must consist of one digit, and so $a = 2$, 3, or 4.

Similar arguments to those above show that in forming b from the three remaining digits, 1 will be juxtaposed with one of the other two digits and the result will be used in an exponentiation. It is also clear that 1 will be the second digit in the juxtaposition. We now have the following possibilities: If $a = 2$, then $b = 3^{41}$, 4^{31}, 41^3, or 31^4; if $a = 3$, then $b = 2^{41}$, 4^{21}, 21^4, or 41^2; if $a = 4$, then $b = 3^{21}$, 2^{31}, 21^3, or 31^2. Using the inequality $3^b < 4^b = 2^{2b}$, it is easy to shorten this list to $a = 2$ and b either 3^{41} or 4^{31}. We now use the inequality

$$4^{31} = 4^{24} \cdot 4^7 = 8^{16} \cdot 16394 < 9^{16} \cdot 19683 = 3^{41},$$

and we are done.

Problem 36

A filing system for newspapers has 366 slots, each corresponding to one day of the year. At some point in a non-leap year, the slot for February 29 is vacant, as it should be; all the other slots are filled, but the papers in them are scrambled.

a. Can the papers always be unscrambled by a series of moves, where in each move one paper is transferred to the current vacant slot? If so, what is the maximum number of moves that might be needed?

b. How would the answer from part (a) change if 366 were replaced by n?

Solution. a. Yes, it is possible to unscramble the papers using this method, but it may take as many as 547 moves. To see why, let's number the slots with papers in them $1, 2, \ldots, 365$, and let 366 denote the empty slot. Also, let's write $m \to n$ if the paper in slot m actually belongs in slot n. Note that for a given n there is one and only one slot, m, such that $m \to n$ (which is to say, n is a one-to-one function of m). Therefore, if we start with m and "follow the arrows" (apply the function repeatedly) we must eventually get back to m. If this happens in k steps, we'll say that the k numbers we encounter along the way form a k-*cycle*. For example, $3 \to 5 \to 182 \to 95 \to 3$ denotes a 4-cycle involving the papers in slots $3, 5, 182, 95$, each of which belongs in the next slot in the cycle. (Note that $3 \to 3$ indicates a 1-cycle, corresponding to a paper that is already in the correct slot.)

If we have a k-cycle with $k \geq 2$, the papers involved can be put in order in $k+1$ moves: Move one of the papers to the vacant slot, then move up the other papers one at a time as their slots become vacant, and finally, move the first paper back from the vacant slot to its proper slot. For example, for the previous 4-cycle, move 95 to 366, then 182 to 95, 5 to 182, 3 to 5, and finally, the paper from slot 366 (namely, 95) to slot 3.

On the other hand, for any k-cycle with $k \geq 2$, getting all k of the papers in order will involve *at least* $k+1$ moves of those papers, because each of them has to move at least once, but none of them can move immediately to its proper location because that slot is occupied (by a paper in the k-cycle).

Now every paper is part of *some* cycle, and in fact, is part of only one cycle. Of course, 1-cycles don't require any moves. So suppose that the cycles with at least two numbers in them consist of a k_1-cycle, a k_2-cycle, $\ldots$, and a k_r-cycle. Then the number of moves required is

$$(k_1 + 1) + (k_2 + 1) + \cdots + (k_r + 1) = (k_1 + \cdots + k_r) + r.$$

Each of the r cycles considered here has at least two numbers in it and all those numbers are distinct, so $2r \leq 365$, or, because r is an integer, $r \leq 182$. Also, $k_1 + \cdots + k_r$ is the total number of numbers in these cycles, so $k_1 + \cdots + k_r \leq 365$. This shows that the total number of moves is at most $365 + 182 = 547$.

On the other hand, we can force 547 moves by dividing $\{1, 2, \ldots, 365\}$ into 181 2-cycles and one 3-cycle, so that every number is in a k-cycle with $k \geq 2$ and there are 182 cycles in all. This completes the argument that 547 is the answer.

b. By essentially the same argument, whenever n is even, so that $n - 1$ is odd, the worst-case scenario has $\frac{1}{2}(n-4)$ 2-cycles and one 3-cycle, and requires $(n-1) + \frac{1}{2}(n-2) = \frac{3}{2}n - 2$ moves. If n is odd, the worst case occurs for $\frac{1}{2}(n-1)$ 2-cycles and requires $(n-1) + \frac{1}{2}(n-1) = \frac{3}{2}(n-1)$ moves.

Problem 37

Babe Ruth's batting performance in the 1921 baseball season is often considered the best in the history of the game. In home games, his batting average was .404; in away games it was .354. Furthermore, his slugging percentage at home was a whopping .929, while in away games it was .772. This was based on a season total of 44 doubles, 16 triples, and 59 home runs. He had 30 more at bats in away games than in home games. What were his overall batting average and his slugging percentage for the year?

Solution. Babe Ruth's batting average was .378; his slugging percentage was .846. To find these powerful numbers, let a_1, h_1, and t_1 be the number of at bats, hits, and total bases (hits + doubles + 2 × triples + 3 × home runs), respectively, in home games, and let a_2, h_2 and t_2 be the corresponding numbers for away games.

We first observe that

$$a_2 = a_1 + 30 \quad \text{and} \quad (t_1 + t_2) - (h_1 + h_2) = 44 + 2 \cdot 16 + 3 \cdot 59 = 253.$$

The given batting averages tell us that

$$.4035\, a_1 \leq h_1 \leq .4045\, a_1 \quad \text{and} \quad .3535\, (a_1 + 30) \leq h_2 \leq .3545\, (a_1 + 30).$$

Adding these inequalities, we get

$$.757\, a_1 + 10.605 \leq h_1 + h_2 \leq .759\, a_1 + 10.635. \qquad (*)$$

Similarly, from the given slugging percentages we find

$$.9285\, a_1 \leq t_1 \leq .9295\, a_1 \quad \text{and} \quad .7715\, (a_1 + 30) \leq t_2 \leq .7725\, (a_1 + 30),$$

hence

$$1.7\, a_1 + 23.145 \leq t_1 + t_2 \leq 1.702\, a_1 + 23.175.$$

Substituting $h_1 + h_2 + 253$ for $t_1 + t_2$, we obtain

$$1.7\, a_1 - 229.855 \leq h_1 + h_2 \leq 1.702\, a_1 - 229.825. \qquad (**)$$

Now we can combine $(*)$ and $(**)$ to get, on the one hand,

$$1.7\, a_1 - 229.855 \leq .759\, a_1 + 10.635,$$

which yields $.941\, a_1 \leq 240.49$, or $a_1 \leq 255.56\ldots$. Since a_1 is an integer, $a_1 \leq 255$. On the other hand,

$$.757\, a_1 + 10.605 \leq 1.702\, a_1 - 229.825,$$

from which we conclude $254.42\ldots \leq a_1$, hence $255 \leq a_1$. Therefore, $a_1 = 255$. With this, (*) easily yields $h_1 + h_2 = 204$, and there is no further difficulty in computing the overall batting average $(h_1 + h_2)/(a_1 + a_2)$ and slugging percentage $(h_1+h_2+253)/(a_1+a_2)$ to be $.378$ and $.846$, respectively.

Comment. From a numerical point of view, Babe Ruth's statistics as given are not very delicate. If we set $h_1 = .404\, a_1$, $h_2 = .354\,(a_1+30)$, $t_1 = .929\, a_1$, $t_2 = .772\,(a_1+30)$, and $t_1+t_2-h_1-h_2 = 253$, and solve this linear system, we obtain $h_1 \approx 103.02$, $h_2 \approx 100.89$, $t_1 \approx 236.89$, $t_2 \approx 220.02$, $a_1 \approx 254.99$. It is then easy to guess the correct answer. However, to justify this, analysis of the round-off error is required. To give a simple example, the system $x+y=2$, $x+1.001y=2$ has $x=2$, $y=0$ as its solution, whereas the system $x+y=2$, $x+1.001y=2.001$ has solution $x=y=1$. In our case, the coefficients have been rounded, as well. For more information, see a text on numerical analysis.

Problem 38

Let C be a circle with center O, and Q a point inside C different from O. Where should a point P be located on the circumference of C to maximize $\angle OPQ$?

Answer. The angle $\angle OPQ$ is a maximum when $\angle OQP$ is a right angle.

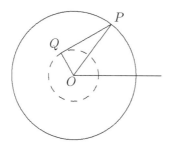

Solution 1. (Jason Colwell, age 13, Edmonton, Canada.) Think of P as fixed, while Q varies along the circle about O with radius OQ. It is then

clear that $\angle OPQ$ is a maximum when PQ is tangent to the small circle, that is, when $\angle OQP = 90°$.

Solution 2. Let $\theta = \angle OPQ$ and $\varphi = \angle OQP$, as shown below.

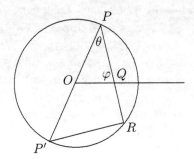

By the law of sines applied to $\triangle OQP$,

$$\frac{OP}{\sin \varphi} = \frac{OQ}{\sin \theta} \quad \text{or} \quad \sin \theta = \frac{OQ}{OP} \sin \varphi \leq \frac{OQ}{OP}$$

with equality if and only if $\varphi = 90°$. Since $\dfrac{OQ}{OP}$ is constant, $\sin \theta$, and hence also θ, is maximized when $\angle OQP$ is a right angle.

Solution 3. Extend PO and PQ to intersect the circle C at P' and R, respectively. For any point P, PP' is a diameter of C, and therefore $\angle PRP'$ is a right angle. Now θ will be maximized when $\cos \theta$ is minimized; $\cos \theta$ is minimized when chord PR is minimized. Finally, PR is minimized when it is perpendicular to OQ, since any other chord through Q passes closer to O, and this completes the proof.

Problem 39

Using technology you can find that $\displaystyle\int_0^1 \frac{\sqrt{x^4 + 2x^2 + 2}}{x^2 + 1}\, dx \approx 1.27798$, while $\dfrac{1}{4}\sqrt{16 + \pi^2} \approx 1.27155$. Clearly, then, $\displaystyle\int_0^1 \frac{\sqrt{x^4 + 2x^2 + 2}}{x^2 + 1}\, dx > \dfrac{1}{4}\sqrt{16 + \pi^2}$. Prove that this is indeed correct, without using any technology.

Solution. Rewrite the integral as

$$\int_0^1 \frac{\sqrt{x^4+2x^2+2}}{x^2+1}\,dx = \int_0^1 \frac{\sqrt{(x^2+1)^2+1}}{x^2+1}\,dx = \int_0^1 \sqrt{1+\frac{1}{(x^2+1)^2}}\,dx,$$

and observe that it is of the form $\int_0^1 \sqrt{1+(f'(x))^2}\,dx$ for $f(x) = \arctan x$, which gives the arc length along the graph of $y = \arctan x$ from $x = 0$ to $x = 1$. Because this arc length is greater than the straight-line distance between the endpoints $(0,0)$ and $(1, \pi/4)$, we have

$$\int_0^1 \frac{\sqrt{x^4+2x^2+2}}{x^2+1}\,dx > \sqrt{1^2+(\pi/4)^2} = \frac{1}{4}\sqrt{16+\pi^2}.$$

Problem 40

Let $f_1(x) = x^2 + 4x + 2$, and for $n \geq 2$, let $f_n(x)$ be the n-fold composition of the polynomial $f_1(x)$ with itself. Let s_n be the sum of the coefficients of the terms of even degree in $f_n(x)$. Find s_{2012}.

Answer. We'll show that $s_{2012} = \frac{1}{2}\left(3^{2^{2012}} - 3\right)$.

Solution. First note that for *any* polynomial

$$P(x) = a_0 + a_1 x + a_2 x^2 + \cdots + a_d x^d = \sum_{i=0}^{d} a_i x^i,$$

we have

$$P(1) = a_0 + a_1 + a_2 + \cdots + a_d = \sum_{i=0}^{d} a_i \quad \text{and}$$

$$P(-1) = a_0 - a_1 + a_2 - \cdots \pm a_d = \sum_{i=0}^{d} (-1)^i a_i,$$

so adding these we get $P(1) + P(-1) = 2a_0 + 2a_2 + \cdots$. Therefore, the sum of the coefficients of the terms of even degree in $P(x)$ is $\frac{1}{2}\bigl(P(1) + P(-1)\bigr)$.

Now we have $f_1(x) = x^2 + 4x + 2 = (x+2)^2 - 2$, so $f_1(x) + 2 = (x+2)^2$. Therefore, $f_2(x) = \bigl(f_1(x) + 2\bigr)^2 - 2 = (x+2)^4 - 2$, so $f_2(x) + 2 = (x+2)^4$;

the pattern continues, because $f_3(x) = f_1(f_2(x)) = \left(f_2(x) + 2\right)^2 - 2 = \left((x+2)^4\right)^2 - 2 = (x+2)^8 - 2$, so $f_3(x) + 2 = (x+2)^8$.

In general, we have $f_n(x) = (x+2)^{2^n} - 2$. To prove this works for all n, use induction: If it's true for n, then $f_{n+1}(x) = f_1(f_n(x)) = \left(f_n(x) + 2\right)^2 - 2 = (x+2)^{2^{n+1}} - 2$ and it's true for $n+1$.

Therefore,

$$S_{2012} = \frac{1}{2}\left(f_{2012}(1) + f_{2012}(-1)\right) = \frac{1}{2}\left(3^{2^{2012}} - 2 + 1 - 2\right) = \frac{1}{2}\left(3^{2^{2012}} - 3\right).$$

Problem 41

Find all real solutions of the equation $\sin(\cos x) = \cos(\sin x)$.

Solution. There are no real solutions.

Note that each side of the equation has period 2π. Therefore, we can assume $-\pi < x \leq \pi$. If x is a solution, then

$$\sin(\cos(-x)) = \sin(\cos x) = \cos(\sin x) = \cos(-\sin x) = \cos(\sin(-x)),$$

hence $-x$ is also a solution. Thus, it suffices to show there are no solutions with $0 \leq x \leq \pi$. On this interval, $0 \leq \sin x \leq 1 < \pi/2$, so we see that $\cos(\sin x) > 0$. In order to have $\sin(\cos x) > 0$, we must have $0 \leq x < \pi/2$. Because the function $x - \sin x$ is increasing and is 0 when $x = 0$, we have $\sin x < x$ for $x > 0$; in particular, $\sin(\cos x) < \cos x$. On the other hand, since the cosine function decreases on $[0, \pi/2)$, we have $\cos x \leq \cos(\sin x)$, so $\sin(\cos x) < \cos(\sin x)$, and we are done.

Problem 42

For what real numbers α does the series

$$\frac{1}{2^\alpha} + \frac{1}{(2 + \frac{1}{2^\alpha})^\alpha} + \frac{1}{\left(2 + \frac{1}{2^\alpha} + \frac{1}{(2+\frac{1}{2^\alpha})^\alpha}\right)^\alpha} + \cdots$$

converge?

Solution. We'll show that the series does not converge for *any* value of α. Suppose that it did, for some α. Then so would the series

$$2 + \frac{1}{2^\alpha} + \frac{1}{(2+\frac{1}{2^\alpha})^\alpha} + \frac{1}{\left(2 + \frac{1}{2^\alpha} + \frac{1}{(2+\frac{1}{2^\alpha})^\alpha}\right)^\alpha} + \cdots.$$

The partial sums

$$2, \quad 2 + \frac{1}{2^\alpha}, \quad 2 + \frac{1}{2^\alpha} + \frac{1}{(2+\frac{1}{2^\alpha})^\alpha}, \quad \ldots$$

of the new series have the form

$$2, \quad f(2), \quad f(f(2)), \quad f(f(f(2))), \quad \ldots$$

where $f(x) = x + \dfrac{1}{x^\alpha}$. If the series converges, this increasing sequence of partial sums has a limit $L \geq 2$. But because f is continuous on $(0, \infty)$, this implies that the sequence $f(2), f(f(2)), f(f(f(2))), \ldots$ would have limit $f(L)$. Since $f(2), f(f(2)), f(f(f(2))), \ldots$ is the same sequence as $2, f(2), f(f(2)), f(f(f(2))), \ldots$ except for a shift, these sequences would have the same limit. That is, $f(L) = L$. But this yields $L + \dfrac{1}{L^\alpha} = L$, $\dfrac{1}{L^\alpha} = 0$, contradiction.

Problem 43

For a natural number $n \geq 2$, let $0 < x_1 \leq x_2 \leq \ldots \leq x_n$ be real numbers whose sum is 1. If $x_n \leq 2/3$, prove that there is some k, $1 \leq k \leq n$, for which $1/3 \leq \sum_{j=1}^{k} x_j < 2/3$.

Solution. (Allen Schwenk, Western Michigan University) We distinguish two cases.

Case 1. $x_n > 1/3$. Then $1/3 \leq 1 - x_n < 2/3$, and since $1 - x_n = \sum_{j=1}^{n-1} x_j$, we can take $k = n - 1$.

Case 2. $x_n \leq 1/3$. Define $s_k = \sum_{j=1}^{k} x_j$. If $1/3 \leq s_k < 2/3$ for any k, we are done. If not, since $s_0 = 0$ and $s_n = 1$, we may select the smallest k with $s_k \geq 2/3$. Then presumably $s_{k-1} < 1/3$. But this forces

$$x_k = s_k - s_{k-1} > 1/3 \geq x_n,$$

a contradiction.

Problem 44

Is there a cubic curve $y = ax^3 + bx^2 + cx + d$, $a \neq 0$, for which the tangent lines at two distinct points coincide?

Answer. No. For each of the solutions below, we assume (x_1, y_1) and (x_2, y_2), $x_1 < x_2$, are two such points on the cubic curve.

Solution 1. By the Mean Value Theorem, there exists a number x_3, with $x_1 < x_3 < x_2$, such that $(y_2 - y_1)/(x_2 - x_1) = y'(x_3)$. Since the tangent lines coincide, we have $y'(x_1) = y'(x_2) = y'(x_3) = M$, say. But then the quadratic polynomial $3ax^2 + 2bx + c - M$ has three distinct roots, x_1, x_2, and x_3, and this is impossible.

Solution 2. If the tangent line through (x_1, y_1) and (x_2, y_2) is $y = Mx + B$, then x_1 and x_2 are multiple roots of the equation

$$ax^3 + bx^2 + (c - M)x + (d - B) = 0.$$

This follows from the fact that r is a multiple root of a polynomial $p(x)$ if and only if $p(r) = p'(r) = 0$. However, a cubic polynomial cannot be divisible by the fourth degree polynomial $(x - x_1)^2(x - x_2)^2$, so we have a contradiction.

(Recall that r is a *multiple root* of a polynomial $p(x)$ if $p(x)$ is divisible by $(x - r)^2$.)

Problem 45

Show that for any positive integer n,

$$\sum_{i=0}^{n-1} \arcsin\left(\frac{i(i+1) + n^2}{\sqrt{i^2 + n^2}\sqrt{(i+1)^2 + n^2}}\right) = \frac{(2n-1)\pi}{4}.$$

Solution 1. For $i = 0, 1, \ldots, n$, define vectors $\mathbf{v}_i = (i, n)$.

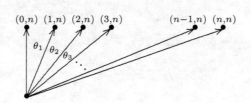

THE SOLUTIONS

Then, letting θ_{i+1} denote the angle between $\mathbf{v}_i$ and $\mathbf{v}_{i+1}$, we have
$$i(i+1) + n^2 = \mathbf{v}_i \cdot \mathbf{v}_{i+1} = \sqrt{i^2 + n^2}\sqrt{(i+1)^2 + n^2}\cos\theta_{i+1},$$
so
$$\theta_{i+1} = \arccos\left(\frac{i(i+1)+n^2}{\sqrt{i^2+n^2}\sqrt{(i+1)^2+n^2}}\right).$$

Also, the sum $\theta_1 + \theta_2 + \cdots + \theta_n$ of all these angles is the angle between the y-axis and the line $y = x$, so $\displaystyle\sum_{i=0}^{n-1}\theta_{i+1} = \frac{\pi}{4}$. Finally,

$$\arcsin\left(\frac{i(i+1)+n^2}{\sqrt{i^2+n^2}\sqrt{(i+1)^2+n^2}}\right) = \frac{\pi}{2} - \arccos\left(\frac{i(i+1)+n^2}{\sqrt{i^2+n^2}\sqrt{(i+1)^2+n^2}}\right).$$

Therefore,
$$\sum_{i=0}^{n-1}\arcsin\left(\frac{i(i+1)+n^2}{\sqrt{i^2+n^2}\sqrt{(i+1)^2+n^2}}\right) = \sum_{i=0}^{n-1}\left(\frac{\pi}{2} - \theta_{i+1}\right)$$
$$= \frac{n\pi}{2} - \frac{\pi}{4} = \frac{(2n-1)\pi}{4}.$$

Solution 2. We have
$$\cos\left(\arccos\frac{i}{\sqrt{i^2+n^2}} - \arccos\frac{i+1}{\sqrt{(i+1)^2+n^2}}\right)$$
$$= \frac{i}{\sqrt{i^2+n^2}} \cdot \frac{i+1}{\sqrt{(i+1)^2+n^2}} + \frac{n}{\sqrt{i^2+n^2}} \cdot \frac{n}{\sqrt{(i+1)^2+n^2}},$$
so
$$\arccos\left(\frac{i(i+1)+n^2}{\sqrt{i^2+n^2}\sqrt{(i+1)^2+n^2}}\right) = \arccos\frac{i}{\sqrt{i^2+n^2}} - \arccos\frac{i+1}{\sqrt{(i+1)^2+n^2}}$$

$\left(\text{note that } \dfrac{i}{\sqrt{i^2+n^2}} < \dfrac{i+1}{\sqrt{(i+1)^2+n^2}}\right)$.

Hence,

$$\sum_{i=0}^{n-1} \arcsin\left(\frac{i(i+1)+n^2}{\sqrt{i^2+n^2}\sqrt{(i+1)^2+n^2}}\right)$$

$$= \sum_{i=0}^{n-1}\left(\frac{\pi}{2} - \arccos\left(\frac{i(i+1)+n^2}{\sqrt{i^2+n^2}\sqrt{(i+1)^2+n^2}}\right)\right)$$

$$= \frac{n\pi}{2} - \sum_{i=0}^{n-1}\arccos\left(\frac{i(i+1)+n^2}{\sqrt{i^2+n^2}\sqrt{(i+1)^2+n^2}}\right)$$

telescopes to

$$\frac{n\pi}{2} - \arccos 0 + \arccos \frac{1}{\sqrt{2}} = \frac{(2n-1)\pi}{4}.$$

Problem 46

A group of friends had n ($n \geq 3$) digital watches between them, each perfectly accurate in the sense that a second was the same according to each watch. The time indicated shifted from one watch to the next, but in such a way that any two watches showed the same time in minutes for part of each minute. Find the largest number x such that at least one pair of watches necessarily showed the same time in minutes for more than the fraction x of each minute. (This is part (b) of the original problem; part (a) will follow by taking $n = 3$.)

Solution. The largest number with the stated property is $x = \dfrac{n-2}{n-1}$; in particular, for $n = 3$ we have $x = 1/2$.

Order the watches so that the first watch is running ahead of (or possibly with) all of the others, the second ahead is of all but the first, ..., and the nth is behind all the rest. Let t_j denote the time in minutes that watch j runs behind watch 1. Our ordering and the fact that all pairs of watches show the same time for some part of every minute implies

$$0 = t_1 \leq t_2 \leq \cdots \leq t_n < 1.$$

Because

$$t_n = \sum_{j=1}^{n-1}(t_{j+1} - t_j) < 1,$$

we must have

$$0 \le t_{j+1} - t_j < \frac{1}{n-1}$$

for some j. For that j, watches j and $j+1$ agree for more than the fraction

$$1 - \frac{1}{n-1} = \frac{n-2}{n-1}$$

of each minute.

Conversely, if

$$t_j = \frac{j-1}{n-1} t_n,$$

then no pair of watches agree for more than $\frac{n-2}{n-1} t_n$. Since t_n can be arbitrarily close to 1, $\frac{n-2}{n-1}$ is the best possible result.

Problem 47

Let C be a circle with center O, and Q a point inside C different from O. Show that the area enclosed by the locus of the centroid of triangle OPQ as P moves about the circumference of C is independent of Q.

Solution. We will show that the locus of the centroid of triangle OQP is a circle whose radius is one-third that of the circle C, as shown in the following figure. Thus, regardless of the location of Q inside C, the area enclosed by the locus of the centroid is one-ninth that of C.

Introduce rectangular coordinates so that O is the origin, C has equation $x^2 + y^2 = r^2$, and $Q = (q, 0)$, $0 < q < r$. Let $P = (x, y)$ be an arbitrary point on C. The midpoint D of PQ has coordinates $D = ((x+q)/2, y/2)$, and the centroid, G, lies two-thirds of the distance along the median line OD, so its coordinates are $G = ((x+q)/3, y/3)$. We have

$$\left(\frac{x+q}{3} - \frac{q}{3}\right)^2 + \left(\frac{y}{3}\right)^2 = \left(\frac{x}{3}\right)^2 + \left(\frac{y}{3}\right)^2 = \left(\frac{r}{3}\right)^2,$$

and we see that the locus of the centroid is the circle with center $(q/3, 0)$ and radius $r/3$.

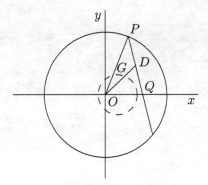

Comment. We can streamline this solution by using complex numbers, as follows. Let O be the origin in the complex plane, let r be the radius of the circle, and let Q correspond to the complex number α.

Any point P on the circle corresponds to $re^{i\theta}$ for some θ; the centroid of triangle OPQ is then given by the complex number

$$z = \frac{0 + re^{i\theta} + \alpha}{3}.$$

Thus, $|z - \alpha/3| = r/3$. This is the equation of a circle of radius $r/3$, centered at $\alpha/3$.

Problem 48

Suppose you have an unlimited supply of identical barrels. To begin with, one of the barrels contains n ounces of liquid, where n is a positive integer, and all the others are empty. You are allowed to redistribute the liquid between the barrels in a series of steps, as follows. If a barrel contains k ounces of liquid and k is even, you may pour exactly half that amount into an empty barrel. If k is odd, you may pour the *largest* integer that is less than half that amount into an empty barrel. No other operations are allowed. Your object is to "isolate" a total of m ounces of liquid, where m is a positive integer less than n.

a. What is the least number of steps (as a function of n) in which this can be done for $m = n - 1$?

b. What is the smallest number of steps in which it can be done regardless of m, as long as m is known in advance?

THE SOLUTIONS 93

Answer. The least number of steps in which this can be done for $m = n-1$ is $\lfloor \log_2 n \rfloor$ (that is, the unique nonnegative integer s with $2^s \leq n < 2^{s+1}$). The answer for (b) is the same as for (a).

Solution. a. Let s be the unique nonnegative integer with $2^s \leq n < 2^{s+1}$. Note that isolating $n-1$ ounces of liquid is equivalent to isolating just 1 ounce of liquid. To isolate 1 ounce of liquid, proceed at each step by continuing with the barrel that you just poured liquid into. We'll show, by induction, that at the end of the kth step (with $k \leq s$) that barrel will contain n_k ounces, with $2^{s-k} \leq n_k < 2^{s-k+1}$. It's true for $k = 0$ by the definition of s. Suppose it's true for $k = j$, so that after j steps we have an amount n_j ounces with $2^{s-j} \leq n_j < 2^{s-j+1}$. If n_j is even, then $n_{j+1} = \frac{1}{2} n_j$, so because $\frac{1}{2} 2^{s-j} \leq \frac{1}{2} n_j < \frac{1}{2} 2^{s-j+1}$, we have $2^{s-(j+1)} \leq n_{j+1} < 2^{s-(j+1)+1}$. If n_j is odd, then at the next step the amount in the new barrel is $\frac{1}{2}(n_j - 1)$. If also $j < s$, 2^{s-j} is even and n_j is odd, so we actually have $2^{s-j} < n_j < 2^{s-j+1}$. Therefore, $2^{s-j} \leq n_j - 1 < 2^{s-j+1}$, so $2^{s-(j+1)} \leq n_{j+1} < 2^{s-(j+1)+1}$. So either way, if the statement is true for $k = j < s$ it's also true for $k = j+1$, and the induction is complete.

In particular, for $k = s$, we have $1 \leq n_s < 2$, so $n_s = 1$, and we've shown we can isolate 1 ounce (and hence $n-1$ ounces) in s steps. On the other hand, it's not hard to see by induction that regardless of which barrel is chosen at each step, after k steps each barrel that isn't empty will contain at least n_k ounces. But since for $k < s$, $n_k \geq 2^{s-k} \geq 2$, it cannot be done in fewer than s steps, which justifies our claim.

b. We'll show, by induction on n, that for any positive integer $m < n$, we can isolate m ounces of liquid in at most $s = \lfloor \log_2 n \rfloor$ steps. Because we've already seen that s steps are needed for $m = n-1$, we'll be done.

It's easy to check that our statement is true for $n = 1, 2$, or 3; suppose it is true for all positive integers less than n. Because isolating m ounces is equivalent to isolating $n - m$ ounces, it's enough to show that for $m \leq n/2$, at most $\lfloor \log_2 n \rfloor$ steps are needed. But if $m \leq n/2$, then after one step there will be at least m ounces in the *new* barrel. By the induction hypothesis, we can then isolate m ounces in at most a further $\lfloor \log_2 (n/2) \rfloor$ steps if n is even, and a further $\lfloor \log_2 ((n-1)/2) \rfloor$ steps if n is odd. Thus in both cases, it will take at most $1 + \lfloor \log_2 (n/2) \rfloor = 1 + \lfloor \log_2 n - 1 \rfloor = \lfloor \log_2 n \rfloor$ steps, and we are done.

Problem 49

Describe the set of points (x, y) in the plane for which

$$\sin(x+y) = \sin x + \sin y.$$

Answer. The set of solutions consists of three infinite families of parallel lines: $x = 2k\pi$, $y = 2k\pi$, $x + y = 2k\pi$, with k an integer.

Solution 1. The relation is certainly true if $x = 2k\pi$ (with k an integer) or if $y = 2k\pi$, so the graph contains the infinitely many horizontal lines $y = 2k\pi$ and vertical lines $x = 2k\pi$.

Now assume that x and y are not (integer) multiples of 2π. We can rewrite the relation as

$$\sin x \cos y + \cos x \sin y = \sin x + \sin y$$

or

$$\sin x (\cos y - 1) = \sin y (1 - \cos x).$$

Since x and y are not multiples of 2π, $1 - \cos x \ne 0$ and $\cos y - 1 \ne 0$, so we can divide by these factors, obtaining

$$\frac{\sin x}{1 - \cos x} = \frac{\sin y}{\cos y - 1} = \frac{-\sin(-y)}{\cos(-y) - 1} = \frac{\sin(-y)}{1 - \cos(-y)}.$$

Let $f(x) = \sin x / (1 - \cos x)$. We have seen that (except for the special cases $x = 2k\pi$, $y = 2k\pi$) the relation is true if and only if $f(x) = f(-y)$.

Now

$$f(x) = \frac{\sin x}{1 - \cos x} = \frac{2\sin(x/2)\cos(x/2)}{2\sin^2(x/2)} = \cot(x/2),$$

and $\cot(x/2)$ has period 2π and decreases from ∞ to $-\infty$ on each interval between successive integer multiples of 2π. So on each such interval, $f(x)$ takes on each value only once. Hence, $f(x) = f(-y)$ if and only if $x = -y + 2k\pi$ for some integer k; that is, when $x + y = 2k\pi$ for some integer k. This completes the proof.

Solution 2. (The late William Firey, Oregon State University) Use trigonometric identities to rewrite the equation as

$$2\sin\left(\frac{x+y}{2}\right)\cos\left(\frac{x+y}{2}\right) = 2\sin\left(\frac{x+y}{2}\right)\cos\left(\frac{x-y}{2}\right).$$

This implies that either
$$\sin\left(\frac{x+y}{2}\right) = 0,$$
resulting in $x + y = 2k\pi$, k an integer, or
$$\cos\left(\frac{x+y}{2}\right) = \cos\left(\frac{x-y}{2}\right),$$
$$\cos(x/2)\cos(y/2) - \sin(x/2)\sin(y/2) = \cos(x/2)\cos(y/2) + \sin(x/2)\sin(y/2),$$
$$\sin(x/2)\sin(y/2) = 0,$$
which implies that $x = 2k\pi$ or $y = 2k\pi$, k an integer.

Problem 50

For n a positive integer, find the smallest positive integer $d = d(n)$ for which there exists a polynomial of degree d whose graph passes through the points $(1, 2)$, $(2, 3)$, ..., $(n, n + 1)$, and $(n + 1, 1)$ in the plane.

Idea. Use the fact that all but one of the points lie on a straight line.

Answer. $d(n) = n$.

Solution. Because $(1, 2)$ and $(2, 1)$ do not lie on a horizontal line, $d(1) = 1$. From now on, assume $n \geq 2$. Let $p(x)$ be a polynomial whose graph passes through the given points, so $p(1) = 2$, $p(2) = 3$, ..., $p(n) = n+1$, $p(n+1) = 1$. Let $q(x) = p(x) - (x+1)$. Then $q(x)$ has roots $1, 2, \ldots, n$, so $q(x)$ must be divisible by $(x-1)(x-2)\cdots(x-n)$. Also, $q(x)$ is not the zero polynomial, because $q(n+1) \neq 0$. Thus $q(x)$, and hence $p(x)$, must have degree at least n.

To achieve degree n, try $p(x) = x+1+\alpha(x-1)(x-2)\cdots(x-n)$ and adjust the constant α so that $p(n+1) = 1$. This leads to
$$p(x) = x + 1 - \frac{n+1}{n!}(x-1)(x-2)\cdots(x-n),$$
a polynomial of degree n whose graph passes through the given points.

Comment. Given any $n+1$ points with distinct x-coordinates, there is a unique polynomial of degree at most n whose graph passes through those points. This *interpolation polynomial* can be written in various forms, perhaps the best known of which is named after Lagrange (although it appears to have been discovered by Waring first).

Problem 51

Show that every 2×2 matrix of determinant 1 is the product of *three* elementary matrices.

Idea. The product of upper triangular matrices is upper triangular; also, the product of lower triangular matrices is lower triangular. Thus, a factorization into a product of elementary matrices must generally involve matrices of type $\begin{pmatrix} 1 & x \\ 0 & 1 \end{pmatrix}$ and of type $\begin{pmatrix} 1 & 0 \\ x & 1 \end{pmatrix}$. On the other hand, if a factorization only had one matrix of type $\begin{pmatrix} y & 0 \\ 0 & 1 \end{pmatrix}$ or $\begin{pmatrix} 1 & 0 \\ 0 & y \end{pmatrix}$, the determinant would be y, so it is probably not worth including any such matrices.

Solution. Let $\begin{pmatrix} a & b \\ c & d \end{pmatrix}$ be a matrix of determinant 1. We first try to write

$$\begin{pmatrix} a & b \\ c & d \end{pmatrix} = \begin{pmatrix} 1 & x \\ 0 & 1 \end{pmatrix} \begin{pmatrix} 1 & 0 \\ y & 1 \end{pmatrix} \begin{pmatrix} 1 & z \\ 0 & 1 \end{pmatrix}.$$

By straightforward computation, this is equivalent to

$$a = 1 + xy, \quad b = z(1 + xy) + x, \quad c = y, \quad d = yz + 1. \qquad (*)$$

If $c \neq 0$, then we can solve $(*)$ for y, x, z in that order using all but the second equation; the result is $y = c$, $x = (a-1)/c$, $z = (d-1)/c$. For these choices of x, y, z the second equation in $(*)$ is satisfied as well, since

$$z(1 + xy) + x = c^{-1}(d-1)(1 + a - 1) + c^{-1}(a - 1) = c^{-1}(ad - 1) = b,$$

where we have used $\det \begin{pmatrix} a & b \\ c & d \end{pmatrix} = 1$. So we are done unless $c = 0$. If $c = 0$, the matrix will be of the form

$$\begin{pmatrix} a & b \\ 0 & a^{-1} \end{pmatrix} = \begin{pmatrix} a & 0 \\ 0 & 1 \end{pmatrix} \begin{pmatrix} 1 & b \\ 0 & 1 \end{pmatrix} \begin{pmatrix} 1 & 0 \\ 0 & a^{-1} \end{pmatrix}.$$

Problem 52

Let $ABCD$ be a convex quadrilateral. Find a necessary and sufficient condition for a point P to exist inside $ABCD$ such that the four triangles ABP, BCP, CDP, DAP all have the same area.

Answer. A necessary and sufficient condition is for one of the diagonals AC and BD to bisect the other.

Solution 1. To prove sufficiency, let Q be the intersection of AC and BD. Without loss of generality, we may assume that $AQ = CQ$. Let P be the midpoint of BD. Furthermore, let R be the foot of the perpendicular from A to BD and let S be the foot of the perpendicular from C to BD.

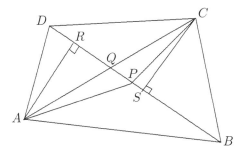

Since $BP = DP$ and since AR is the altitude for triangles ABP and DAP, these triangles have equal area. Similarly, the areas of triangles BCP and CDP are equal. Triangles ARQ and CSQ are congruent; therefore, $AR = CS$ and, hence, the areas of triangles ABP and BCP are equal. We conclude that the four triangles ABP, BCP, CDP, and DAP all have the same area.

To prove the necessity, assume that P is a point in the interior of $ABCD$ for which triangles ABP, BCP, CDP, and DAP have the same area (see next figure). Since triangles ABP and BCP have the same area, vertices A and C are equidistant from line BP. Because the quadrilateral $ABCD$ is convex, A and C lie on opposite sides of the line BP. An argument using congruent triangles (similar to the one in the sufficiency proof above) shows that the line BP must actually pass through the midpoint of the line segment AC. Similarly, the line DP must also pass through the midpoint of AC. If B, P, and D are not collinear, then P must be the midpoint of AC. But then, since BCP and CDP have the same area, B and D are equidistant from the line AC and thus AC bisects the segment BD. If B, P, and D are collinear, then P must be the midpoint of BD. In this case,

A and C are equidistant from the line BD, hence BD bisects the segment AC, completing the proof.

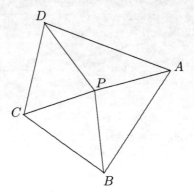

Solution 2. We prove the sufficiency as above. To prove the necessity, let θ_1, θ_2, θ_3, and θ_4 be the measures of the angles APB, BPC, CPD, and DPA, respectively.

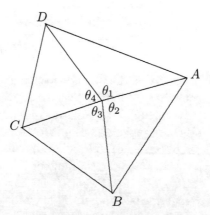

The area of an arbitrary triangle is half the product of two sides and the sine of their included angle. The product of the areas of triangles APB and CPD equals the product of the areas of BPC and DPA. Thus,

$$\sin\theta_1 \sin\theta_3 = \sin\theta_2 \sin\theta_4,$$

or, from the formula for the cosine of a sum,

$$\cos(\theta_1 - \theta_3) - \cos(\theta_1 + \theta_3) = \cos(\theta_2 - \theta_4) - \cos(\theta_2 + \theta_4).$$

Since $\theta_1 + \theta_2 + \theta_3 + \theta_4 = 2\pi$, this implies
$$\cos(\theta_1 - \theta_3) = \cos(\theta_2 - \theta_4).$$
There is no loss of generality in assuming
$$\theta_3 \leq \theta_1 < \pi \quad \text{and} \quad \theta_4 \leq \theta_2 < \pi$$
(if necessary, rename the vertices and/or reflect the quadrilateral). Under these conditions, $\theta_1 - \theta_3 = \theta_2 - \theta_4$, hence $\theta_1 + \theta_4 = \theta_2 + \theta_3$, from which we conclude that B, P, and D are collinear. As in the first solution, this implies that BD bisects AC.

Problem 53

Let k be a positive integer. Find the largest power of 3 which divides $10^k - 1$.

Solution. If $k = 3^m n$, where n is not divisible by 3, we prove that $10^k - 1$ is divisible by 3^{m+2}, but not by 3^{m+3}.

The proof is by induction on m, and uses the fact that an integer is divisible by 3 or 9 if and only if the sum of its digits is divisible by 3 or 9, respectively. For $m = 0$, k is not divisible by 3. We have
$$10^k - 1 = 9 \cdot \underbrace{111 \cdots 1}_{k \text{ digits}},$$
and the sum of the digits of $111\cdots 1$ is k, so $10^k - 1$ is divisible by 3^2, but not by 3^3. We now assume the claim holds for m. Let $k = 3^{m+1}n$ where n is not divisible by 3. Then
$$10^k - 1 = (10^{3^m n})^3 - 1 = (10^{3^m n} - 1)(10^{2 \cdot 3^m n} + 10^{3^m n} + 1).$$
By the inductive hypothesis, the largest power of 3 dividing the first factor is 3^{m+2}. The sum of the digits of the second factor is 3, which is divisible by 3, but not by 3^2. Therefore, the highest power of 3 dividing $10^k - 1$ is 3^{m+3}, and the proof is complete.

Problem 54

a. Show that there is always at least one normal line, and that there are at most three normal lines, to $y = x^2$ that pass through any given point P.

b. Show that there are exactly two normal lines to $y = x^2$ that pass through $Q = (4, 7/2)$, and find and sketch the set of all points Q with this property.

c. Are there any such points Q for which *both* coordinates of Q are integers?

Solution. Let P be the point (p, q) in the plane, and let X be the point (a, a^2) on the parabola. The equation of the normal line at X is

$$y - a^2 = -\frac{1}{2a}(x - a),$$
$$2a^3 + (1 - 2y)a - x = 0.$$

(The latter form includes the case $a = 0$, where the normal line is vertical.) Thus, point P will be on this line if and only if

$$2a^3 + (1 - 2q)a - p = 0. \qquad (*)$$

a. Because a cubic equation has at least one real root and at most three real roots, $(*)$ has at least one and at most three solutions for a, so there are at least one and at most three points X on the parabola such that the normal line at X passes through P.

b. For $p = 4, q = 7/2$, equation $(*)$ becomes $2a^3 - 6a - 4 = 0$, which can be written as $(a+1)^2(a-2) = 0$. This has exactly two solutions $a = -1, a = 2$, and so there are exactly two normal lines through Q.

To find the set of all points Q with this property, we need to find all values of the coordinates (p, q) for which $(*)$ has exactly two (real) solutions for a. Now the polynomial $f(a) = 2a^3 + (1 - 2q)a - p$ will always factor completely over the complex numbers, say as

$$f(a) = 2(a - a_1)(a - a_2)(a - a_3).$$

Comparing coefficients of a^2, we see that $0 = -2(a_1 + a_2 + a_3)$. In particular, because two of the roots must be real, all three must be real, so to have exactly two distinct roots there must be a double root but no triple root. If a_1 is the double root, then $f(a_1) = f'(a_1) = 0$, so

$$2a_1^3 + (1 - 2q)a_1 - p = 0 \quad \text{and} \quad 6a_1^2 + (1 - 2q) = 0.$$

Multiplying the second of these by $a_1/3$ and subtracting, we get

$$\frac{2}{3}(1 - 2q)a_1 - p = 0.$$

Now if $1-2q = 0$, then $p = 0$ and $a = 0$ is a triple root of $f(a)$, contradiction. So $1 - 2q \neq 0$ and $a_1 = \dfrac{3p}{2(1-2q)}$. Substituting this into $6a_1^2 + (1-2q) = 0$, we find

$$6\left(\frac{9p^2}{4(1-2q)^2}\right) + (1-2q) = 0$$
$$27p^2 + 2(1-2q)^3 = 0, \qquad 1 - 2q \neq 0.$$

So the desired set is the curve $27x^2 + 2(1-2y)^3 = 0$, *except* for the single point $(0, 1/2)$.

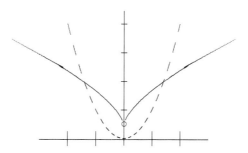

Points above the curve will have three normal lines whereas points below the graph will have only one normal line, as illustrated in the following figures.

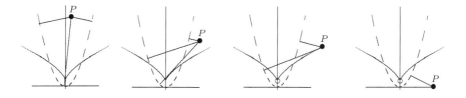

c. No, p and q can't both be integers, for suppose they are. Then from $27p^2 + 2(1-2q)^3 = 0$ we see that $27p^2$ is even, so p is even. But then $27p^2$ is divisible by 4, so $2(1-2q)^3$ is divisible by 4, and thus $1 - 2q$ is even, contradiction.

Problem 55

Suppose that n dancers, n even, are arranged in a circle so that partners are directly opposite each other. During the dance, two dancers who are next to each other change places while all others stay in the same place; this is repeated with different pairs of adjacent dancers until, in the ending position, the two dancers in each couple are once again opposite each other, but in the opposite of the starting position (that is, every dancer is halfway around the circle from her/his original position). What is the least number of interchanges (of two adjacent dancers) necessary to do this?

Solution. The least number of interchanges required is $n^2/4$.

If the dancers are numbered $1, 2, \ldots, n$ as we move clockwise around the circle, first begin by interchanging dancer number $n/2$ with dancer $n/2+1$, then with dancer $n/2+2, \ldots$, and finally with dancer n. We have made $n/2$ interchanges and dancer $n/2$ is in the correct position. Next, interchange dancer $n/2-1$ with dancers $n/2+1, n/2+2, \ldots, n$. Continuing this process until dancer 1 has switched with dancers $n/2+1, n/2+2, \ldots, n$, we arrive at the desired position, having made $(n/2)^2 = n^2/4$ interchanges. On the other hand, each dancer starts $n/2$ places away from his or her final position, and thus must participate in at least $n/2$ interchanges. Each interchange is between two dancers; therefore, at least $(n \cdot n/2)/2 = n^2/4$ interchanges are required, completing the proof.

Comments. Another systematic way to move each dancer halfway around the circle using $n^2/4$ interchanges is as follows. As a first step, interchange the dancers in positions 1 and 2, those in positions 3 and 4, etc., for $n/2$ interchanges in all. In the second step, interchange the dancers now in positions 2 and 3, those in positions 4 and 5, etc. The third step is like the first, the fourth is like the second, etc. After $n/2$ steps, or $n^2/4$ interchanges, all dancers will have shifted either $n/2$ places forward, or $n/2$ places backward, so they are halfway around the circle from their original positions.

To the best of our knowledge, $n^2/4$ is the largest number of interchanges required to arrive at an arrangement of dancers from any other arrangement.

Problem 56

Suppose that n points in the plane are such that there are just two different distances between them. What is the largest possible value of n?

Answer. The largest possible value of n is 5.

Solution. To see that $n = 5$ is possible, take the five vertices of a regular pentagon; the distance between any two such vertices is either the length of a side or the length of a diagonal, so because all sides have the same length and all diagonals do also, there are just two different distances between the five points.

To show that no larger value of n is possible, it's enough to show that $n = 6$ is impossible. Suppose we did have six points with just two distances a, b between them. Pick any of the six points and connect it to the five others. Of these five line segments, at least three must have the same length, say x. Consider the three "other" endpoints of those three line segments. If any two of these are at distance x, we have an equilateral triangle (of side length x) formed by those points and the original chosen point. If not, all three of the "other" endpoints have the "other" distance to each other, and so they form an equilateral triangle. (Note: This is the standard argument that shows that the Ramsey number $R(3,3)$ is at most 6.)

In either case, three of the six points must form an equilateral triangle, and let's assume that this triangle, call it ABC, is normalized so that it has side length $a = 1$. We have three other points, D, E, F. For each of these, say D, there are several cases: (i) D has distance 1 to two of A, B, C, say B and C; (ii) D has distance b to exactly two of A, B, C, say B and C; (iii) D has distance b to all three of A, B, C.

The following figures show the possibilities in each case. In case (ii), there are two possibilities for the configuration: D must be on the perpendicular bisector of BC, but there are two ways its distance to A can equal 1.

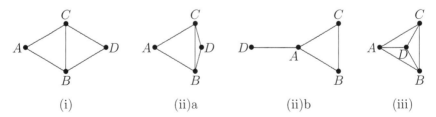

(i) (ii)a (ii)b (iii)

In case (i), AD is twice the altitude of the equilateral triangle ABC, so

$$AD = 2 \cdot \frac{\sqrt{3}}{2} = \sqrt{3}.$$

In case (ii)a, $\angle DAB = 30°$, so by the law of cosines,

$$BD = \sqrt{1^2 + 1^2 - 2 \cdot 1 \cdot 1 \cdot \cos 30°} = \sqrt{2 - \sqrt{3}}.$$

In case (ii)b, again by the law of cosines,

$$BD = \sqrt{1^2 + 1^2 - 2 \cdot 1 \cdot 1 \cos 150°} = \sqrt{2 + \sqrt{3}}.$$

In case (iii), AD is two-thirds of the altitude of triangle ABC, so

$$AD = \frac{2}{3} \cdot \frac{\sqrt{3}}{2} = \frac{1}{\sqrt{3}}.$$

All four of these distances are distinct (look at their squares), so E and F must fall into exactly the same case as D. Case (iii) is immediately ruled out because there is only one possible location for D. In cases (i), (ii)a, and (ii)b, by the rotational symmetry of $\triangle ABC$ about its center, $\triangle DEF$ must be equilateral.

Then in cases (i) and (ii)b, the distance from D to E is clearly greater than the two distances we have already, a contradiction. In case (ii)a, say E is on the perpendicular bisector of AC and F is on the perpendicular bisector of AB, as shown.

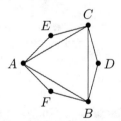

For there to be just two distances between points, ED must equal AC, so $\triangle ECD$ and $\triangle AEC$ are congruent. Then by symmetry, all the angles of hexagon $AECDBF$ are equal. However, $\angle BDC$ actually equals $150°$, not $120°$, and we have a contradiction.

Problem 57

Let L be a line in the plane; let A and B be points on L which are a distance 2 apart. If C is any point in the plane, there may or may not (depending on C) be a point X on the line L for which the distance from X to C is equal to the average of the distances from X to A and B. Give a precise description of the set of all points C in the plane for which there is no such point X on the line.

THE SOLUTIONS

Answer. The set of all points C in the plane for which there is no such point X consists of those points on the perpendicular bisector of AB whose distance from AB is greater than 1 (see the first figure below).

Solution 1. Let M denote the midpoint of AB. First note that if X is a point on the line L, the average of the distances from X to A and B is the distance from X to M if X is outside the line segment AB, and 1 if X is on the line segment AB. Therefore, for the distance from X to C to equal this average, we must have X at the same distance to M and C if X is outside AB, while we must have X at distance 1 to C if X is on AB.

It is easy to show that the set of points C for which there is a point X on AB with distance 1 to C is the closed set Γ shown in the second figure. (The boundary of Γ consists of two line segments parallel to AB on either side of AB

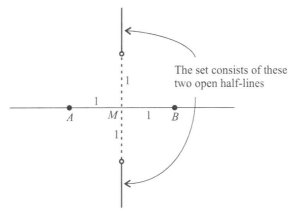

with distance 1 to AB and the same length as AB, and two semicircles. The semicircles are centered at A and at B; they have radius 1, and the diameter connecting the end points of each semicircle is perpendicular to AB.) Thus, for any point C in Γ, there is a point X on AB whose distance to C is equal to the average of the distances from X to A and B.

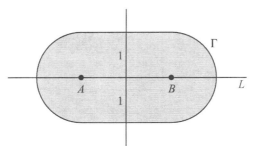

Suppose that C is not in Γ and that MC is not perpendicular to AB. Then the perpendicular bisector of MC is not parallel to AB; let X be the intersection point of this perpendicular bisector with L.

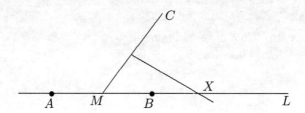

If X were in the line segment AB, we would have $CX = MX \leq 1$ and thus C would be in Γ. Therefore, X is outside the line segment AB, and since the distance from X to C equals the distance from X to M, this distance equals the average of the distances from X to A and B.

Thus, the only points C in the plane for which there *might* not be a suitable point X on the line are the points C on the two rays consisting of those points on the perpendicular bisector of AB which are not in Γ. If C is on one of these two open rays, there is no point X: X cannot be on AB because then C would have to be in Γ, and X cannot be outside AB because then X would have to be on the perpendicular bisector of MC, which is parallel to L and thus does not intersect L.

Solution 2. (Bruce Reznick, University of Illinois) We can assume that L is the x-axis, $A = (-1, 0)$ and $B = (1, 0)$. Let $C = (a, b)$, and by symmetry, suppose $a \geq 0$ and $b \geq 0$. The average of the distances of a point $X = (u, 0)$ on the x-axis to A and B is $\max\{1, |u|\}$; this equals the distance from X to C if and only if $(a - u)^2 + b^2 = \max\{1, |u|^2\}$.

There are two cases. First, consider the equation $(a - u)^2 + b^2 = u^2$. If $a \neq 0$, it has the unique solution $u = (a^2 + b^2)/2a$. This gives a valid X if $|u| \geq 1$; that is, if $a^2 + b^2 \geq 2a$, or $(a - 1)^2 + b^2 \geq 1$. If $a = 0$, the equation has a solution with $|u| \geq 1$ if and only if $b = 0$. Thus, we can find X for C unless C is on the positive y-axis or inside the circle with center $(1, 0)$ and radius 1.

Next, consider the equation $(a - u)^2 + b^2 = 1$. It has a solution if and only if $b \leq 1$, in which case its solutions are $u = a \pm \sqrt{1 - b^2}$. The solution with the smaller absolute value is $a - \sqrt{1 - b^2}$. This leads to an X for C provided $a - \sqrt{1 - b^2} \leq 1$. If $a \leq 1$, this holds for all $b \leq 1$; in particular, it holds if C is on the positive y-axis with $b \leq 1$. If $a > 1$, this holds if and only if $(a-1)^2 + b^2 \leq 1$; that is, if C is inside the circle with center $(1,0)$ and radius 1.

We conclude that the only points C for which no X exists are the points $(0, b)$, $|b| > 1$.

Problem 58

Show that there exists a positive number λ such that
$$\int_0^\pi x^\lambda \sin x \, dx = 3.$$

Solution 1. We first observe that
$$\int_0^\pi \sin x \, dx = 2 \quad (\lambda = 0) \quad \text{and} \quad \int_0^\pi x \sin x \, dx = \pi \quad (\lambda = 1),$$

the latter integral computed using integration by parts. Because $2 < 3 < \pi$, the Intermediate Value Theorem will imply the existence of a λ between 0 and 1 for which
$$\int_0^\pi x^\lambda \sin x \, dx = 3,$$
provided we prove that
$$f(\alpha) = \int_0^\pi x^\alpha \sin x \, dx$$
defines a continuous function on $[0, 1]$.

Suppose $\beta \geq \alpha \geq 0$. Note that $\sin x \geq 0$ on $[0, \pi]$. Further noting that $x^\beta \leq x^\alpha$ for $x \leq 1$, while $x^\beta \geq x^\alpha$ for $x \geq 1$, we have

$$|f(\beta) - f(\alpha)| = \left| \int_0^\pi (x^\beta - x^\alpha) \sin x \, dx \right|$$
$$\leq \int_0^\pi |x^\beta - x^\alpha| \sin x \, dx$$
$$= \int_0^1 (x^\alpha - x^\beta) \sin x \, dx + \int_1^\pi (x^\beta - x^\alpha) \sin x \, dx.$$

Because $|\sin x| \leq 1$, we obtain

$$|f(\beta) - f(\alpha)| \leq \int_0^1 (x^\alpha - x^\beta)\, dx + \int_1^\pi (x^\beta - x^\alpha)\, dx$$
$$= \frac{\pi^{\beta+1} - 2}{\beta + 1} - \frac{\pi^{\alpha+1} - 2}{\alpha + 1}.$$

This shows that, whether or not $\beta \geq \alpha$, we have

$$|f(\beta) - f(\alpha)| \leq \left| \frac{\pi^{\beta+1} - 2}{\beta + 1} - \frac{\pi^{\alpha+1} - 2}{\alpha + 1} \right|.$$

Because the right-hand side of the above tends to 0 as $\beta \to \alpha$, we have

$$\lim_{\beta \to \alpha} f(\beta) = f(\alpha),$$

hence f is continuous on $[0, 1]$, as claimed. The desired λ must therefore exist.

Solution 2. This and the next solution provide two alternative proofs that the function f defined above is continuous on $[0, 1]$.

Note that $\sin x \leq x$ for $x \geq 0$ (this can be shown by looking at the derivative of $x - \sin x$). Therefore,

$$|f(\beta) - f(\alpha)| = \left| \int_0^\pi (x^\beta - x^\alpha) \sin x\, dx \right|$$
$$\leq \int_0^\pi |x^\beta - x^\alpha| \sin x\, dx$$
$$\leq \int_0^\pi |x^{\beta+1} - x^{\alpha+1}|\, dx.$$

For any x with $0 < x \leq \pi$, the Mean Value Theorem applied to the function $g(y) = x^{y+1}$ implies that there exists a γ (which depends on x) between α and β for which

$$|x^{\beta+1} - x^{\alpha+1}| = x^{\gamma+1} |\ln x|\, |\beta - \alpha|.$$

It is well known that $\lim_{x \to 0^+} x \ln x = 0$ (one can use l'Hôpital's rule). This is enough to imply that $x^{\gamma+1} |\ln x|$ is bounded for $0 < x \leq \pi$, $0 \leq \gamma \leq 1$, and hence, for some constant C,

$$|x^{\beta+1} - x^{\alpha+1}| \leq C |\beta - \alpha|.$$

Therefore,
$$|f(\beta) - f(\alpha)| \leq \int_0^\pi C|\beta - \alpha|\, dx = C\pi|\beta - \alpha|.$$
This clearly implies $\lim_{\beta \to \alpha} f(\beta) = f(\alpha)$, hence the continuity of f.

Solution 3. For a third proof of the continuity of f, again begin with
$$|f(\beta) - f(\alpha)| \leq \int_0^\pi |x^{\beta+1} - x^{\alpha+1}|\, dx.$$

Because x^{y+1} is continuous on the closed, bounded rectangle $0 \leq x \leq \pi$, $0 \leq y \leq 1$, it is uniformly continuous on this rectangle. Given $\varepsilon > 0$, there exists a $\delta > 0$ such that for points (x_1, β) and (x_2, α) in the rectangle with distance between them less than δ, $|x_1^{\beta+1} - x_2^{\alpha+1}| < \varepsilon/\pi$. In particular, for the case $x_1 = x_2 = x$, we see that if $|\beta - \alpha| < \delta$, then
$$|f(\beta) - f(\alpha)| < \int_0^\pi \frac{\varepsilon}{\pi}\, dx = \varepsilon,$$
proving the continuity of f.

Problem 59

The equation $x^5 - 5x^4 + 8x^3 - 6x^2 + 3x + 3 = 0$ has two solutions that sum to 2. Using this information, find all solutions.

Answer. The roots of the quintic polynomial are $1 \pm \sqrt{2}$, $1 + \sqrt[3]{2}$, and $1 + (-1 \pm i\sqrt{3})/\sqrt[3]{4}$.

Solution. Let r_1 and r_2 be roots of the quintic polynomial for which $r_1 + r_2 = 2$. Then
$$(x - r_1)(x - r_2) = x^2 - 2x + p,$$
where $p = r_1 r_2$, must be a factor of the quintic. On the other hand, long division yields
$$x^5 - 5x^4 + 8x^3 - 6x^2 + 3x + 3$$
$$= \left(x^3 - 3x^2 + (2-p)x + (p-2)\right)\left(x^2 - 2x + p\right)$$
$$+ (p^2 - 1)x + (-p^2 + 2p + 3).$$

It follows that the remainder, $(p^2 - 1)x + (-p^2 + 2p + 3)$, must be zero, so

$$p^2 - 1 = 0 \quad \text{and} \quad -p^2 + 2p + 3 = 0.$$

Adding these, we get $2p + 2 = 0$, $p = -1$. So r_1 and r_2 must be the roots of $x^2 - 2x - 1 = 0$, namely $1 \pm \sqrt{2}$.

The other factor of the quintic is

$$x^3 - 3x^2 + 3x - 3 = (x-1)^3 - 2,$$

so the remaining roots will satisfy $(x-1)^3 = 2$. This yields the final real root $x = 1 + \sqrt[3]{2}$ along with two complex roots $x = 1 + \sqrt[3]{2}\omega$, $x = 1 + \sqrt[3]{2}\omega^2$, where

$$\omega = e^{2\pi i/3} = \cos(2\pi/3) + i\sin(2\pi/3) = -\frac{1}{2} + \frac{1}{2}i\sqrt{3}$$

is a primitive cube root of unity.

Problem 60

One regular n-gon is inscribed in another regular n-gon and the area of the large n-gon is twice the area of the small one.

a. What are the possibilities for n?

b. What are the possibilities for the angles that the sides of the large n-gon make with the sides of the small one?

Solution. a. We may suppose the larger n-gon is inscribed in a circle of radius 1. Let O denote the center of this circle, let A and B be adjacent vertices of the larger n-gon and let A' and B' be adjacent vertices of the smaller n-gon, with A' on AB. Take x to be the distance from A to A' along AB.

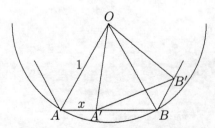

Because $\triangle AOB$ and $\triangle A'OB'$ are similar we will have the desired proportion, Area $\triangle AOB = 2 \cdot$ Area $\triangle A'OB'$, if and only if $OA^2 = 2(OA')^2$.

THE SOLUTIONS

Applying the law of cosines in $\triangle OAA'$ we have

$$OA' = \sqrt{1+x^2 - 2x\cos(\pi/2 - \pi/n)} = \sqrt{1+x^2 - 2x\sin(\pi/n)}\,.$$

Therefore, we want

$$1 = 2\Big(1 + x^2 - 2x\sin(\pi/n)\Big),$$
$$x^2 - 2x\sin(\pi/n) + 1/2 = 0,$$
$$x = \frac{2\sin(\pi/n) \pm \sqrt{4\sin^2(\pi/n) - 2}}{2}.$$

For x to be real, we need $\sin^2(\pi/n) \geq 1/2$. This means that $n \leq 4$.

For $n = 4$, we get

$$x = \frac{2(\sqrt{2}/2) \pm \sqrt{4(1/2) - 2}}{2} = \sqrt{2}/2,$$

while for $n = 3$,

$$x = \frac{2(\sqrt{3}/2) \pm \sqrt{4(3/4) - 2}}{2} = \frac{\sqrt{3} \pm 1}{2}.$$

Note that in each case, $0 \leq x \leq AB$ (for $n = 4$, $AB = \sqrt{2}$, and for $n = 3$, $AB = \sqrt{3}$), so the construction is really possible.

b. As we've seen, for $n = 4$, $AB = \sqrt{2}$ and $x = \sqrt{2}/2$, so that the corners of the inscribed square are at the midpoints of the outer square, and the angle between the sides of the squares is $\pi/4$.

For $n = 3$, $AB = \sqrt{3}$ and $x = (\sqrt{3} \pm 1)/2$. Let $\theta = \angle B'A'B$. For the smaller x, $BB' = AA' = (\sqrt{3}-1)/2$ and $A'B = \sqrt{3}-(\sqrt{3}-1)/2 = (\sqrt{3}+1)/2$, as shown.

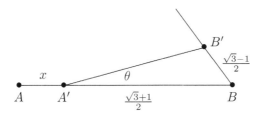

Because $A'B' = \sqrt{3/2}$, by the law of sines

$$\sin\theta = \frac{BB'}{A'B'}\sin\angle B = \frac{(\sqrt{3}-1)/2}{\sqrt{3/2}}\sin(\pi/3) = \frac{\sqrt{3}-1}{2\sqrt{2}}\,.$$

By the double angle formula

$$\cos 2\theta = 1 - 2\sin^2\theta = \frac{\sqrt{3}}{2},$$

so $2\theta = \pi/6$, or $\theta = \pi/12$. Because the equilateral triangles have angles $\pi/3$, the sides of the large triangle make angles $\pi/12$, $7\pi/12$ with the sides of the small one.

The case where $x = (\sqrt{3}+1)/2$ is reflected from the previous case, so the angles between the sides are again $\pi/12$, $7\pi/12$.

Problem 61

Consider a rectangular array of numbers, extending infinitely to the left and right, top and bottom. Start with all the numbers equal to 0 except for a single 1. Then, expanding outward from this nonzero entry, go through a series of steps, where at each step each number gets replaced by the sum of its four neighbors.

a. After n steps, what will be the sum of all the numbers in the array, and why?

b. After n steps, what will be the number in the center of the array (at the position of the original 1)?

c. Can you describe the various nonzero numbers that will occur in the array after n steps?

Solution. a. Each number, say k, that is in the array at one step, contributes k to each of its neighbors for the next step, for a total contribution of $4k$. As a result, the sum of all the numbers is multiplied by 4 at each step. Because the sum is 1 after 0 steps (and therefore 4 after 1 step, 16 after 2 steps, and so forth), it will be 4^n after n steps.

b. If you "color" the array in a checkerboard pattern with a white square at the center, then since every white square is surrounded by black squares and vice versa, after any odd number of steps all the numbers on the white squares will be zero, and after any even number of steps all the numbers on the black squares will be zero. In particular, if n is *odd* there will be a 0 at the center of the array after n steps. If n is *even*, after n steps the number at the center of the array will be $\binom{n}{n/2}^2$, which we'll show in part (c) when we describe *all* the nonzero numbers in the array.

THE SOLUTIONS 113

c. As we've seen, the nonzero numbers will always be interspersed with zeros in a checkerboard pattern. Also, it's easy to see that they will keep forming a "diamond" shape. If we omit the zeros and turn the pattern through 45°, then for the first few values of n, we see (for reference, the center of the array is circled):

$$
\begin{array}{ccccc}
& & & & 1\ 4\ 6\ 4\ 1 \\
& & 1\ 3\ 3\ 1 & & 4\ 16\ 24\ 16\ 4 \\
& 1\ 2\ 1 & 3\ 9\ 9\ 3 & & 6\ 24\ \textcircled{36}\ 24\ 6 \\
\textcircled{1}\ 1 & 2\ \textcircled{4}\ 2 & 3\ 9\ 9\ 3 & & 4\ 16\ 24\ 16\ 4 \\
\textcircled{1} & 1\ 1 & 1\ 2\ 1 & 1\ 3\ 3\ 1 & 1\ 4\ 6\ 4\ 1 \\
n=0 & n=1 & n=2 & n=3 & n=4
\end{array}
$$

It looks like the numbers along the outside of each square are the numbers in the n-th row of Pascal's triangle, and once you notice this it's easy to see why, because the outside of each square is formed from the outside of the previous square by adding adjacent (nonzero) numbers, which is exactly how each row of Pascal's triangle is formed from the previous row. Also, the number in the k-th row and the ℓ-th column in each of the squares above seems to be the product of the number at the beginning of that row and the number at the beginning of that column. That is, if we number the rows and columns of the square starting from 0 then it seems like the number in the k-th row and ℓ-th column is always $\binom{n}{k} \cdot \binom{n}{\ell}$. If this is true, then for even n the number at the center, which is row $n/2$ and column $n/2$, will be $\binom{n}{n/2}^2$, as claimed in part (b).

To see that this pattern persists, suppose it works for n, and look at the $(k-1)$-st and k-th rows and the $(\ell-1)$-st and ℓ-th columns of the square for n. Where they intersect, the entries are

$$\binom{n}{k-1}\binom{n}{\ell-1} \qquad \binom{n}{k-1}\binom{n}{\ell}$$

$$\binom{n}{k}\binom{n}{\ell-1} \qquad \binom{n}{k}\binom{n}{\ell}$$

(As usual, a binomial coefficient $\binom{n}{i}$ with $i < 0$ or $i > n$ should be interpreted as 0.) In the actual array, these numbers are interspersed with zeros and turned 45°, so the corresponding part of the actual array looks like

$$\begin{array}{ccc} 0 & \binom{n}{k-1}\binom{n}{\ell-1} & 0 \\ \binom{n}{k}\binom{n}{\ell-1} & 0 & \binom{n}{k-1}\binom{n}{\ell} \\ 0 & \binom{n}{k}\binom{n}{\ell} & 0 \end{array}$$

At the next step, the zero that is now surrounded by the four nonzero numbers will be replaced by their sum, which is

$$\binom{n}{k-1}\binom{n}{\ell-1} + \binom{n}{k-1}\binom{n}{\ell} + \binom{n}{k}\binom{n}{\ell-1} + \binom{n}{k}\binom{n}{\ell}$$
$$= \left[\binom{n}{k-1} + \binom{n}{k}\right]\left[\binom{n}{\ell-1} + \binom{n}{\ell}\right]$$
$$= \binom{n+1}{k}\binom{n+1}{\ell}.$$

This shows that the pattern persists at the next step, and our proof by induction is complete.

Problem 62

What is the fifth digit from the end (the ten thousands digit) of the number $5^{5^{5^{5^5}}}$?

Idea. Remainder problems generally exhibit some sort of periodic behavior.

Answer. The fifth digit from the end is 0.

Solution 1. We are interested in the remainder when $5^{5^{5^{5^5}}}$ is divided by $100{,}000 = 2^5 5^5$. Clearly 5^r is divisible by 5^5 for $r \geq 5$. We determine those r for which $5^r - 5^5$ is divisible by 2^5 as well, implying that 5^r and 5^5 have the same last five digits. This will be the case if and only if $5^{r-5} - 1$ is divisible by 32. The binomial expansion of $(1+4)^{r-5}$ starts out

$$1 + (r-5)4 + \frac{(r-5)(r-6)}{2}4^2 + \cdots,$$

and all further terms are divisible by 32. Thus, $5^{r-5} - 1$ is divisible by 32 if and only if

$$(r-5) + \frac{(r-5)(r-6)}{2} 4$$

is divisible by 8. This certainly occurs for r of the form $8k+5$. Since, for any m, $(2m+1)^2 = 4m(m+1) + 1$, 5 to any even power is of the form $8k+1$ and 5 to any odd power is of the form $8k+5$. In particular, $5^{5^{5^5}}$ is of the form $8k+5$, so it has the same last five digits as $5^5 = 3125$. Thus, the fifth digit from the end is 0.

Solution 2. We make repeated use of Euler's theorem, which states that if a is an integer relatively prime to $n > 1$, then $a^{\phi(n)} \equiv 1 \pmod{n}$, where $\phi(n)$ is the number of positive integers which are less than or equal to n and relatively prime to n.

As in Solution 1, we are interested in the remainder when $5^{5^{5^{5^5}}}$ is divided by $100{,}000 = 2^5 5^5$. Because it is divisible by 5^5, we need only determine the remainder upon division by 2^5. Since $\phi(2^5) = 16$, we want the remainder when $5^{5^{5^5}}$ is divided by 16. Since $\phi(16) = 8$, we want the remainder when 5^{5^5} is divided by 8. We now note that every odd square is congruent to 1 (mod 8), hence every odd power of 5 is congruent to 5 modulo 8. In particular,

$$5^{5^5} \equiv 5 \pmod{8}.$$

Working backward, we see that

$$5^{5^{5^5}} \equiv 5^5 = 3125 \equiv 5 \pmod{16}.$$

Finally, we conclude

$$5^{5^{5^{5^5}}} \equiv 5^5 \pmod{32}.$$

Rather than computing the remainder upon dividing 5^5 by 32, we observe that

$$5^{5^{5^{5^5}}} \equiv 5^5 \pmod{2^5} \quad \text{and} \quad \pmod{5^5},$$

hence (mod 10^5). Because $5^5 = 3125$, the fifth digit from the end of $5^{5^{5^{5^5}}}$ is a 0.

Problem 63

On a particular bridge deal, West had to play the very first card (on the first trick). One of her options was to play the king of hearts, and it turned out that if West did so, the East-West partnership would take all thirteen tricks, *no matter what*. On the other hand, if West started the play with any of her other twelve cards, the North-South partnership would take all thirteen tricks, *no matter what*! Given that North had the two of spades and the jack of clubs, who had the five of diamonds?

Answer. South had the five of diamonds.

Solution. First, consider the heart suit. Who had the ace? West can't have had it, because then she could have led it and taken the first trick. North and South can't have had it, because if either of them did they could have won the first trick with it if West led the king of hearts. So East had the ace of hearts. However, East cannot have had any other hearts, because if East had a second heart, it could be played "under" the king of hearts on the first trick. That would keep West "on lead" (first to play) for the second trick, and West could now play another card which North or South could win (since they would have won any original lead except the king of hearts). So of the hearts, East had only the ace. Similarly, West had only the king. (If West had another heart to lead, it could have been led to East's ace at the first trick.) Therefore, North/South had all the other 11 hearts.

Now suppose West led something other than the king of hearts and (as given) either North or South won the trick. If that player (North or South) had any heart, (s)he could legally lead it at trick 2, and East would win with the ace (for East cannot have had to play that ace on the first trick), contradicting the given. Conclusion: One of North/South must have had all eleven hearts except the ace and king, and that player must never have been able to win a trick. Suppose that player were South. Then North would have to win every trick after West led any of her cards other than the king of hearts. But North had the two of spades and could legally play it at the second trick; North could not possibly win that trick unless (s)he had *all* the remaining spades, but it is not hard to see that in that case North couldn't have been obliged to win the first trick and/or other players couldn't have been obliged to discard spades. So the player with eleven hearts must have been North, and we now know North's entire hand: Spades: 2; Hearts: Q, J, 10, 9, 8, 7, 6, 5, 4, 3, 2; Clubs: J.

Now, who had the missing aces? West can't have had any, because otherwise West could have led an ace to start play and win the trick. South must have had at least one, because otherwise East would have all four and could win the first trick. In fact, if South had only one ace (s)he could lose a trick after taking one by leading a suit in which (s)he didn't have the ace (South must have had at least two suits for otherwise (s)he could never take the first trick, because no one else could lead a card (s)he could take). On the other hand, East cannot have had a card in a suit in which South had the ace, because then East could lose a trick to South after taking the first trick with the ace of hearts. So all twelve of East's cards except the ace of hearts must have been *in the same suit* (since (s)he had no other hearts and South had aces in the other two suits). Also, they all had to be winners no matter what, so the twelve cards must be the top twelve of the suit (else they would have included the 2 and East could lead the 2 and lose the trick to the thirteenth card of the suit). In particular, the suit can't have been clubs since North had the jack of clubs. Suppose the suit were diamonds. Who had the 2 of diamonds? North didn't, and West didn't (else West could lead it and East would win). But if South had it, (s)he could have led it after winning the first trick, and East would then get all the other tricks, contradiction. So East's long suit can't have been diamonds; it must have been spades. Now we know East's entire hand: Spades: A, K, Q, J, 10, 9, 8, 7, 6, 5, 4, 3; Hearts: A.

So what's left for West and South? West had the king of hearts, but otherwise they had only clubs and diamonds. Every diamond West had was lower than every diamond South had, else there would be a legal way for West to win the first trick with a diamond. Since North must never have been able to win a trick, either West or South must have had no clubs lower than the jack. But if West had no clubs lower than the jack, West could have led a club higher than the jack and won the trick when South played a lower one – unless West had no clubs at all. But if West had no clubs at all, she would have had 12 diamonds, leaving South with only the ace and with 12 clubs; after winning the ace of diamonds, South could play a low club to North's jack. So it must have been South who had no clubs below the jack; in other words, South had at most three clubs (A, K, Q) and thus at least ten diamonds. South's diamonds were all higher than the only others, because West had the only others, and so South had at least the top ten diamonds. But the tenth diamond from the top is the 5, so South had the five of diamonds, as claimed. One possibility for the actual hand is:

```
                    ♠: 2
                    ♡: Q J 10 9 8 7 6 5 4 3 2
                    ◇: —
                    ♣: J
                          North

♠: —                                          ♠: A K Q J 10 9 8 7 6 5 4 3
♡: K                                           ♡: A
◇: 4 3 2      West            East             ◇: —
♣: 10 9 8 7 6 5 4 3 2                          ♣: —

                          South
                    ♠: —
                    ♡: —
                    ◇: A K Q J 10 9 8 7 6 5
                    ♣: A K Q
```

[For bridge players: The bidding, with a few comments:

North: (dealer)	4 Hearts	(A bit conservative, perhaps. About half the time, you'll make this without *any* help from partner.)
East:	7 Spades	(Not subtle, but surely you'll make.)
South:	7 No Trump	(East is known as a solid bidder, so if I trust my partner, it sounds like East has 13 spades, in which case I'll make 7 No Trump, assuming my partner has the ace of hearts.)
West:	Pass	
North:	Pass	
East:	Double	(More than flesh and blood can bear . . .)
	All Pass	

Did West find the killing lead of the king of hearts despite North's bid? Or did West opt for the "safe" ten of clubs? You can choose your own ending to the story . . .   .]

Problem 64

Describe the set of all points P in the plane such that exactly two tangent lines to the curve $y = x^3$ pass through P.

Solution. We will show that all points P on exactly two tangent lines are of the form $(a, 0)$ or (a, a^3), where $a \neq 0$.

Let $P = (a, b)$. If a tangent to the curve $y = x^3$ passes through P, then the point of tangency (x, x^3) satisfies

$$x^3 - b = 3x^2(x - a),$$

or

$$2x^3 - 3ax^2 + b = 0.$$

If the tangent lines through distinct points (x_1, x_1^3) and (x_2, x_2^3) have the same slope, then we must have $x_1 = -x_2$. However, the tangent line through (x_1, x_1^3) is

$$y = 3x_1^2 x - 2x_1^3,$$

while the one through $(-x_1, -x_1^3)$ is

$$y = 3x_1^2 x + 2x_1^3.$$

Thus, no line is tangent to $y = x^3$ at two points. We conclude that there are exactly two tangents to the curve which pass through P if and only if

$$p(x) = 2x^3 - 3ax^2 + b$$

has exactly two real roots.

Since a cubic polynomial with two real roots must have three, not necessarily distinct, real roots, $p(x)$ must have a multiple root. This occurs if and only if $p(x)$ and $p'(x)$ have a common root. Since $p'(x) = 6x^2 - 6ax$, this common root must be either 0 or a. If $p(0) = 0$, then $b = 0$. The third root of $p(x)$ is then $3a/2$, which is a distinct real root provided $a \neq 0$. On the other hand, if $p(a) = 0$, then $b = a^3$. The third root of $p(x)$ is now $-a/2$, hence again $a \neq 0$. Thus, the only points on exactly two tangent lines to $y = x^3$ are of the form $(a, 0)$ or (a, a^3), where $a \neq 0$.

Problem 65

Show that for any $n \geq 3$, there exist two disjoint sets of size n, whose elements are positive integers, such that the two sets have the same sum and the same product.

Solution 1. The case $n = 3$ is given in the statement of the problem. For $n = 4$ we can take $A = \{1, 8, 12, 24\}$ and $B = \{3, 4, 6, 32\}$, where the common sum is 45 and the common product is $2^8 3^2$. For $n = 5$ an example is $A = \{1, 8, 12, 24, 36\}$ and $B = \{2, 6, 9, 16, 48\}$, where the common sum is 81 and the common product is $2^{10} 3^4$.

Now suppose the result is true for all $k < n$, and choose disjoint sets A and B of $n - 3$ distinct integers which have the same sum, say S, and the same product, say P. Pick an integer m larger than any number in A or B (or a prime number that doesn't occur as a factor in any element of A or B) and define sets

$$A^* = A \cup \{m, 8m, 12m\} \quad \text{and} \quad B^* = B \cup \{2m, 3m, 16m\}$$

Then A^* and B^* are disjoint and their elements are distinct. The elements of A^* add to $S + 21m$, as do the elements of B^*. The respective products are $96Pm^3$. Therefore A^* and B^* are sets with n elements that satisfy the desired conditions, so the result follows by induction.

Solution 2. Such a pair of sets with four elements is $A = \{1, 8, 12, 24\}$, $B = \{3, 4, 6, 32\}$. We will proceed by induction. Suppose that $A = \{a_1, a_2, \ldots, a_n\}$ and $B = \{b_1, b_2, \ldots, b_n\}$ are disjoint sets of n distinct elements with the same sum and the same product. Let

$$A^* = \{2^{a_1}, 2^{a_2}, \ldots, 2^{a_n}, \ 2^{a_1} + \cdots + 2^{a_n}, \ 2(2^{b_1} + \cdots + 2^{b_n})\} \quad \text{and}$$
$$B^* = \{2^{b_1}, 2^{b_2}, \ldots, 2^{b_n}, \ 2^{b_1} + \cdots + 2^{b_n}, \ 2(2^{a_1} + \cdots + 2^{a_n})\}.$$

Using the fact that every positive integer has a unique binary representation, we see that A^* and B^* are disjoint sets of $n+2$ elements; it is then straightforward to check that they have the same sum and the same product. The result follows by induction.

Observe that this solution doesn't use the part of the induction hypothesis that says the products are equal. So the solution could be tweaked to give an explicit, noninductive solution; for example, for $n \geq 3$,

$$A = \{2^1, 2^3, \ldots, 2^{2n-5}, 2^{3n-5}, 2^1 + 2^3 + \cdots + 2^{2n-5} + 2^{3n-5},$$
$$2(2^2 + 2^4 + \cdots + 2^{2n-4} + 2^{2n-3})\}$$
$$B = \{2^2, 2^4, \ldots, 2^{2n-4}, 2^{2n-3}, 2^2 + 2^4 + \cdots + 2^{2n-4} + 2^{2n-3},$$
$$2(2^1 + 2^3 + \cdots + 2^{2n-5} + 2^{3n-5})\}.$$

Problem 66

Let f and g be odd functions that are infinitely differentiable at $x = 0$, and assume that $f'(0) = g'(0) = 1$. Let $F = f \circ g$ and $G = g \circ f$.

a. Show that $F'(0) = G'(0)$, $F^{(3)}(0) = G^{(3)}(0)$, and $F^{(5)}(0) = G^{(5)}(0)$.

b. Show that for all even n, $F^{(n)}(0) = G^{(n)}(0)$.

c. Is it always true that for all odd n, $F^{(n)}(0) = G^{(n)}(0)$? If so, prove it; if not, give a counterexample.

Solution. a. Because $F(x) = f(g(x))$, we have $F'(x) = f'(g(x))g'(x)$, so $F'(0) = f'(g(0))g'(0)$. But g is an odd function, so $g(0) = 0$, so $F'(0) = f'(0)g'(0) = 1$ by the givens. Interchanging the roles of f and g, we also have $G'(0) = 1$, so $F'(0) = G'(0)$.

From $F'(x) = f'(g(x))g'(x)$ we get

$$F''(x) = f''(g(x))[g'(x)]^2 + f'(g(x))g''(x),$$
$$F^{(3)}(x) = f^{(3)}(g(x))[g'(x)]^3 + 3f''(g(x))g'(x)g''(x) + f'(g(x))g^{(3)}(x), \text{ so}$$
$$F^{(3)}(0) = f^{(3)}(0) + 3f''(0)g''(0) + g^{(3)}(0).$$

But this expression is symmetric in f and g, so if we interchange f and g (which will change F to G) we'll get the same answer (since the givens we used were also symmetric in f and g). Thus, $F^{(3)}(0) = G^{(3)}(0)$.

In the same way we find the fourth and fifth derivatives (you might prefer using a computer algebra system, such as *Mathematica* or *MAPLE*, to do this for you):

$$F^{(4)}(x) = f^{(4)}(g(x))[g'(x)]^4 + 6f^{(3)}(g(x))[g'(x)]^2 g''(x) + 3f''(g(x))[g''(x)]^2$$
$$+ 4f''(g(x))g'(x)g^{(3)}(x) + f'(g(x))g^{(4)}(x)$$
$$F^{(5)}(x) = f^{(5)}(g(x))[g'(x)]^5 + 10f^{(4)}(g(x))[g'(x)]^3 g''(x)$$
$$+ 15f^{(3)}(g(x))g'(x)[g''(x)]^2 + 10f^{(3)}(g(x))[g'(x)]^2 g^{(3)}(x)$$
$$+ 10f''(g(x))g''(x)g^{(3)}(x) + 5f''(g(x))g'(x)g^{(4)}(x) + f'(g(x))g^{(5)}(x).$$

Substituting $x = 0$ and using $g(0) = 0$ and $f'(0) = g'(0) = 1$, we get

$$F^{(5)}(0) = f^{(5)}(0) + 10f^{(4)}(0)g''(0) + 15f^{(3)}(0)[g''(0)]^2 + 10f^{(3)}(0)g^{(3)}(0)$$
$$+ 10f''(0)g''(0)g^{(3)}(0) + 5f''(0)g^{(4)}(0) + g^{(5)}(0).$$

Now, because $f(-x) = -f(x)$ for all x, $-f'(-x) = -f'(x)$ and $f''(-x) = -f''(x)$ for all x. Thus, $f''(x)$ is an odd function. In particular, $f''(0) = 0$;

similarly, $g''(0) = 0$, so the expression for $F^{(5)}(0)$ simplifies to

$$F^{(5)}(0) = f^{(5)}(0) + 10f^{(3)}(0)g^{(3)}(0) + g^{(5)}(0).$$

This is symmetric in f and g, so $F^{(5)}(0) = G^{(5)}(0)$.

Comment. There is a simpler solution in the case that f and g are analytic functions (functions that can be represented by their Maclaurin series). For suppose that

$$f(x) = a_0 + a_1 x + a_2 x^2 + a_3 x^3 + \cdots$$

where $a_k = f^{(k)}(0)/k!$ for $k = 0, 1, \ldots$. As previously pointed out, f'' is an odd function, so $f''(0) = 0$. Similarly, $f^{(4)}$ is an odd function, so $f^{(4)}(0) = 0$, and by induction $f^{(n)}(0) = 0$ for all even n. Also $a_1 = 1$ because $f'(0) = 1$. Therefore f has the form

$$f(x) = x + a_3 x^3 + a_5 x^5 + \cdots.$$

In the same manner

$$g(x) = x + b_3 x^3 + b_5 x^5 + \cdots$$

where $b_k = g^{(k)}(0)/k!$. It follows that

$$F(x) = f(g(x)) = (x + b_3 x^3 + b_5 x^5 + \cdots) + a_3 (x + b_3 x^3 + \cdots)^3$$
$$+ a_5 (x + b_3 x^3 + \cdots)^5 + \cdots$$
$$= x + (b_3 + a_3) x^3 + (b_5 + 3a_3 b_3 + a_5) x^5 + \cdots$$

and

$$G(x) = g(f(x)) = x + (a_3 + b_3) x^3 + (a_5 + 3b_3 a_3 + b_5) x^5 + \cdots.$$

From these Maclaurin series we read off that $F'(0) = G'(0) = 1$, $F^{(3)}(0) = G^{(3)}(0) = (a_3 + b_3) \cdot 3!$ and $F^5(0) = G^5(0) = (a_5 + 3a_3 b_3 + b_5) \cdot 5!$.

One can adapt this argument to our problem by writing f and g as Taylor polynomials of degree 6 with remainder; that is, write

$$f(x) = x + a_3 x^3 + a_5 x^5 + R_6(x)$$

where $R_6(x) = \int_0^x \frac{(x-t)^6}{6!} f^{(7)}(t) dt$, and similarly for g. Proceed to write

$$F(x) = f(g(x)) = x + (b_3 + a_3) x^3 + (b_5 + 3a_3 b_3 + a_5) x^5 + \text{terms involving}$$

the remainder terms and/or x^7 and higher powers of x

and similarly for $G(x)$. Then take five derivatives of these compositions making use of the fact (explained in the note below) that

$$R_6'(x) = \int_0^x \frac{6(x-t)^5}{6!} f^{(7)}(t)\, dt$$

so $R_6'(0) = 0$. Similarly, $R_6^{(k)}(0) = 0$ for $k = 2, 3, 4, 5$. The same holds for g and its remainder term.

Note. Write $R_6(u, v) = \int_0^u \frac{(v-t)^6}{6!} f^{(7)}(t)\, dt$. Then

$$R_6'(x) = \left.\frac{\partial R_6(u,v)}{\partial u}\right|_{u=v=x} + \left.\frac{\partial R_6(u,v)}{\partial v}\right|_{u=v=x}$$

$$= \frac{(x-x)^6}{6!} f^{(7)}(x) + \int_0^x \frac{6(x-t)^5}{6!} f^{(7)}(t)\, dt\,.$$

b. The composition of the odd functions f and g is odd because

$$F(-x) = f(g(-x)) = f(-g(x)) = -f(g(x)) = -F(x);$$

as previously shown, the second derivative of an odd function is odd. So

$$F''(0) = 0 = G''(0),\ F^{(4)}(0) = 0 = G^{(4)}(0),$$

and for all even n, $F^{(n)}(0) = 0 = G^{(n)}(0)$.

c. It is not necessarily true that $F^{(n)}(0) = G^{(n)}(0)$ for all n. As a counterexample, take $f(x) = x + x^3$ and $g(x) = x + 2x^3$. Then f and g are odd functions, $f'(0) = g'(0) = 1$, and

$$f(g(x)) = (x + 2x^3) + (x + 2x^3)^3 = x + 3x^3 + 6x^5 + 12x^7 + 8x^9,$$

$$g(f(x)) = (x + x^3) + 2(x + x^3)^3 = x + 3x^3 + 6x^5 + 6x^7 + 2x^9.$$

Therefore, $F^{(7)}(0) = 12 \cdot 7!$ whereas $G^{(7)}(0) = 6 \cdot 7!$.

Problem 67

In how many ways can the integers $1, 2, \ldots, n$, $n \geq 2$, be listed (once each) so that as you go through the list, there is exactly one integer which is immediately followed by a smaller integer?

Answer. The number of ways of listing the integers $1, 2, \ldots, n$ so that only one integer is immediately followed by a smaller integer is $2^n - (n+1)$.

Solution 1. Suppose the answer is $f(n)$. For instance, $f(2) = 1$ since the only possibility is 2 1, while $f(3) = 4$, with the possibilities 1 3 2, 2 1 3, 2 3 1, 3 1 2.

Suppose we have listed the numbers $1, 2, \ldots, n, n+1$ so there is exactly one decrease. If the number $n+1$ is at the end of the list, then if we leave it off we'll have a listing of $1, 2, \ldots, n$ with exactly one decrease. Conversely, for any listing of $1, 2, \ldots, n$ with exactly one decrease we can get a listing of $1, 2, \ldots, n, n+1$ with exactly one decrease by putting $n+1$ at the end of the list. Now suppose $n+1$ is not at the end of the list. Then since $n+1$ is the largest number on the list, the one decrease must be immediately after $n+1$. Then if we leave $n+1$ out, we'll get a list of the numbers $1, 2, \ldots, n$ which may or may not still have one decrease. Specifically, if the number before $n+1$ is x and the number after $n+1$ is y and $x > y$, then there will still be a decrease when we omit $n+1$. On the other hand, if $x < y$ or if there is no x (because $n+1$ is at the beginning of the list), then the new list of numbers $1, 2, \ldots, n$ will have no decrease. Now the only way to list the numbers $1, 2, \ldots, n$ with no decrease is to list them in order. Conversely, if we have these numbers in order we can insert $n+1$ immediately before any of $1, 2, \ldots, n$ and get a list of $1, 2, \ldots, n, n+1$ with exactly one decrease. Also, if we have a listing of $1, 2, \ldots, n$ with exactly one decrease, say from x to y, then we can get a listing of $1, 2, \ldots, n, n+1$ with one decrease by inserting $n+1$ between x and y. The upshot of all this is that

$$f(n+1) = f(n) + f(n) + n = 2f(n) + n$$

where the first $f(n)$ on the right is the number of ways to list $1, 2, \ldots, n$ with one decrease, then add $n+1$ to the end of the list, the second $f(n)$ is the number of ways to list $1, 2, \ldots, n$ with one decrease, then insert $n+1$ where the decrease occurs, and the final term, n, is the number of ways to insert $n+1$ before one of $1, 2, \ldots, n$ (in order). That is, $f(n+1) = 2\,f(n) + n$. So, from $f(2) = 1$, we get $f(3) = 2\,f(2) + 2 = 4$, $f(4) = 2\,f(3) + 3 = 11$, $f(5) = 2\,f(4) + 4 = 26$, and so forth.

If we just had $f(n+1) = 2\,f(n)$, the values of f would be powers of 2 (given that $f(2) = 1$), so perhaps we should look at how close we are to powers of 2. So let $g(n) = 2^n - f(n)$, and we get $g(2) = 4 - 1 = 3$, $g(3) = 8 - 4 = 4$, $g(4) = 16 - 11 = 5$, $g(6) = 32 - 26 = 6$, which certainly suggests $g(n) = n+1$, that is $f(n) = 2^n - (n+1)$. This is now easily shown by induction on n: It works for $n = 2$, and if it works for n, then $f(n+1) = 2\,f(n) + n = 2(2^n - n - 1) + n = 2^{n+1} - (n+2)$, as desired.

Solution 2. Let $f(n)$ denote the number of ways of listing $1, 2, \ldots, n$ so that only one number in the list is followed by a smaller number. In particular, $f(2) = 1$, since 2, 1 is the only way to do it when $n = 2$. Now suppose that $n > 2$. There are $f(n-1)$ ways to make such a list of $1, 2, \ldots, n$ where the first element of the list is 1, because the second element of such a list is necessarily larger than 1 so the decrease will be in the list of numbers $2, 3, \ldots, n$. If A is any non-empty subset of $\{2, 3, \ldots, n\}$ we can get a listing of $1, 2, \ldots, n$ with one decrease, by listing the numbers in A in increasing order, then 1 (a decrease), followed by the list of numbers in the complement of A (which may be empty) in increasing order. Conversely, any listing of $1, 2, \ldots, n$ with one decrease where 1 is not the first number in the list will correspond to a non-empty subset of $\{2, 3, \ldots, n\}$, namely, the subset of numbers in the list that precede 1. We conclude that the number of lists of $1, 2, \ldots, n$ with a single decrease, where 1 is not the first number in the list, is the same as the number of non-empty subsets of $\{2, 3, \ldots, n\}$, namely, $2^{n-1} - 1$. Therefore, $f(n) = f(n-1) + 2^{n-1} - 1$. Using this recurrence we find

$$\begin{aligned} f(n) &= f(n) - f(1) \\ &= \Big(f(n) - f(n-1)\Big) + \Big(f(n-1) - f(n-2)\Big) + \cdots + \Big(f(2) - f(1)\Big) \\ &= (2^{n-1} - 1) + (2^{n-2} - 1) + \cdots + (2^1 - 1) \\ &= 2^{n-1} + 2^{n-2} + \cdots + 2^1 + 1 - n = \frac{2^n - 1}{2 - 1} - n = 2^n - (n+1). \end{aligned}$$

Solution 3. Let $\mathcal{S}$ denote the set of all subsets of $\{1, 2, \ldots, n\}$ except for the subsets $\emptyset, \{1\}, \{1, 2\}, \ldots, \{1, 2, \ldots, n\}$. There are $2^n - (n+1)$ subsets in $\mathcal{S}$. We will show that there is a one-to-one correspondence between the subsets of S and the listings of $1, 2, \ldots, n$ which have exactly one decrease.

Suppose A is a subset in $\mathcal{S}$ (note that A is non-empty). We get a listing of $1, 2, \ldots, n$ with one decrease by listing the elements of A (in increasing order) followed by the numbers in the complement of A (in increasing order). There will be a number in the complement that is smaller than the largest number in A, so there will be exactly one decrease.

Conversely, suppose we have a listing of $1, 2, \ldots, n$ with one decrease. Let A be the set of numbers that occur before the decrease. Then A is one of the subsets in $\mathcal{S}$.

Clearly, the correspondence is one-to-one so the number of listings of $1, 2, \ldots, n$ with one decrease is $2^n - (n+1)$.

Solution 4. As in the previous solution, suppose the drop is after k, for $k = 2, \ldots, n$. Then the list is completely determined by which nonempty subset of $\{1, \ldots, k-1\}$ follows k. Thus, the number of such lists is

$$\sum_{k=2}^{n}(2^{k-1} - 1) = (2^n - 2) - (n-1) = 2^n - (n+1).$$

Problem 68

A two-player game is played as follows. The players take turns changing a positive integer to a smaller one and then passing that smaller integer back to their opponent. If the integer is even, the two legal moves are (i) subtracting 1 from the integer and (ii) halving the integer. If the integer is odd, the two legal moves are (i) subtracting 1 from the integer and (ii) subtracting 1 and then halving the result. The game ends when the integer reaches 0, and the player making the last move wins.

a. Given best play, if the starting integer is 1000, should the first or second player win? How about if the starting integer is N?

b. If we take a starting integer at random from all integers from 1 to n inclusive, what will be the limit of the probability that the second player should win, as $n \to \infty$?

Idea. Analyze systematically what happens for small values of N.

Answer. If the starting integer is $1000 = 2^3 \cdot 125$ the second player should win. If the starting integer N has an even number of factors 2 (in particular, if N is odd) the first player should win; if N has an odd number of factors 2 the second player should win.

As $n \to \infty$, the probability that the second player should win approaches $1/3$.

Solution. a. We use strong induction on N. For $N = 1$ the claim is correct because the first player wins by moving from 1 to 0. Suppose the claim is correct for all positive integers $N \leq k$, and consider $N = k + 1$.

Case 1. $k + 1$ odd. In this case the legal moves are from $k + 1$ to the even number k and to $k/2$. Because k has one more factor 2 than $k/2$ (note that $k > 0$) either k or $k/2$ will have an odd number of factors 2, so the first player can move to that integer. Then by the induction hypothesis the second player (who is now the first to play) should lose. That is, the first player should win.

Case 2a. $k+1$ is even with an even number of factors 2. In this case the first player will win by moving to $(k+1)/2$, which has an odd number of factors 2. By the induction hypothesis the second player is now in a losing position, so the first player will win.

Case 2b. $k+1$ is even with an odd number of factors 2. If the first player moves to k, the induction hypothesis says that the second player (now the first to play) should win. If the first player moves to $(k+1)/2$ (which has an even number of factors 2), the induction hypothesis again says that the second player is in a winning position. So no matter what the first player does, the second player should win. This completes the induction step and the proof of the claim.

b. Using the result of part (a), the second player should win if and only if the starting integer N has an odd number of factors 2. The probability of this occurring is the sum of the probability that N has exactly one factor 2, the probability that N has exactly three factors 2, the probability that N has exactly five factors 2, and so forth. For a given n, when N is chosen at random between 1 and n, only finitely many of these probabilities will be nonzero; however, in the limit they will all contribute. As n goes to infinity the probability that N has exactly one factor 2 will be $1/4$ (half of the even numbers qualify: $2, 6, 10, 14, \ldots$), the probability that N has exactly three factors 2 will be $1/16$ (the possibilities are $8, 24, 40, 56, \ldots$), the probabilty that N has exactly five factors 2 will be $1/64$ ($N = 32, 96, \ldots$), and so forth. So the probability that the second player should win as $n \to \infty$ will be

$$\frac{1}{4} + \frac{1}{4^2} + \frac{1}{4^3} + \cdots = \frac{1}{4}\left(\frac{1}{1-1/4}\right) = \frac{1}{3}.$$

Problem 69

Find a solution to the system of simultaneous equations
$$\begin{cases} x^4 - 6x^2y^2 + y^4 = 1 \\ 4x^3y - 4xy^3 = 1, \end{cases}$$
where x and y are real numbers.

Idea. The coefficients and the powers of x and y in the equations are reminiscent of the binomial expansion

$$(x+y)^4 = x^4 + 4x^3 y + 6x^2 y^2 + 4xy^3 + y^4.$$

Solution. One solution is

$$x = \frac{\sqrt{2+\sqrt{2+\sqrt{2}}}}{2^{7/8}}, \quad y = \frac{\sqrt{2-\sqrt{2+\sqrt{2}}}}{2^{7/8}}.$$

If we put $z = x + iy$, where $i^2 = -1$, we have

$$z^4 = x^4 + 4x^3(iy) + 6x^2(iy)^2 + 4x(iy)^3 + (iy)^4$$
$$= (x^4 - 6x^2 y^2 + y^4) + (4x^3 y - 4xy^3)i.$$

Hence, the two given equations are equivalent to the one equation $z^4 = 1 + i$ in the complex unknown z. Now write z and $1+i$ in polar form:

$$z = r(\cos\theta + i\sin\theta), \quad 1 + i = \sqrt{2}\left(\cos(\pi/4) + i\sin(\pi/4)\right).$$

Using the formula $(\cos\theta + i\sin\theta)^n = \cos n\theta + i\sin n\theta$ (de Moivre's theorem), our equation becomes

$$r^4\left(\cos 4\theta + i\sin 4\theta\right) = \sqrt{2}\left(\cos(\pi/4) + i\sin(\pi/4)\right).$$

We get a solution by taking $r = \sqrt[4]{\sqrt{2}}$, $4\theta = \pi/4$, or equivalently, $r = \sqrt[8]{2}$, $\theta = \pi/16$, and therefore $x = \sqrt[8]{2}\cos(\pi/16)$, $y = \sqrt[8]{2}\sin(\pi/16)$. Using half-angle formulas, this solution can be written as shown above.

Comments. There are exactly three other solutions, obtained by taking

$$4\theta = \pi/4 + 2\pi, \quad 4\theta = \pi/4 + 4\pi, \quad 4\theta = \pi/4 + 6\pi.$$

For a more elementary (but laborious) approach, rewrite the equations as

$$(x^2 - y^2)^2 - 4x^2 y^2 = 1,$$
$$4xy(x^2 - y^2) = 1.$$

Then let $xy = u$, $x^2 - y^2 = v$, so the system becomes $v^2 - 4u^2 = 1$, $4uv = 1$. Although this reduces to a quadratic equation in u^2, considerable computation is needed to eventually recover x and y.

Problem 70

Show that if $p(x)$ is a polynomial of odd degree greater than 1, then through any point P in the plane, there will be at least one tangent line to the curve $y = p(x)$. Is this still true if $p(x)$ is of even degree?

Solution. If $y = p(x)$, where $p(x)$ is a polynomial of odd degree $d > 1$, then $P = (a, b)$ is on some tangent to the curve if and only if the equation

$$p(x) - b = p'(x)(x - a)$$

has a real solution. Thus, we are looking for a real root of

$$xp'(x) - p(x) - ap'(x) + b,$$

which has degree d (with leading coefficient $(d-1)$ times the leading coefficient of $p(x)$). Since, by the Intermediate Value Theorem, any real polynomial of odd degree has a real root, there is a real x_0 for which the tangent to $y = p(x)$ at $(x_0, p(x_0))$ passes through $P = (a, b)$.

The result does not hold for polynomials of even degree. For instance, since $y = x^2$ is concave up, no tangent line can pass through any point above this parabola.

Comment. For any polynomial $p(x)$ of even degree, one can find points in the plane through which no tangent line to $y = p(x)$ passes. In fact, any point "inside" the curve which is sufficiently far from the x-axis will do. To show this, use the concavity of the curve for $|x|$ large and the boundedness of $p'(x)$ on a finite interval.

Problem 71

For each positive integer n, let $N(n) = \left\lceil \dfrac{n}{2} \right\rceil + \left\lceil \dfrac{n}{4} \right\rceil + \left\lceil \dfrac{n}{8} \right\rceil + \cdots + \left\lceil \dfrac{n}{2^k} \right\rceil$, where k is the unique integer such that $2^{k-1} \leq n < 2^k$, and $\lceil x \rceil$ denotes the smallest integer greater than or equal to x. For which numbers n is $N(n) = n$?

Answer. $N(n) = n$ if and only if the binary expansion of n consists of a string of 1's followed by a string of 0's (the string of 0's might be empty), or equivalently, if and only if $n = 2^k - 2^\ell$ for some k, ℓ, $k > \ell$.

Solution. Note that $N(2n) = n + N(n)$, so $N(2n) = 2n$ if and only if $N(n) = n$. Thus it's enough to consider the case of odd n.

If $n = 2^k - 1$, then

$$N(n) = \left[\frac{n}{2}\right] + \left[\frac{n}{4}\right] + \cdots + \left[\frac{n}{2^k}\right]$$

$$= \frac{n+1}{2} + \frac{n+1}{4} + \cdots + \frac{n+1}{2^k} \quad (n+1 \text{ is divisible by } 2, 4, \ldots, 2^k)$$

$$= (n+1)\left(1 - \frac{1}{2^k}\right) = \frac{(n+1)(2^k - 1)}{2^k} = \frac{(n+1)n}{n+1} = n.$$

On the other hand, if $2^{k-1} \leq n < 2^k - 1$ and n is odd, then

$$N(n) = \left[\frac{n}{2}\right] + \left[\frac{n}{4}\right] + \cdots + \left[\frac{n}{2^k}\right]$$

$$\geq \frac{n+1}{2} + \frac{n+1}{4} + \cdots + \frac{n+1}{2^{k-1}} + \frac{n+3}{2^k}$$

$$= (n+1)\left(1 - \frac{1}{2^k}\right) + \frac{2}{2^k}$$

$$= n + 1 - \frac{n+1}{2^k} + \frac{2}{2^k}$$

$$> n + \frac{2}{2^k} \quad \text{because } n + 1 < 2^k$$

$$> n.$$

Problem 72

Call a convex pentagon "parallel" if each diagonal is parallel to the side with which it does not have a vertex in common. That is, $ABCDE$ is parallel if the diagonal AC is parallel to the side DE and similarly for the other four diagonals. It is easy to see that a regular pentagon is parallel, but is a parallel pentagon necessarily regular?

Answer. No, a parallel pentagon need not be a regular pentagon.

Solution 1. A one-to-one linear transformation of the plane onto itself takes parallel lines to parallel lines. So, start with a regular pentagon, say with vertices

$$\mathbf{v}_k = \begin{pmatrix} \cos(2k\pi/5) \\ \sin(2k\pi/5) \end{pmatrix}, \quad k = 0, 1, 2, 3, 4,$$

THE SOLUTIONS

and now simply change the scale on the x- and y-axes. For example, take the specific linear transformation $A = \begin{pmatrix} 2 & 0 \\ 0 & 1 \end{pmatrix}$. This stretches the x-coordinate by a factor of 2 and leaves the y-coordinate unchanged. The resulting figure is a non-regular parallel pentagon.

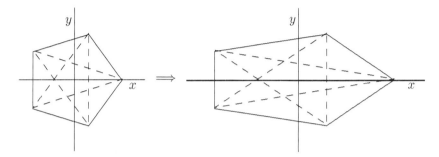

Solution 2. Another way to construct a non-regular parallel pentagon is as follows: Start with a square $ABXE$, say of side 1. Extend EX by x units to C and BX by x units to D, where $x > 0$ will be determined later. We will choose x so that $ABCDE$ is a parallel pentagon; obviously it will be non-regular.

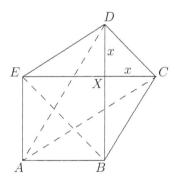

Regardless of the value of $x > 0$, $EC \parallel AB$ and $BD \parallel AE$. Also, since $DX = CX = x$, DXC and BXE are both isosceles right triangles, so $BE \parallel CD$. This leaves us to choose x (if possible) so that $AC \parallel DE$ and $AD \parallel BE$. By symmetry about AX, it is enough to find x so that $AC \parallel DE$. For this, it is sufficient to choose x so that triangles DEX and ACE are similar, or equivalently, so that the ratios of corresponding legs of the right triangles are equal to each other. Thus, we want

$$\frac{x}{1} = \frac{DX}{EX} = \frac{AE}{CE} = \frac{1}{1+x}.$$

This holds, for $x > 0$, when $x = (-1+\sqrt{5})/2$, and thus our construction can be carried out.

Problem 73

Given a permutation of the integers $1, 2, \ldots, n$, define the *total fluctuation* of that permutation to be the sum of all the differences between successive numbers along the permutation, where all differences are counted positively. What is the greatest possible total fluctuation, as a function of n, for permutations of $1, 2, \ldots, n$?

Answer. The maximum total fluctuation is $\frac{1}{2}n^2 - 1$ if n is even and $\frac{1}{2}(n^2 - 3)$ if n is odd (except 0 if $n = 1$).

Solution 1. Let's experiment with trying to construct a permutation with maximal total fluctuation. One idea is to work from the middle outward, making adjacent differences as large as possible. Here are some possibilities for the first few cases.

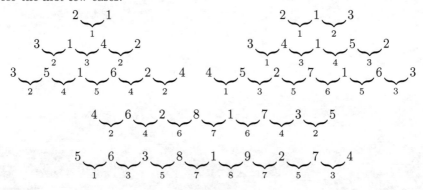

Now these may not have maximal fluctuation; for example, in the last instance if we exchange 6 and 9, or 1 and 3, we obtain

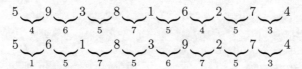

which have the same fluctuation ($= 39$) but don't have the unimodal pattern of differences (with the greater differences in the middle). This suggests that some other arrangement might have greater total fluctuation.

THE SOLUTIONS 133

One feature common to these examples is that the sequence of "rises" and "falls" alternates. It seems reasonable to expect that this will be the case for a *maximal* permutation (that is, one with maximal total fluctuation). So, let's assume that $a_1, a_2, \ldots, a_n$ is a maximal permutation with two rises or two falls in a row, and suppose this happens for the *first* time (counting from the left) at a_i, a_{i+1}, a_{i+2}. We may suppose that $a_i < a_{i+1} < a_{i+2}$ because if they were falls we could replace our permutation with its *complement*, $b_1, b_2, \ldots, b_n$ where $b_i = n + 1 - a_i$, which has the same total fluctuation but where falls are replaced with rises and vice versa. If $i + 2 = n$ then switching a_{i+1} and a_{i+2} definitely *increases* the total fluctuation, contrary to our assumption about having a maximal permutation. So $i + 2 < n$. We want to keep track of what happens to the total fluctuation when we switch a_{i+1} and a_{i+2}, and for this we have to consider a_{i+3}. There are three cases.

Case 1. $a_i < a_{i+1} < a_{i+2} < a_{i+3}$. Switching a_{i+1} and a_{i+2} clearly increases the fluctuation, contradicting that we had a maximal permutation.

Case 2. $a_i < a_{i+1} < a_{i+3} < a_{i+2}$. The contribution to the total fluctuation from $a_i, a_{i+1}, a_{i+2}, a_{i+3}$ goes from

$$(a_{i+1} - a_i) + (a_{i+2} - a_{i+1}) + (a_{i+2} - a_{i+3})$$

to

$$(a_{i+2} - a_i) + (a_{i+2} - a_{i+1}) + (a_{i+3} - a_{i+1}),$$

an *increase* of $2(a_{i+3} - a_{i+1}) > 0$, contradiction.

Case 3. $a_i < a_{i+1} < a_{i+2}$, $a_{i+3} < a_{i+1}$. Here the contribution changes from

$$(a_{i+1} - a_i) + (a_{i+2} - a_{i+1}) + (a_{i+2} - a_{i+3})$$

to

$$(a_{i+2} - a_i) + (a_{i+2} - a_{i+1}) + (a_{i+1} - a_{i+3}),$$

which is in fact no change, meaning that after the switch we will still have a maximal permutation and this new permutation will not have two rises (or falls) in a row until later (if at all). So whereas there might be a maximal permutation with two consecutive rises or falls, by repeating this process we will eventually find a maximal permutation whose rises and falls alternate. Moreover, by replacing a permutation with its complement if necessary, we may assume that our maximal permutation begins with a rise.

Looking at our examples that begin with a rise, we notice that for even n, the odd-numbered terms are the numbers $1, 2, 3, \ldots, \frac{1}{2}n$ in some order, and that for odd n, they are the numbers $1, 2, \ldots, \frac{1}{2}(n+1)$ in some order. We claim that this will always be the case for a maximal permutation of the kind we're considering. Specifically, suppose $a_1, \ldots, a_n$ is a maximal permutation and that $a_1 < a_2, a_2 > a_3, a_3 < a_4, a_4 > a_5$, etc., and suppose that $a_i > a_j$ for some i odd and j even. Then, if we interchanged a_i and a_j, the total fluctuation would increase (because a_i is less than its neighbors and its replacement a_j is even less, and at the same time, a_j is more than its neighbors and its replacement a_i would be even more), contradiction. This establishes the claim.

We are now ready to calculate the total fluctuation of a maximal permutation $a_1, a_2, \ldots, a_n$. As we've seen we may assume that the permutation starts with a rise, that the rises and falls alternate, and that the odd-numbered terms are $1, 2, \ldots, \frac{1}{2}n$ in some order for even n and $1, 2, \ldots, \frac{1}{2}(n+1)$ for odd n.

Case 1. n even. The total fluctuation T is

$$T = (a_2 - a_1) + (a_2 - a_3) + (a_4 - a_3) + (a_4 - a_5) + \cdots + (a_n - a_{n-1})$$
$$= 2a_2 + 2a_4 + \cdots + 2a_{n-2} + a_n - a_1 - 2a_3 - 2a_5 - \cdots - 2a_{n-1}$$
$$= 2(a_2 + a_4 + \cdots + a_n) - a_n - 2(a_1 + a_3 + \cdots + a_{n-1}) + a_1$$
$$= 2\big((\tfrac{1}{2}n+1) + (\tfrac{1}{2}n+2) + \cdots + (\tfrac{1}{2}n+\tfrac{1}{2}n)\big) - a_n - 2(1+2+\cdots+\tfrac{1}{2}n) + a_1$$
$$= 2(\tfrac{1}{2}n + \tfrac{1}{2}n + \cdots + \tfrac{1}{2}n) - a_n + a_1 = \tfrac{1}{2}n^2 - a_n + a_1.$$

To get a maximal permutation we should take $a_n = \tfrac{1}{2}n + 1$ and $a_1 = \tfrac{1}{2}n$, and this yields

$$T = \tfrac{1}{2}n^2 - (\tfrac{1}{2}n + 1) + \tfrac{1}{2}n = \tfrac{1}{2}n^2 - 1.$$

Case 2. n odd. In this case

$$T = (a_2 - a_1) + (a_2 - a_3) + \cdots + (a_{n-1} - a_n)$$
$$= 2(a_2 + a_4 + \cdots + a_{n-1}) - a_1 - 2(a_3 + a_5 + \cdots + a_{n-2}) - a_n$$
$$= 2(a_2 + a_4 + \cdots + a_{n-1}) - 2(a_1 + a_3 + \cdots + a_n) + a_1 + a_n$$
$$= 2\Big(\tfrac{1}{2}(n+3) + \cdots + n\Big) - 2\Big(1 + 2 + \cdots + \tfrac{1}{2}(n+1)\Big) + a_1 + a_n$$
$$= 2(1 + 2 + \cdots + n) - 4\Big(1 + 2 + \cdots + \tfrac{1}{2}(n+1)\Big) + a_1 + a_n$$
$$= n(n+1) - \tfrac{1}{2}(n+1)(n+3) + a_1 + a_n.$$

THE SOLUTIONS

For a maximal permutation we must make $a_1 + a_n$ as large as possible, so a_1 and a_n should be $\frac{1}{2}(n-1)$, $\frac{1}{2}(n+1)$ in some order. This yields

$$T = n^2 + n - \tfrac{1}{2}n^2 - 2n - \tfrac{3}{2} + \tfrac{1}{2}(n-1) + \tfrac{1}{2}(n+1) = \tfrac{1}{2}(n^2 - 3).$$

Solution 2. (Robert Boucher, Mathcamp student, 2006) Let $M = \frac{n+1}{2}$ be the average of the integers $1, 2, \ldots, n$, and let $a_1, a_2, \ldots, a_n$ be any permutation of those integers. Then the total fluctuation of that permutation is

$$\sum_{i=1}^{n-1} |a_{i+1} - a_i| = \sum_{i=1}^{n-1} |a_{i+1} - M + M - a_i|$$

$$\leq \sum_{i=1}^{n-1} \left(|a_{i+1} - M| + |M - a_i| \right)$$

$$= \sum_{i=1}^{n-1} \left(|M - a_{i+1}| + |M - a_i| \right)$$

$$= |M - a_1| + 2|M - a_2| + \cdots + 2|M - a_{n-1}| + |M - a_n|$$

$$= -|M - a_1| - |M - a_n| + 2 \sum_{i=1}^{n} |M - a_i|$$

$$= -|M - a_1| - |M - a_n| + 2 \sum_{i=1}^{n} |M - i|,$$

because the $|M - a_i|$ are just the $|M - i|$ in some other order. The inequality above is strict if the integers a_i, a_{i+1} are ever on the same side of M (both larger than M or both smaller than M); if this never happens, the inequality is an equality. Note that $2\sum_{i=1}^{n} |M - i|$ is independent of the permutation. Thus to make the total fluctuation as large as possible, if we can we should both choose the a_i on alternating sides of M (to get equality; this can be done since there is an equal number of the integers $1, 2, \ldots, n$ to either side of M) and choose $|M - a_1| + |M - a_n|$ to be as small as possible.

If n is odd, M is an integer, and $|M - a_1| + |M - a_n|$ is as small as possible for $a_1 = M$, $a_n = M + 1$ (and similar cases). If n is even, $|M - a_1| + |M - a_n|$ is as small as possible for $a_1 = M - \frac{1}{2}$, $a_n = M + \frac{1}{2}$ (or vice versa). In either case, the minimal value of $|M - a_1| + |M - a_n|$ is 1 and can be achieved with the a_i on alternating sides of M, so the greatest possible total fluctuation is

$$-1 + 2 \sum_{i=1}^{n} |M - i|.$$

If n is odd,

$$-1 + 2\sum_{i=1}^{n}|M-i| = -1 + 4\sum_{i=1}^{(n-1)/2}\left(\frac{n+1}{2} - i\right)$$

$$= -1 + 4\left(\frac{n+1}{2} \cdot \frac{n-1}{2} - \frac{1}{2} \cdot \frac{n-1}{2} \cdot \frac{n+1}{2}\right)$$

$$= -1 + \frac{n^2 - 1}{2} = \frac{1}{2}(n^2 - 3),$$

whereas if n is even,

$$-1 + 2\sum_{i=1}^{n}|M-i| = -1 + 4\sum_{i=1}^{n/2}\left(\frac{n+1}{2} - i\right)$$

$$= -1 + 4\left(\frac{n+1}{2} \cdot \frac{n}{2} - \frac{1}{2} \cdot \frac{n}{2}\left(\frac{n}{2} + 1\right)\right)$$

$$= -1 + 4\left(\frac{n^2}{4} - \frac{n^2}{8}\right) = \frac{1}{2}n^2 - 1.$$

Problem 74

Find the sum of the infinite series

$$\sum_{n=1}^{\infty}\frac{1}{2n^2 - n} = 1 + \frac{1}{6} + \frac{1}{15} + \frac{1}{28} + \cdots.$$

Answer. The series sums to $2\ln 2$.

Solution 1. We begin with the partial fraction decomposition

$$\frac{1}{2n^2 - n} = \frac{2}{2n-1} - \frac{1}{n} = \frac{2}{2n-1} - \frac{2}{2n}.$$

Thus,

$$\sum_{n=1}^{\infty}\frac{1}{2n^2 - n} = 2\sum_{n=1}^{\infty}\left(\frac{1}{2n-1} - \frac{1}{2n}\right) = 2\left(1 - \frac{1}{2} + \frac{1}{3} - \frac{1}{4} + \cdots\right).$$

The last step is legitimate because the alternating series on the right is convergent. We now recall the well-known Taylor series expansion

$$\ln(1+x) = \sum_{n=1}^{\infty}(-1)^{n+1}\frac{x^n}{n},$$

THE SOLUTIONS

which is valid for $-1 < x \leq 1$. In particular, setting $x = 1$ yields

$$\sum_{n=1}^{\infty} \frac{1}{2n^2 - n} = 2\ln 2.$$

Solution 2. We express the series as a definite integral we can evaluate. Observing that

$$\frac{1}{2n^2 - n} = \frac{1}{2n - 1} \cdot \frac{1}{n},$$

we have

$$\sum_{n=1}^{\infty} \frac{1}{2n^2 - n} = \sum_{n=1}^{\infty} \frac{1}{n} \int_0^1 x^{2n-2} \, dx.$$

At this point we would like to switch summation and integration. However, we cannot be too cavalier, because $\sum_{n=1}^{\infty} x^{2n-2}/n$ diverges for $x = 1$, making the resulting integral improper. To avoid this difficulty, we can argue as follows.

By the continuity of a power series on its interval of convergence (Abel's Limit Theorem),

$$\lim_{b \to 1^-} \sum_{n=1}^{\infty} \frac{b^{2n-1}}{2n^2 - n} = \sum_{n=1}^{\infty} \frac{1}{2n^2 - n},$$

and therefore

$$\sum_{n=1}^{\infty} \frac{1}{2n^2 - n} = \lim_{b \to 1^-} \sum_{n=1}^{\infty} \frac{1}{n} \int_0^b x^{2n-2} \, dx = \lim_{b \to 1^-} \int_0^b \sum_{n=1}^{\infty} \frac{x^{2n-2}}{n} \, dx.$$

Now, from the Taylor series for $\ln(1+x)$, we see that

$$\sum_{n=1}^{\infty} \frac{x^{2n-2}}{n} = \frac{1}{x^2} \sum_{n=1}^{\infty} \frac{x^{2n}}{n} = -\frac{\ln(1 - x^2)}{x^2}, \qquad 0 < x < 1,$$

and so we have

$$\sum_{n=1}^{\infty} \frac{1}{2n^2 - n} = \lim_{b \to 1^-} \int_0^b -\frac{\ln(1 - x^2)}{x^2} \, dx = \int_0^1 -\frac{\ln(1 - x^2)}{x^2} \, dx.$$

An integration by parts followed by a partial fraction decomposition yields

$$\int -\frac{\ln(1 - x^2)}{x^2} \, dx = \frac{\ln(1 - x^2)}{x} - \ln(1 - x) + \ln(1 + x) + C$$

$$= \frac{(1 - x)\ln(1 - x)}{x} + \frac{(1 + x)\ln(1 + x)}{x} + C.$$

The well-known limits

$$\lim_{c \to 0^+} c \ln c = 0 \quad \text{and} \quad \lim_{c \to 0} \frac{\ln(1+c)}{c} = 1$$

(provable using l'Hôpital's rule) imply

$$\sum_{n=1}^{\infty} \frac{1}{2n^2 - n} = \lim_{b \to 1^-} \left[\frac{(1-b)\ln(1-b)}{b} + \frac{(1+b)\ln(1+b)}{b} \right]$$

$$- \lim_{a \to 0^+} \left[\frac{(1-a)\ln(1-a)}{a} + \frac{(1+a)\ln(1+a)}{a} \right]$$

$$= (0 + 2\ln 2) - (-1 + 1) = 2\ln 2.$$

Comment. The second solution is more difficult, but it illustrates a useful method.

Problem 75

Can one get any integer that is not itself a cube by adding and/or subtracting distinct cubes of integers?

Solution 1. Yes. To see why, consider four successive cubes $n^3, (n+1)^3, (n+2)^3, (n+3)^3$ and the differences

$$(n+3)^3 - (n+2)^3 = 3(n+2)^2 + 3(n+2) + 1,$$

$$(n+1)^3 - n^3 = 3n^2 + 3n + 1.$$

Subtract the second from the first to get

$$(n+3)^3 - (n+2)^3 - (n+1)^3 + n^3 = 12n + 18.$$

We can replace n by $n + 3$ to get

$$(n+6)^3 - (n+5)^3 - (n+4)^3 + (n+3)^3 = 12(n+3) + 18$$

and then subtract the previous equation from this one to get

$$(n+6)^3 - (n+5)^3 - (n+4)^3 + (n+2)^3 + (n+1)^3 - n^3 = 36. \quad (*)$$

This is true for *any* n. Now suppose we can write every positive integer up to and including 18 as a sum/difference of distinct cubes, and consider an arbitrary integer. Write this integer in the form $36q + r$ with $-18 < r \le 18$, write $36q$ as $36 + 36 + \cdots + 36$ (q times), and then apply (∗) repeatedly (q times), with a different n for each term 36 and with each n larger than the ones used for r and at least 6 greater than the previous one so no repetition of cubes occurs. Then we will have $36q + r$ in terms of distinct cubes and we will be done.

It remains to show that every positive integer through 18 can be written as a sum/difference of distinct cubes. We start with

$$1 = 1^3,$$
$$2 = 7^3 - 6^3 - 5^3,$$
$$7 = 2^3 - 1^3,$$
$$8 = 2^3,$$
$$9 = 2^3 + 1^3,$$
$$18 = 3^3 - 2^3 - 1^3.$$

Adding, subtracting, and multiplying these expressions yields

$$3 = 2 + 1 = 7^3 - 6^3 - 5^3 + 1^3,$$
$$4 = 28 - 24 = 3^3 + 1^3 - 2^3 \cdot 3$$
$$\quad = 3^3 + 1^3 - 14^3 + 12^3 + 10^3 - 2^3,$$
$$5 = 2^3 - 3 = 2^3 - 7^3 + 6^3 + 5^3 - 1^3,$$
$$6 = 2^3 - 2 = 2^3 - 7^3 + 6^3 + 5^3,$$
$$10 = 2^3 + 2 = 2^3 + 7^3 - 6^3 - 5^3,$$
$$11 = 2^3 + 3 = 2^3 + 7^3 - 6^3 - 5^3 + 1^3,$$
$$12 = 28 - 2^3 \cdot 2 = 3^3 + 1^3 - 14^3 + 12^3 + 10^3,$$
$$13 = 2 \cdot 7 - 1^3 = 14^3 - 12^3 - 10^3 - 7^3 + 6^3 + 5^3 - 1^3,$$
$$14 = 2 \cdot 7 = 14^3 - 12^3 - 10^3 - 7^3 + 6^3 + 5^3,$$
$$15 = 2 \cdot 7 + 1^3 = 14^3 - 12^3 - 10^3 - 7^3 + 6^3 + 5^3 + 1^3,$$
$$16 = 2^3 \cdot 2 = 14^3 - 12^3 - 10^3,$$
$$17 = 3^3 - 10 = 3^3 - 2^3 - 7^3 + 6^3 + 5^3.$$

Eugene Luks, University of Oregon, used a similar approach, considering polynomials of the form

$$f(x) = \sum_{i=1}^{k}(x + a_k)^3$$

for any k, with distinct integers $a_1, \ldots, a_k$. The idea is to find two such polynomials f_1, f_2 that differ by a constant c, for then we can express any multiple of c as a sum/difference of arbitrarily large cubes (as above). It is not hard to show that c must be divisible by 6; with the help of *Mathematica*,

$c = 6$ is attainable, for example with

$$f_1(x) = x^3 + (x+6)^3 + (x+12)^3 + (x+16)^3 + (x+20)^3$$
$$= 5x^3 + 162x^2 + 2508x + 14040,$$

$$f_2(x) = (x+1)^3 + (x+4)^3 + (x+13)^3 + (x+17)^3 + (x+19)^3$$
$$= 5x^3 + 162x^2 + 2508x + 14034.$$

Hence, following the idea above, it suffices to find specific expressions for just $1, 2, 3$.

Solution 2. Suppose we are able to represent each of the numbers $-3, -2, -1, 1, 2, 3, 4$ as a sum/difference of cubes of distinct odd integers. Then we can use those numbers as "octal digits", as follows. For any integer n, write n as

$$n = 8^k a_k + 8^{k-1} a_{k-1} + \cdots + 8 a_1 + a_0, \qquad a_i \in \{0, \pm 1, \pm 2, \pm 3, \pm 4\}.$$

Substituting the representation for each a_i and distributing the $8^i = (2^i)^3$, we get an expression for n as a sum/difference of distinct cubes.

It's enough to find representations for $2, 3$, and 4. To do this, calculate the cubes of the first 25 or so odd integers, $1, 3, 5, \ldots, 51$, and look for pairs of disjoint subsets of these cubes whose elements add to approximately the same number, and take their difference. By combining numbers obtained in this way, we are able to find the desired representations. Specifically,

$$2 = [(17^3 + 15^3 + 9^3 + 7^3) - 21^3] - [(5^3) - (3^3 + 1^3)] = 99 - 97,$$

$$3 = [7^3 - (5^3 + 1^3)] - [(47^3 + 43^3 + 41^3) - (51^3 + 45^3 + 25^3 + 21^3 + 15^3)]$$
$$= 217 - 214,$$

$$4 = [7^3 - 5^3] - [(47^3 + 43^3 + 41^3) - (51^3 + 45^3 + 25^3 + 21^3 + 15^3)]$$
$$= 218 - 214.$$

Problem 76

For any vector $\mathbf{v} = (x_1, \ldots, x_n)$ in $\mathbb{R}^n$ and any permutation σ of $1, 2, \ldots, n$, define $\sigma(\mathbf{v}) = (x_{\sigma(1)}, \ldots, x_{\sigma(n)})$. Now fix $\mathbf{v}$ and let V be the span of $\{\sigma(\mathbf{v}) \mid \sigma \text{ is a permutation of } 1, 2, \ldots, n\}$. What are the possibilities for the dimension of V?

Solution. The possibilities for $\dim V$ are 0, 1, $n-1$, and n.

To see this, first consider the case when all coordinates of $\mathbf{v}$ are equal: $\mathbf{v} = (z, z, \ldots, z)$. Then $\sigma(\mathbf{v}) = \mathbf{v}$ for every permutation σ, so V is just the span of $\mathbf{v}$, which has dimension 0 or 1 according to whether $\mathbf{v}$ is $\mathbf{0}$ or not.

Now suppose not all coordinates of $\mathbf{v}$ are equal; let x and y, with $x \neq y$, be among the coordinates of $\mathbf{v}$. Then we can find permutations σ_1 and σ_2 such that $\sigma_1(\mathbf{v}) = (x, y, a_3, \ldots, a_n)$ and $\sigma_2(\mathbf{v}) = (y, x, a_3, \ldots, a_n)$ for some $a_3, \ldots, a_n \in \mathbb{R}$. Therefore,

$$\frac{1}{y-x}\left(\sigma_1(\mathbf{v}) - \sigma_2(\mathbf{v})\right) = (-1, 1, 0, \ldots, 0)$$

is in V. That is, $\mathbf{e}_2 - \mathbf{e}_1 \in V$, where $\mathbf{e}_1, \mathbf{e}_2, \ldots, \mathbf{e}_n$ is the standard basis for $\mathbb{R}^n$. Similarly, $\mathbf{e}_3 - \mathbf{e}_1, \ldots, \mathbf{e}_n - \mathbf{e}_1$ are all in V. It is easy to see that the vectors $\mathbf{e}_2 - \mathbf{e}_1, \mathbf{e}_3 - \mathbf{e}_1, \ldots, \mathbf{e}_n - \mathbf{e}_1$ are linearly independent, so $\dim V \geq n-1$.

Finally, we can write

$$\mathbf{v} = x_1 \mathbf{e}_1 + x_2 \mathbf{e}_2 + \cdots + x_n \mathbf{e}_n$$
$$= (x_1 + x_2 + \cdots + x_n)\mathbf{e}_1 + x_2(\mathbf{e}_2 - \mathbf{e}_1) + \cdots + x_n(\mathbf{e}_n - \mathbf{e}_1). \quad (*)$$

This shows that if $x_1 + x_2 + \cdots + x_n = 0$, then $\mathbf{v}$ is in the span of $\mathbf{e}_2 - \mathbf{e}_1, \ldots, \mathbf{e}_n - \mathbf{e}_1$; similarly, each $\sigma(\mathbf{v})$ will be in this span, so V will equal this span and $\dim V = n-1$. On the other hand, if $x_1 + x_2 + \cdots + x_n \neq 0$, then $(*)$ shows that $\mathbf{e}_1 \in V$ and thus also $\mathbf{e}_2, \ldots, \mathbf{e}_n \in V$, so $V = \mathbb{R}^n$ and $\dim V = n$.

Problem 77

Say a positive integer n is *suitable* if there is a way to write the set $\{1, 2, \ldots, n\}$ as the union of sets S and T which have no elements in common and such that the average of all the elements of S is an element of T and vice versa. Show that, with one exception, composite numbers are suitable and prime numbers are not suitable.

Solution. We'll make repeated use of the fact that the sum of a finite arithmetic series is equal to the product of the number of terms in the series and the average of the first and last terms.

First we'll show that all positive even integers are suitable except for $n = 2$ and $n = 6$. It is obvious that 2 is not suitable. Suppose that $n = 6$ is suitable and let S and T be two disjoint sets each of whose averages belongs to the other. Clearly the sets S and T have at least two elements each, so suppose that one of the sets, say S, has exactly two elements. Those two elements would have to have the same parity, else their average would not be an integer. If both elements are even, then T consists of one even and three odd numbers; if both elements are odd, then T consists of one odd and three even numbers. Either way, the sum of the four elements of T is odd, so the average of the numbers in T is not an integer, contradiction. Therefore S and T each have exactly three elements. Consider the set which contains the number 3, say $3 \in S$. Then $6 \notin S$, because otherwise the sum of the three numbers in S would not be divisible by 3. For the same reason, S must consist of one of $1, 2, 3$; $1, 3, 5$; $2, 3, 4$; $3, 4, 5$ (for all other combinations, the average is not an integer). But in each of these cases, the average of the three integers in S is itself in S, contradiction. Therefore $n = 6$ is not suitable.

Now suppose that n is divisible by 4, say $n = 4k$. Then take

$$S = \{1, 3, 5, \ldots, 4k-1\} \quad \text{and} \quad T = \{2, 4, 6, \ldots, 4k\}.$$

The average of the elements of S is $\left(\frac{1}{2} \cdot 2k \cdot 4k\right)/(2k) = 2k$, which is in T, while the average of the elements of T is $2k + 1$ (each element of T is 1 larger than the corresponding element of S), which is in S.

Now suppose that n is even but not divisible by 4, say $n = 4k + 2$, with $k \geq 2$. Then take

$$S = \{\underbrace{k, \ k+1, \ \ldots, 2k-2}_{k-1 \text{ integers}}, \ 2k, \ 2k+1, \ \underbrace{2k+5, \ 2k+6, \ \ldots, 3k+4}_{k \text{ integers}}\}$$

and let T be the complementary set. The sum of the integers in S (broken into three subsets as shown in the display) is

$$\tfrac{1}{2}(k-1)(3k-2) + (4k+1) + \tfrac{1}{2}k(5k+9)$$
$$= \tfrac{1}{2}(3k^2 - 5k + 2 + 8k + 2 + 5k^2 + 9k)$$
$$= 4k^2 + 6k + 2 = (2k+1)(2k+2),$$

so the average is $2k + 2$, which is in T. The sum of the elements of T is the sum of $1, 2, \ldots, 4k+2$ minus the sum of the elements of S, which gives

$$\tfrac{1}{2}(4k+2)(4k+3) - (2k+1)(2k+2) = (2k+1)(4k+3-2k-2) = (2k+1)^2,$$

so the average of the elements of T is $2k + 1$, which is in S.

Next we will show that odd composite numbers are suitable (solution by Christie Chiu, Mathcamp student, 2007). Let n be an odd composite integer, and let p be the smallest prime factor of n. Let S be the set

$\{1, 2, \ldots, p-1, \frac{n+p}{2}\}$, and let T be the complementary set. The average of the integers in S is

$$\frac{1}{p}\left(1 + 2 + \cdots + (p-1) + \frac{n+p}{2}\right) = \frac{1}{p}\left(\frac{p(p-1)}{2} + \frac{n}{2} + \frac{p}{2}\right)$$
$$= \frac{p-1}{2} + \frac{n}{2p} + \frac{1}{2} = \frac{1}{2}\left(p + \frac{n}{p}\right),$$

which is an integer because p and n/p are odd integers. Also,

$$\frac{1}{2}\left(p + \frac{n}{p}\right) \geq \frac{1}{2}(p+p) = p \quad \text{and} \quad \frac{1}{2}\left(p + \frac{n}{p}\right) < \frac{1}{2}(p+n) = \frac{n+p}{2},$$

so $\frac{1}{2}\left(p + \frac{n}{p}\right)$ cannot be in S. Therefore, the average of the integers in S must be in T.

On the other hand, the average of the integers in T is

$$\frac{1}{n-p}\left(p + (p+1) + \cdots + n - \frac{n+p}{2}\right)$$
$$= \frac{1}{n-p}\left(\frac{1}{2}(n-p+1)(n+p) - \frac{n+p}{2}\right) = \frac{n+p}{2},$$

which is in S, and we are done.

Alternatively, consider a number of the form $n = km$ where k and m are odd, $3 \leq k \leq m$. Let S_0 consist of the middle k numbers of $\{1, 2, \ldots, km\}$; that is,

$$S_0 = \{M{-}d, \ldots, M, \ldots, M{+}d\},$$

where $M = (km{+}1)/2$ and $d = (k{-}1)/2$.

By symmetry, both S_0 and the $k(m{-}1)$ numbers in its complement average to M. We adjust S_0 slightly to achieve our goal. Let $s = (m{-}1)/2$ and define

$$S = \left(S_0 - \{M, M+1\}\right) \cup \{M + sk, M + sk + 1\}, \quad \text{(see figure)}$$

and take T to be the complement of S. (Note that $sk + 1 = \left(\frac{m-1}{2}\right)k + 1 = \frac{mk+1}{2} - \frac{k-1}{2} = M{-}d$.)

We have replaced two numbers in S_0 with numbers from its complement, increasing each by ks, so the average of the numbers in S is

$$M + 2ks/k = M + (m-1),$$

which lies in T because

$$M + d = M + \frac{k-1}{2} \leq M + \frac{m-1}{2} < M + (m-1) < M + \frac{k(m-1)}{2} = M + ks.$$

Also, the average of the elements of T is $M - 2ks/(km-k) = M - 1$, which is in S.

Finally, we show that prime numbers are not suitable. Suppose an odd prime p is suitable. Let a and b denote the averages of the elements of S and T, respectively. Because S and T are disjoint sets, $a \neq b$, so suppose, say, that $a < b$, and set $c = b - a$. Then if we add all the elements of S and all the elements of T together, we get $|S|a + |T|b$ (where $|S|$ denotes the number of elements of S, etc.) and we also get $1 + 2 + \ldots + p = \frac{p(p+1)}{2}$. In particular,

$$\frac{p(p+1)}{2} = |S|a + |T|b = |S|a + |T|(a+c) = \Big(|S| + |T|\Big)a + |T|c = pa + |T|c$$

is divisible by p. But then $|T|c$ is divisible by the prime p. Because both $|T|$ and c are positive integers less than p, this is impossible.

Problem 78

Suppose three circles, each of radius 1, go through the same point in the plane. Let A be the set of points which lie inside at least two of the circles. What is the smallest area A can have?

Solution. The smallest area is $\pi - \frac{3}{2}\sqrt{3}$.

Let C_1, C_2, C_3 be the circles, let O_1, O_2, O_3 be their centers, and let P be their common point. Let A_1, A_2, A_3 be the sets of points inside both C_2 and C_3, inside both C_1 and C_3, and inside both C_1 and C_2, respectively, so that A is the (not necessarily disjoint) union of A_1, A_2 and A_3. Note that the three centers of the circles all have distance 1 to P and therefore lie on a fourth circle with center at P. We may assume that O_2 follows O_1 as one moves counterclockwise around this circle. Let α, β, γ be the counterclockwise angles O_2PO_3, O_3PO_1, O_1PO_2, respectively. Then $\alpha + \beta + \gamma = 2\pi$, and we may assume $\alpha \geq \beta$, $\alpha \geq \gamma$. There are then two cases, as illustrated in the following pair of figures.

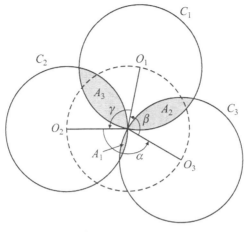

Case 1. $\alpha \leq \pi$.

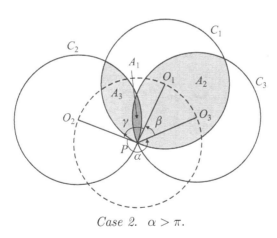

Case 2. $\alpha > \pi$.

In Case 1, there is no overlap between the regions A_1, A_2, A_3, so the area of A is the sum of the areas of A_1, A_2, and A_3. In Case 2, on the other hand, A_1 is the intersection of A_2 and A_3, so the area of A equals the sum of the areas of A_2 and A_3 minus the area of A_1.

To find the areas of the individual A_i, it is enough to find how the area of A_2 depends on β. By symmetry, the result of this computation will immediately give us all the other areas, since the region A_1 in Case 2 (for the angle $\alpha > \pi$) is congruent to the region A_1 in Case 1 for the angle $2\pi - \alpha$.

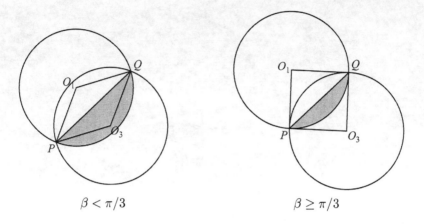

$\beta < \pi/3$ \qquad\qquad $\beta \geq \pi/3$

The area of A_2 is twice the area of the shaded region shown in the preceding figure. That region is obtained by removing triangle O_1PQ from circular sector O_1PQ, where Q is the second intersection point of C_1 and C_3. Now the area of the circular sector O_1PQ is $(\pi - \beta)/2$ and the area of triangle O_1PQ is

$$\frac{1}{2}\sin(\pi - \beta) = \frac{1}{2}\sin\beta.$$

Thus, the area of the shaded region is $(\pi - \beta - \sin\beta)/2$ and the area of A_2 is $\pi - \beta - \sin\beta$.

Combining the areas as indicated above, we find that in both cases, the area we wish to minimize is

$$\pi - \sin\alpha - \sin\beta - \sin\gamma, \qquad \alpha, \beta, \gamma \geq 0, \quad \alpha + \beta + \gamma = 2\pi.$$

Suppose the angles are not all equal. Then $\alpha > 2\pi/3$ and at least one of β and γ is less than $2\pi/3$. Suppose $\beta < 2\pi/3$. Now fix γ and note that

$$\sin\alpha + \sin\beta = 2\sin\left(\frac{\alpha+\beta}{2}\right)\cos\left(\frac{\alpha-\beta}{2}\right)$$
$$= 2\sin\left(\pi - \frac{\gamma}{2}\right)\cos\left(\frac{\alpha-\beta}{2}\right).$$

Since $0 \leq \gamma \leq 2\pi$, $\sin(\pi - \frac{\gamma}{2}) \geq 0$. Now steadily decrease α and increase β, in such a way that their sum remains constant, until one or the other is equal to $2\pi/3$. As the difference $\alpha - \beta$ decreases, the value of $\sin\alpha + \sin\beta$

increases, so the area decreases. Thus we can assume that one of the angles is equal to $2\pi/3$. Suppose $\beta = 2\pi/3$ (the other case is similar). Then

$$\sin \alpha + \sin \gamma = 2 \sin\left(\pi - \frac{\beta}{2}\right) \cos\left(\frac{\alpha - \gamma}{2}\right) = \sqrt{3} \cos\left(\frac{\alpha - \gamma}{2}\right).$$

This sum is maximized when $\cos(\frac{\alpha-\gamma}{2}) = 1$, or equivalently, when

$$\alpha = \gamma = 2\pi/3,$$

as well. Thus, under the given constraints, $\sin \alpha + \sin \beta + \sin \gamma$ is maximized, and the area of A is minimized, when α, β, γ all equal $2\pi/3$, and the minimal area is $\pi - 3\sin(2\pi/3) = \pi - \frac{3}{2}\sqrt{3}$.

Comments. As we have seen, the problem reduces to that of finding the maximum value of

$$\sin \alpha + \sin \beta + \sin \gamma, \qquad \alpha, \beta, \gamma > 0, \quad \alpha + \beta + \gamma = 2\pi.$$

The following figure shows how this can be reformulated geometrically as: Given a circle of radius 1, find the maximum area of an inscribed triangle. It is known that the area of such a triangle is maximized when the triangle is equilateral, that is, when $\alpha = \beta = \gamma = 2\pi/3$. On the other hand, the problem can also be solved using Lagrange multipliers.

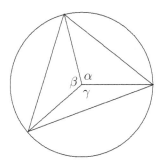

Problem 79

Do there exist two different positive integers (written as usual in base 10) with an equal number of digits so that the square of each of the integers starts with the other? If two such integers exist, give an example; if not, show why not.

Answer. Yes, two such integers exist; one is the pair $2154435, 4641590$, for which $2154435^2 = 4641590169225$ and $4641590^2 = 21544357728100$.

Solution. Even using technology, a search by trial and error alone would be quite laborious; here is a good way to get started. Suppose the integers are a and b with $a < b$. If a and b have $n+1$ digits, then

$$10^n \leq a < b < 10^{n+1}$$

and consequently,

$$10^{2n} \leq a^2 < b^2 < 10^{2n+2}.$$

Thus a^2 and b^2 have either $2n+1$ or $2n+2$ digits.

Because $a^2 < b \cdot 10^{n+1}$, yet a^2 starts with b, a^2 has only $2n+1$ digits. Also, using again that a^2 starts with b,

$$(a+1) \cdot 10^n \leq b \cdot 10^n \leq a^2 < b^2,$$

yet b^2 starts with a, so b^2 has $2n+2$ digits. It follows that

$$b \cdot 10^n \leq a^2 < (b+1) \cdot 10^n \quad \text{and} \quad a \cdot 10^{n+1} \leq b^2 < (a+1) \cdot 10^{n+1}.$$

One way to think about this is that we have reasonable approximations $a^2 \approx b \cdot 10^n$ and $b^2 \approx a \cdot 10^{n+1}$. Combining these approximations, we have $a^4 = (a^2)^2 \approx (b \cdot 10^n)^2 = b^2 \cdot 10^{2n} \approx a \cdot 10^{n+1} \cdot 10^{2n}$, so that $a^3 \approx 10^{3n+1}$ and $a \approx 10^n \cdot \sqrt[3]{10}$. So good candidates for the number a are integer approximations to $10^n \cdot \sqrt[3]{10}$ (which is indeed between 10^n and 10^{n+1}, as required). Now $\sqrt[3]{10} = 2.15443469\ldots$ and the approximation 2154435 to $10^6 \cdot \sqrt[3]{10}$ turns out to be good enough.

Problem 80

How many real solutions does the equation $\sqrt[7]{x} - \sqrt[5]{x} = \sqrt[3]{x} - \sqrt{x}$ have?

Answer. There are three solutions to the given equation.

Solution 1. We wish to find the number of solutions of

$$\sqrt{x} - \sqrt[3]{x} - \sqrt[5]{x} + \sqrt[7]{x} = 0.$$

THE SOLUTIONS

Letting $x = y^{210}$, it is enough to determine the number of nonnegative roots of

$$y^{105} - y^{70} - y^{42} + y^{30},$$

which factors as

$$y^{30}(y-1)\left[y^{40}\left(y^{34} + y^{33} + \cdots + y + 1\right) - \left(y^{11} + y^{10} + \cdots + y + 1\right)\right].$$

If we let

$$p(y) = y^{40}\left(y^{34} + y^{33} + \cdots + y + 1\right) - \left(y^{11} + y^{10} + \cdots + y + 1\right),$$

then $p(0) = -1$, while $p(1) = 23$. Thus, by the Intermediate Value Theorem, there must be a root of $p(y)$ between 0 and 1. On the other hand, the coefficients of $p(y)$ change sign exactly once, hence, by Descartes' rule of signs, $p(y)$ can have at most one positive root. Therefore, there are three values of x, namely 0, 1, and some number between 0 and 1, which solve the given equation.

Solution 2. Let

$$f(x) = \sqrt[7]{x} - \sqrt[5]{x} - \sqrt[3]{x} + \sqrt{x},$$

which has domain $x \geq 0$. By inspection, $f(0) = f(1) = 0$. Differentiating f, we obtain

$$f'(x) = \frac{1}{7}x^{-6/7} - \frac{1}{5}x^{-4/5} - \frac{1}{3}x^{-2/3} + \frac{1}{2}x^{-1/2}.$$

In particular, $f'(1) = 23/210 > 0$, hence $f(x) < 0$ for x less than 1 and sufficiently close to 1. Since

$$\lim_{x \to 0^+} x^{6/7} f'(x) = \frac{1}{7},$$

$f'(x) > 0$ for sufficiently small positive x. For such x, the Mean Value Theorem implies $f(x) = f'(c)x$ for some c, $0 < c < x$, hence $f(x) > 0$. The continuity of f now implies the existence of a zero of f in the open interval $(0, 1)$.

If we can show that $f'(x)$, or equivalently, $x^{6/7}f'(x)$, has at most 2 positive zeros, then Rolle's theorem will imply that $f(x)$ has at most 3 nonnegative zeros. We prove the more general result that

$$g(x) = \sum_{j=1}^{n} a_j x^{r_j}, \qquad r_1 > r_2 > \cdots > r_n, \quad a_j \neq 0,$$

can have at most $n-1$ positive roots. This is clear for $n=1$. Assume the result for a sum of $n-1$ terms. Differentiating $x^{-r_n}g(x)$ yields

$$\sum_{j=1}^{n-1}(r_j-r_n)a_j x^{r_j-r_n-1},$$

which has at most $n-2$ positive roots by the inductive hypothesis. Rolle's theorem implies $x^{-r_n}g(x)$, and therefore also $g(x)$, has at most $n-1$ positive roots, proving the generalization.

It follows that $f(x)$ has exactly three zeros.

Problem 81

Suppose you draw n parabolas in the plane. What is the largest number of regions that the plane may be divided into by those parabolas?

Answer. The maximum number of regions is $2n^2-n+1$.

Solution. First of all, for two parabolas we can get 7 regions as shown.

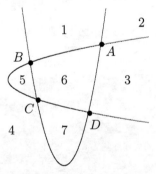

Note that the two parabolas shown intersect in four points. If we started with the "upward" parabola (which divides the plane into two regions, one with the numbers $1,6,7$ and one with the numbers $2,3,4,5$) and we then draw the second parabola, say counterclockwise so as to go through the intersection points in order A, B, C, D, then each of the five segments in which the second parabola is drawn will add one to the original number of two regions, as follows:

The segment "before" (to the right of A) will divide the "outside" region $2,3,4,5$ into 2 and $3,4,5$. The segment between A and B will divide the "inside" region into 1 and $6,7$. The segment between B and C will divide the 3, 4, 5 region into 5 and 3, 4. The segment between C and D will separate 6 and 7, and finally, the segment "beyond" D will separate 3 and 4.

Now let $F(n)$ be the number we are looking for, that is, the largest number of regions into which the plane can be divided by n parabolas. Suppose we have already drawn $n-1$ parabolas and we now draw an nth. Because any two parabolas can intersect in at most four points, there is a total of at most $4(n-1)$ intersection points of the new parabola with the $n-1$ previous parabolas. Therefore, when we draw the new parabola it will come in at most $4(n-1) + 1 = 4n - 3$ segments. Because each segment adds 1 to the count of regions, we end up with at most $4n - 3 + F(n-1)$ regions. That is,

$$F(n) \leq 4n - 3 + F(n-1).$$

Thus, from $F(1) = 2$ we get $F(2) \leq (4 \cdot 2 - 3) + 2 = 7$, and then $F(3) \leq (4 \cdot 3 - 3) + 7 = 16$, $F(4) \leq (4 \cdot 4 - 3) + 16 = 29$, and so forth. In general,

$$\begin{aligned} F(n) &\leq \Big(4n - 3\Big) + \Big(4(n-1) - 3\Big) + \cdots + \Big(4 \cdot 2 - 3\Big) + 2 \\ &= 4\Big(n + (n-1) + \cdots + 2\Big) - \underbrace{(3 + 3 + \cdots + 3)}_{(n-1) \text{ terms}} + 2 \\ &= 4 \cdot (n-1) \cdot \left(\frac{n+2}{2}\right) - 3(n-1) + 2 = 2n^2 - n + 1. \end{aligned}$$

To show that we can, in fact, get $2n^2 - n + 1$ regions, it will be enough to show that we can always arrange the n parabolas so any two of them do in fact intersect in four points and all the intersection points are distinct from each other, because then tracing the kth parabola will (for $k \geq 2$) add $4(k-1) + 1$ regions, as we've seen. This can be done by arranging the vertices $V_1, V_2, V_3, \ldots$ of the parabolas along a quarter circle, with the parabolas opening "inward" so that their axes are radii of the circle; the fact that any two axes meet on the "inside" guarantees four intersection points for each pair of parabolas, and it's easy to adjust the "widths" of the parabolas so no two intersection points coincide.

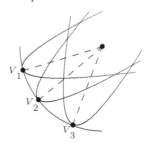

Problem 82

A particle starts somewhere in the plane and moves 1 unit in a straight line. Then it makes a "shallow right turn," abruptly changing direction by an acute angle α, and moves 1 unit in a straight line in the new direction. Then it again changes direction by α (to the right) and moves 1 unit, and so forth. In all, the particle takes 9 steps of 1 unit each, with each direction at an angle α to the previous direction.

a. For which value(s) of α does the particle end up exactly at its starting point?

b. For how many values of the acute angle α does the particle end up at a point whose (straight-line) distance to the starting point is exactly 1 unit?

Answer. The particle will end up at the starting point after 9 steps if and only if the acute angle α is either $2\pi/9$ or $4\pi/9$. The particle will end up 1 unit from the starting point after 9 steps for exactly three different acute angles α.

Solution 1. (Amy Becker '11, Carleton College) Let the starting point be P_0, and let $P_1, P_2, P_3, \ldots$ be the points the particle reaches at the end of the various steps. Let C be the intersection point of the perpendicular bisectors of P_0P_1 and P_1P_2. Then C is an equal distance from P_0, P_1, and P_2, so the isosceles triangles P_0P_1C and P_1P_2C are congruent, with base angles $\frac{1}{2}(\pi - \alpha)$. But then

$$\angle CP_2P_3 = \angle P_1P_2P_3 - \angle CP_2P_1$$
$$= \pi - \alpha - \tfrac{1}{2}(\pi - \alpha) = \tfrac{1}{2}(\pi - \alpha)$$

also, so triangles P_3P_2C and P_1P_2C are congruent, and therefore P_3 is the same distance to C as P_0, P_1, and P_2 are.

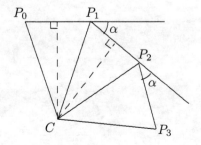

THE SOLUTIONS

Continuing in this way, we see that all the points $P_0, P_1, P_2, \ldots$ lie on the same circle centered at C. Also, the angles $\angle P_0 C P_1, \angle P_1 C P_2, \ldots$ at the center all equal α.

a. To have $P_9 = P_0$, 9α needs to be a multiple of 2π. Given that α is acute, the possibilities are $\alpha = 2\pi/9$ and $\alpha = 4\pi/9$.

b. Because α is acute, taking four steps is not enough to go around the circle once, so the circumference of the circle is greater than 4 and hence its diameter is greater than $4/\pi$, so certainly greater than 1. Therefore, there are exactly two points on the circle that have distance 1 to P_0. For P_9 to be one of those points, we must have either $P_8 = P_0$ (and $P_9 = P_1$) or $P_{10} = P_0$. Therefore, either 8α or 10α must be a multiple of 2π. Given that α is acute, the possibilities are $\alpha = \pi/4$, $\alpha = \pi/5$, and $\alpha = 2\pi/5$.

Solution 2. We may assume that the starting point is $P_0 = (0,0)$ and after one step the particle is at $P_1 = (1,0)$. After two steps (see figure) the particle is at $P_2 = P_1 + (\cos\alpha, -\sin\alpha)$, and after three steps it is at $P_3 = P_2 + (\cos 2\alpha, -\sin 2\alpha)$.

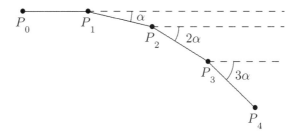

Continuing in this way, we see that after 9 steps the particle is at

$$P_9 = (1 + \cos\alpha + \cos 2\alpha + \cdots + \cos 8\alpha, \ -\sin\alpha - \sin 2\alpha - \cdots - \sin 8\alpha).$$

Now, switching to the complex plane and using complex exponentials ($e^{i\theta} = \cos\theta + i\sin\theta$),

$$P_9 = 1 + e^{-i\alpha} + e^{-2i\alpha} + \cdots + e^{-8i\alpha} = \frac{e^{-9i\alpha} - 1}{e^{-i\alpha} - 1},$$

where in the last step we have applied the formula for the sum of a finite geometric series (assuming $e^{-i\alpha} \neq 1$; if $e^{-i\alpha} = 1$, then $P_n = (n, 0)$ and moves along the positive x-axis to infinity).

a. We need to find the acute angles α for which $e^{-9i\alpha} = 1$ but $e^{-i\alpha} \neq 1$, so $e^{9i\alpha} = 1$ but $e^{i\alpha} \neq 1$. $e^{9i\alpha} = 1$ implies that 9α is a multiple of 2π, and so there are exactly two angles that qualify: $\alpha = 2\pi/9$, $\alpha = 4\pi/9$.

b. We need to find the acute angles α for which $\left| \dfrac{e^{-9i\alpha} - 1}{e^{-i\alpha} - 1} \right| = 1$. For this, note that

$$\left| \frac{e^{-9i\alpha} - 1}{e^{-i\alpha} - 1} \right| = 1 \iff \left(\frac{e^{9i\alpha} - 1}{e^{i\alpha} - 1} \right) \left(\frac{e^{-9i\alpha} - 1}{e^{-i\alpha} - 1} \right) = 1 \quad \text{(because } |z|^2 = \bar{z}z\text{)}$$

$$\iff 1 - e^{9i\alpha} - e^{-9i\alpha} + 1 = 1 - e^{i\alpha} - e^{-i\alpha} + 1$$

$$\iff 2 - 2\cos 9\alpha = 2 - 2\cos \alpha$$

$$\iff \frac{1 - \cos 9\alpha}{2} = \frac{1 - \cos \alpha}{2} \iff \sin^2 \left(\frac{9\alpha}{2} \right) = \sin^2 \left(\frac{\alpha}{2} \right)$$

$$\iff \sin \left(\frac{9\alpha}{2} \right) = \pm \sin \left(\frac{\alpha}{2} \right).$$

Now the graph of $\sin \left(\frac{9\alpha}{2} \right)$ has amplitude 1, period $4\pi/9$, and goes through the origin, while the graphs of $\pm \sin \left(\frac{\alpha}{2} \right)$ have amplitude 1, period 4π, and go through the origin. The graphs make it clear that there are exactly three values of α between 0 and $\pi/2$ for which $\sin \left(\frac{9\alpha}{2} \right) = \pm \sin \left(\frac{\alpha}{2} \right)$ (that is, for which the straight-line distance from starting point to end point is exactly one unit).

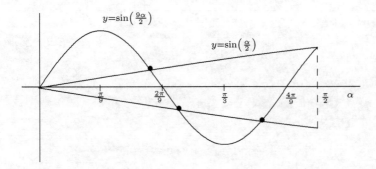

In fact, those values of α are $\pi/5$, $\pi/4$, and $2\pi/5$. (See Solution 1 for a systematic way to find them.)

Problem 83

Let $A \neq 0$ and B_1, B_2, B_3, B_4 be 2×2 matrices (with real entries) such that

$$\det(A + B_i) = \det A + \det B_i \quad \text{for } i = 1, 2, 3, 4.$$

Show that there exist real numbers k_1, k_2, k_3, k_4, not all zero, such that

$$k_1 B_1 + k_2 B_2 + k_3 B_3 + k_4 B_4 = 0.$$

Solution. Let

$$A = \begin{pmatrix} a & b \\ c & d \end{pmatrix} \quad \text{and} \quad B_i = \begin{pmatrix} w_i & x_i \\ y_i & z_i \end{pmatrix}, \quad i = 1, 2, 3, 4.$$

If $\det(A + B_i) = \det A + \det B_i$, then

$$dw_i - cx_i - by_i + az_i = 0.$$

Because $A \neq 0$, the solution space to $dw - cx - by + az = 0$ is a 3-dimensional vector space. Since any four vectors in a 3-dimensional space are linearly dependent, there must exist k_1, k_2, k_3, k_4, not all 0, for which

$$k_1 B_1 + k_2 B_2 + k_3 B_3 + k_4 B_4 = 0.$$

Comment. The "obvious" analog of this statement for nine 3×3 matrices $B_1, \ldots, B_9$ is false.

Problem 84

Let g be a continuous function defined on the positive real numbers. Define a sequence (f_n) of functions as follows. Let $f_0(x) = 1$, and for $n \geq 0$ and $x > 0$, let

$$f_{n+1}(x) = \int_1^x f_n(t) g(t) \, dt.$$

Suppose that for all $x > 0$, $\sum_{n=0}^{\infty} f_n(x) = x$. Find the function g.

Answer. The function g is defined by $g(x) = \frac{1}{x}$, $x > 0$.

Solution. By the Fundamental Theorem of Calculus,
$$f_1'(x) = \frac{d}{dx}\int_1^x f_0(t)g(t)\,dt = \frac{d}{dx}\int_1^x g(t)\,dt = g(x).$$

Also, note that $f_1(1) = \int_1^1 g(t)\,dt = 0$. Then
$$f_2(x) = \int_1^x f_1(t)g(t)\,dt = \int_1^x f_1(t)f_1'(t)\,dt = \frac{1}{2}\left(f_1(t)\right)^2\Big|_1^x = \frac{1}{2}\left(f_1(x)\right)^2.$$

Next,
$$f_3(x) = \int_1^x f_2(t)g(t)\,dt = \int_1^x \frac{1}{2}\left(f_1(t)\right)^2 f_1'(t)\,dt = \frac{1}{6}\left(f_1(t)\right)^3\Big|_1^x = \frac{1}{6}\left(f_1(x)\right)^3,$$

and continuing in this way we see that $f_n(x) = \frac{1}{n!}\left(f_1(x)\right)^n$ for all positive integers n. This also holds for $n = 0$, because $f_0(x) = 1$. Therefore,
$$x = \sum_{n=0}^\infty f_n(x) = \sum_{n=0}^\infty \frac{1}{n!}\left(f_1(x)\right)^n = e^{f_1(x)},$$

so we have
$$f_1(x) = \ln x,$$
$$g(x) = f_1'(x) = \frac{1}{x}.$$

Problem 85

Let r and s be specific positive integers. Let F be a function from the set of all positive integers to itself with the following properties:
  (i) F is one-to-one and onto;
  (ii) For every positive integer n, either $F(n) = n + r$ or $F(n) = n - s$.

a. Show that there exists a positive integer k such that the k-fold composition of F with itself is the identity function.

b. Find the smallest such positive integer k.

Answer. The smallest positive integer k such that $F^{(k)}(n) = F(n)$ for all positive integers n is $k = \dfrac{r+s}{\gcd(r,s)}$.

THE SOLUTIONS 157

Solution. a. First note that for any integer n with $1 \leq n \leq s$, $n - s$ is not positive, so for those integers n, we must have $F(n) = n + r$. On the other hand, F is onto, so there is an integer n with $F(n) = 1$ and for this n we must have $F(n) = n - s = 1$ which means $n = s + 1$; that is, $F(s+1) = 1$. Similarly, $F(s+2) = 2, \ldots, F(s+r) = r$, and we now know exactly what $F(n)$ is for each integer n in the interval $[1, r+s]$:

$$F(n) = \begin{cases} n + r & \text{for } 1 \leq n \leq s, \\ n - s & \text{for } s + 1 \leq n \leq r + s; \end{cases}$$

in particular, note that F actually maps the interval (of integers) $[1, r+s]$ onto itself.

Once we know this, we can use similar arguments to see that F must map each of the intervals $[r+s+1, 2(r+s)], [2(r+s)+1, 3(r+s)], \ldots$ to itself. For example, for $r+s+1 \leq n \leq r+2s$, $n-s$ is in the interval $[r+1, r+s]$ which is already "taken", so $F(n)$ cannot equal $n-s$ and must equal $n+r$. Continuing in this way, we see that

$$F(n) = \begin{cases} n + r & \text{for } r+s+1 \leq n \leq r+2s, \\ n - s & \text{for } r+2s+1 \leq n \leq 2(r+s), \end{cases}$$

and, proceeding by induction, that F acts in the same way on each of the intervals $[1, r+s], [r+s+1, 2(r+s)], \ldots, [i(r+s)+1, (i+1)(r+s)], \ldots$:

$$F(n) = \begin{cases} n + r & \text{for } i(r+s)+1 \leq n \leq ir+(i+1)s, \\ n - s & \text{for } ir+(i+1)s+1 \leq n \leq (i+1)(r+s). \end{cases}$$

Because F maps each of these intervals to itself, the same will be true for the k-fold composition of F with itself. Also, because F gives a permutation of the finite set $\{1, 2, \ldots, r+s\}$ of integers in the first interval, there is some finite positive integer k such that the k-fold composition of F with itself is the identity. And once the k-fold composition of F with itself is the identity on $\{1, 2, \ldots, r+s\}$, the same will be true for all the other intervals $[i(r+s)+1, (i+1)(r+s)]$, and we're done.

b. As we've seen, it is enough to look at what happens to the integers in the interval $[1, r+s]$. If we apply F repeatedly to such an integer m, it will go up a number of times, say u times, and down a number of times, say d times, before returning to m, so $ur = ds$. Now there are exactly s integers in $[1, r+s]$ that go up when we apply F and exactly r that go down. Therefore, $u \leq s$ and $d \leq r$.

Let $g = \gcd(r, s)$. Then at each step the integer must go up or down by a multiple of g, so there are at most $(r+s)/g$ integers that can be in the cycle for m. On the other hand, $ur = ds$ implies that $u(r/g) = d(s/g)$, and because r/g and s/g are relatively prime, u is divisible by s/g and d is divisible by r/g. Since u and d are positive, $u+d \geq (s/g)+(r/g) = (r+s)/g$. But we had seen that we could encounter at most $(r+s)/g$ integers in the cycle, so there can be at most that many steps, and $u + d = (r + s)/g$. Since this applies to any starting integer m, the smallest integer k must be $(r + s)/g$. This completes the proof.

Problem 86

Two ice fishermen have set up their ice houses on a perfectly circular lake, in exactly opposite directions from the center, two-thirds of the way from the center to the lakeshore. Can a third ice house be put on the lake in such a way that the three regions, each consisting of all points on the lake for which one of the three ice houses is closest, all have the same area?

Idea. If a third ice house were placed near the edge of the lake, it could not "control" one-third of the lake; if it were placed in the center of the lake, it would control more than one-third of the lake. There should be an intermediate spot somewhere between these extremes where the third ice house will control exactly one third of the area.

Solution. Yes, it is possible to put a third ice house on the lake in such a way that the three areas described in the problem are all equal. To see why, suppose that the center of the lake is at C and that the diameter on which the first two ice houses are located is AB; let H_1, H_2 be the locations of these ice houses, so we are given that $H_1 C = \frac{2}{3} AC$ and $CH_2 = \frac{2}{3} CB$.

If the third ice house is placed at the center of the lake, the three areas controlled by the three ice houses will be parallel strips of the circle bounded by the perpendicular bisectors of the line segments $H_1 C$ and CH_2, and it is easy to see that the third ice house controls more than one-third of the area of the circle.

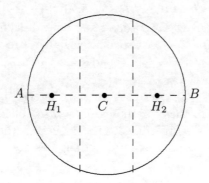

Suppose we place the third ice house at the same distance from C as H_1 and H_2, but along a radius at right angles to AB. The shaded sector in the following figure is the region of points for which the closest ice house is H_3, and it is easy to see that the sector is one quarter of the circle.

Now imagine moving the third ice house along the radius toward the center of the lake. The three areas (controlled by each of the ice houses) will change continuously as we do this. By symmetry, the regions consisting of points for which H_1 is closest and of points for which H_2 is closest will always have the same area. If we look at the difference between this area and the area of points for which H_3 is closest, we see that this difference is originally positive (specifically, $3/8 - 1/4 = 1/8$ of the area of the circle) but eventually becomes negative (by the time H_3 reaches the center of the circle). Therefore, by the Intermediate Value Theorem, the difference must be zero at some intermediate point. To get all three areas equal, put the third ice house at that point.

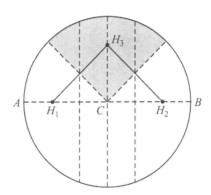

Problem 87

a. Let a_n be the result of writing down the first n odd integers in order and b_n be the result of writing down the first n even integers in order. Evaluate $\lim_{n\to\infty} \frac{a_n}{b_n}$.

b. Now suppose we do the same thing, but we write all the odd and even integers in base B (and we interpret the fractions a_n/b_n in base B). Show that for any base $B \geq 2$, $\lim_{n\to\infty} \frac{a_n}{b_n}$ exists. For what values of B will the limit be the same as for $B = 10$?

Answer. $\lim_{n\to\infty} \dfrac{a_n}{b_n} = 0$ if and only if B is even, in particular if $B = 10$.

Solution. a. The first few terms of the sequence are

$$\frac{1}{2}, \quad \frac{13}{24}, \quad \frac{135}{246}, \quad \frac{1357}{2468}, \quad \frac{13579}{246810}$$

which are increasing until the fifth term. For that term, though, the numerator has one more digit than the denominator, so that fraction is smaller than $\frac{1}{10}$. What's more, as we now go through the two-digit numbers beyond 10, successively extending the numerator by $11, 13, 15, \ldots$ and the denominator by $12, 14, 16, \ldots$, all the fractions we get will still be smaller than $\frac{1}{10}$ (for the same reason). At the end of the two-digit numbers, when we extend the numerator by 99 and the denominator by 100, the numerator starts having *two* fewer digits than the denominator, so the new fraction will be smaller than $\frac{1}{100}$. At the end of the three-digit numbers the numerator will be extended by 999 and the denominator by 1000, and the fraction and all subsequent ones will be smaller than $\frac{1}{1000}$. In general, we can make sure that $\dfrac{a_n}{b_n} < \dfrac{1}{10^k}$ by taking n large enough so that the first n odd integers include all the k-digit odd integers. Because $\lim_{k\to\infty} \dfrac{1}{10^k} = 0$, it follows that $\lim_{n\to\infty} \dfrac{a_n}{b_n} = 0$.

b. For any *even* base B, the previous argument still works. For example, the 1-digit numbers in base B are $1, 2, \ldots, B-1$, so the last 1-digit number is odd and the corresponding extension of the denominator will be by the 2-digit number 10. After that, the 2-digit numbers alternate between odd and even:

$$11 \,(= B+1, \text{ odd}), \; 12, \; \ldots, \; (B{-}1)0, \; (B{-}1)1, \; \ldots, \; (B{-}1)(B{-}1)$$

(this last one represents $(B-1)B + (B-1) = B^2 - 1$, odd). So the last 2-digit number is matched with the even 3-digit number 100 $(= B^2)$, and again the denominator gets further "ahead." In general, we can make sure that b_n has at least k more base B digits than a_n by taking n large enough so that the first n odd integers include all the k-digit ones in base B. Therefore, $\lim_{n\to\infty} \dfrac{a_n}{b_n} = 0$.

For *odd* bases B, the transition from k-digit to $(k+1)$-digit integers always occurs right after an even number (specifically, after $B^k - 1$, which in base B is written with k digits as $(B{-}1)(B{-}1)\cdots(B{-}1)$). So the number of base B digits in a_n is always the same as in b_n. Because a_n starts with 1 and b_n starts with 2, we have $a_n/b_n < 1$ for all n. We'll show that the

THE SOLUTIONS

sequence (a_n/b_n) is increasing. Then because it is also bounded above (by 1), it must have a limit, but the limit cannot be 0.

So we need to show that $\dfrac{a_n}{b_n} < \dfrac{a_{n+1}}{b_{n+1}}$ for all n. This will follow in the course of proving that

$$\frac{a_n}{b_n} < \frac{2n+1}{2n+2} \qquad (*)$$

for all n. Clearly, $(*)$ is true for $n = 1$ because $1/2 < 3/4$. So suppose $(*)$ holds for a fixed n. Note that if the $(n+1)$st odd number, which is $2n+1$, has k digits in base B, then

$$a_{n+1} = a_n B^k + 2n + 1 \quad \text{and} \quad b_{n+1} = b_n B^k + 2n + 2.$$

It follows that

$$\frac{a_n}{b_n} = \frac{a_n B^k}{b_n B^k} < \frac{a_n B^k + 2n + 1}{b_n B^k + 2n + 2} = \frac{a_{n+1}}{b_{n+1}} < \frac{2n+1}{2n+2}, \qquad (**)$$

where we are using the fact that for positive numbers a, b, c, d, if $\dfrac{a}{b} < \dfrac{c}{d}$ then $\dfrac{a}{b} < \dfrac{a+c}{b+d} < \dfrac{c}{d}$, which is easily proved by cross-multiplication. From $(**)$ we see that $\dfrac{a_n}{b_n} < \dfrac{a_{n+1}}{b_{n+1}}$ and also that $\dfrac{a_{n+1}}{b_{n+1}} < \dfrac{2n+1}{2n+2} < \dfrac{2n+3}{2n+4}$; therefore $(*)$ follows by induction.

Problem 88

Note that the integers $2, -3$, and 5 have the property that the difference of any two of them is an integer times the third. Suppose three distinct integers a, b, c have this property.

a. Show that a, b, c cannot all be positive.
b. Now suppose that a, b, c, in addition to having the above property, have no common factors (except $1, -1$). Is it true that one of the three integers has to be either $1, 2, -1$, or -2?

Solution. Without loss of generality, we may assume $a < b < c$.

a. We prove the stronger result that one of the three, in our case a, must be negative. Suppose $0 \leq a$. Then $0 < b - a \leq b < c$, hence $0 < (b-a)/c < 1$, a contradiction.

b. One of a, b, c must be ± 1 or ± 2.

Replacing (a, b, c) by $(-c, -b, -a)$ if necessary, we may assume $0 \leq b$. Since $c - a$ is a nonzero multiple of b, we must actually have $a < 0 < b < c$. The divisibility condition implies the existence of positive integers m and n such that
$$b - c = na,$$
$$b - a = mc.$$
Then $c - a = mc - na$ (by subtraction), so $(m - 1)c = (n - 1)a$. This, together with $c > 0$ and $a < 0$, implies $m = 1, n = 1$. Thus, $b - c = a$, or $c + a = b$. Using this, and the supposition that $c - a = kb$ for some integer k, we have
$$2c = (c + a) + (c - a) = b + kb = (k + 1)b.$$
If b and c are relatively prime, then b divides 2, and we are done. If not, then the equation $a = b - c$ implies a, b, and c have a nontrivial common factor, a contradiction.

Problem 89

Start with four numbers arranged in a circle. Form the average of each pair of adjacent numbers, and put these averages in the circle between the original numbers; then delete the original numbers so that once again there are four numbers in the circle. Repeat. Suppose that after twenty steps, you find the numbers $1, 2, 3, 4$ in *some* order. What numbers did you have after one step? Can you recover the original numbers from this information?

Idea. Although the numbers themselves keep changing at every step, is there anything about them that doesn't change?

Solution 1. If the original numbers are a_0, b_0, c_0, d_0, then after one step we have
$$a_1 = \frac{a_0 + b_0}{2}, \quad b_1 = \frac{b_0 + c_0}{2}, \quad c_1 = \frac{c_0 + d_0}{2}, \quad d_1 = \frac{d_0 + a_0}{2}$$
around the circle.

In the next step we replace a_1, b_1, c_1, d_1 by
$$a_2 = \frac{a_1 + b_1}{2}, \quad b_2 = \frac{b_1 + c_1}{2}, \quad c_2 = \frac{c_1 + d_1}{2}, \quad d_2 = \frac{d_1 + a_1}{2}.$$

THE SOLUTIONS

Note that at each step the *average* of the four numbers doesn't change. For example,

$$\frac{a_2+b_2+c_2+d_2}{4} = \frac{\frac{a_1+b_1}{2}+\frac{b_1+c_1}{2}+\frac{c_1+d_1}{2}+\frac{d_1+a_1}{2}}{4} = \frac{a_1+b_1+c_1+d_1}{4}.$$

If this average is A and we replace a_1, b_1, c_1, d_1 by $a_1 - A, b_1 - A, c_1 - A, d_1 - A$ (which will make their average 0), then a_2, b_2, c_2, d_2 will similarly be replaced by $a_2 - A, b_2 - A, c_2 - A, d_2 - A$. Thus, in studying our process we can first look at the special case in which the average is 0. Then for the general case, subtract A from each number to make the average 0, carry out the process, and then add A back in again.

Note that $a_1 + c_1 = b_1 + d_1 = (a_0 + b_0 + c_0 + d_0)/2$; suppose our original numbers are chosen so that their average is 0. Then $(a_1 + c_1) = (b_1 + d_1) = 0$, $b_1 = -d_1$ and $a_1 = -c_1$. So our numbers after the first step are

$$a_1, \quad b_1, \quad -a_1, \quad -b_1.$$

One step later we have

$$a_2 = \frac{a_1+b_1}{2}, \quad b_2 = \frac{b_1-a_1}{2}, \quad c_2 = \frac{-a_1-b_1}{2} = -a_2, \quad d_2 = \frac{a_1-b_1}{2} = -b_2,$$

and if we take one more step we find

$$a_3 = \frac{a_2+b_2}{2} = \frac{b_1}{2}, \quad b_3 = -\frac{a_1}{2}, \quad c_3 = -\frac{b_1}{2}, \quad d_3 = \frac{a_1}{2}.$$

Because only the order around the circle is important, we can list these as

$$\frac{a_1}{2}, \quad \frac{b_1}{2}, \quad -\frac{a_1}{2}, \quad -\frac{b_1}{2}.$$

That is, in two steps all the numbers have been halved (from $a_1, b_1, -a_1, -b_1$). But now we know how to reverse the steps after the first: To take two steps backward, double all the numbers; to take one step back, first take two steps back and then one forward, going from $a_2, b_2, c_2, d_2 = a_2, b_2, -a_2, -b_2$ (say) to $2a_2, 2b_2, -2a_2, -2b_2$ to $a_2 + b_2, b_2 - a_2, -a_2 - b_2, a_2 - b_2$.

Now for the specific numbers in our problem, where it is given that $a_{20}, b_{20}, c_{20}, d_{20}$ (the numbers after 20 steps) are $1, 2, 3, 4$ in some order. The average is $A = (1+2+3+4)/4 = 5/2$. Subtracting A from $a_{20}, b_{20}, c_{20}, d_{20}$ yields $-\frac{3}{2}, -\frac{1}{2}, \frac{1}{2}, \frac{3}{2}$. But that order is wrong; we saw above that "opposite" numbers (a_i and c_i, or b_i and d_i) are literally opposite in our case $A = 0$ once at least one step has been carried out. So we must have

$$-\frac{3}{2}, \quad -\frac{1}{2}, \quad \frac{3}{2}, \quad \frac{1}{2}$$

(or the reverse order) in that order around the circle. We want to go back 19 steps; so we will go back 18 steps by doubling the numbers nine times to

$$-\frac{3}{2} \cdot 2^9 = -768, \quad -\frac{1}{2} \cdot 2^9 = -256, \quad \frac{3}{2} \cdot 2^9 = 768, \quad \frac{1}{2} \cdot 2^9 = 256,$$

and then go back one more step (see above) to

$$-768-256 = -1024, \ -256+768 = 512, \ 768+256 = 1024, \ 256-728 = -512.$$

But now we have to add $A = \frac{5}{2}$ back in, so we have

$$-1021\tfrac{1}{2}, \quad 514\tfrac{1}{2}, \quad 1026\tfrac{1}{2}, \quad -509\tfrac{1}{2}$$

as the actual numbers after one step.

We cannot recover the original numbers from the knowledge of a_1, b_1, c_1, d_1; in fact, if we started with $a_0 + x, b_0 - x, c_0 + x, d_0 - x$ instead of a_0, b_0, c_0, d_0, the resulting averages a_1, b_1, c_1, d_1 after one step would be the same.

Solution 2. As in Solution 1, we note that the average of the numbers around the circle doesn't change during the process, so it suffices to consider the case in which the average is 0. Also, as in Solution 1, we may assume the numbers around the circle have the form

$$a_n, \quad b_n, \quad -a_n, \quad -b_n,$$

in that order. Because the numbers after 20 steps are $1, 2, 3, 4$ in some order, we may assume, once their average has been subtracted away, that $a_{20} = 3/2$, $b_{20} = 1/2$.

From the given recurrence

$$\begin{pmatrix} a_{n+1} \\ b_{n+1} \end{pmatrix} = \begin{pmatrix} 1/2 & 1/2 \\ -1/2 & 1/2 \end{pmatrix} \begin{pmatrix} a_n \\ b_n \end{pmatrix},$$

we then have

$$\begin{pmatrix} a_n \\ b_n \end{pmatrix} = \begin{pmatrix} 1/2 & 1/2 \\ -1/2 & 1/2 \end{pmatrix}^{-1} \begin{pmatrix} a_{n+1} \\ b_{n+1} \end{pmatrix} = \begin{pmatrix} 1 & -1 \\ 1 & 1 \end{pmatrix} \begin{pmatrix} a_{n+1} \\ b_{n+1} \end{pmatrix},$$

and consequently

$$\begin{pmatrix} a_1 \\ b_1 \end{pmatrix} = \begin{pmatrix} 1 & -1 \\ 1 & 1 \end{pmatrix}^{19} \begin{pmatrix} a_{20} \\ b_{20} \end{pmatrix}.$$

To compute this we can diagonalize the matrix by finding the eigenvalues $1 \pm i$ and corresponding eigenvectors $\begin{pmatrix} \pm i \\ 1 \end{pmatrix}$; we then have

$$\begin{pmatrix} a_1 \\ b_1 \end{pmatrix} = \begin{pmatrix} 1 & -1 \\ 1 & 1 \end{pmatrix}^{19} \begin{pmatrix} 3/2 \\ 1/2 \end{pmatrix}$$

$$= \left(\begin{pmatrix} i & -i \\ 1 & 1 \end{pmatrix} \begin{pmatrix} 1+i & 0 \\ 0 & 1-i \end{pmatrix} \begin{pmatrix} i & -i \\ 1 & 1 \end{pmatrix}^{-1} \right)^{19} \begin{pmatrix} 3/2 \\ 1/2 \end{pmatrix}$$

$$= \begin{pmatrix} i & -i \\ 1 & 1 \end{pmatrix} \begin{pmatrix} 1+i & 0 \\ 0 & 1-i \end{pmatrix}^{19} \begin{pmatrix} -i/2 & 1/2 \\ i/2 & 1/2 \end{pmatrix} \begin{pmatrix} 3/2 \\ 1/2 \end{pmatrix}$$

$$= \begin{pmatrix} i & -i \\ 1 & 1 \end{pmatrix} \begin{pmatrix} 2^9(i-1) & 0 \\ 0 & 2^9(-i-1) \end{pmatrix} \begin{pmatrix} -i/2 & 1/2 \\ i/2 & 1/2 \end{pmatrix} \begin{pmatrix} 3/2 \\ 1/2 \end{pmatrix}$$

$$= 2^9 \begin{pmatrix} -1 & -1 \\ 1 & -1 \end{pmatrix} \begin{pmatrix} 3/2 \\ 1/2 \end{pmatrix} = 512 \begin{pmatrix} -2 \\ 1 \end{pmatrix} = \begin{pmatrix} -1024 \\ 512 \end{pmatrix}.$$

We now get the numbers after one step by adding the average $A = \frac{5}{2}$ back in:

$\frac{5}{2} + a_1 = -1021\frac{1}{2}$, $\quad \frac{5}{2} + b_1 = 514\frac{1}{2}$, $\quad \frac{5}{2} - a_1 = 1026\frac{1}{2}$, $\quad \frac{5}{2} - b_1 = -509\frac{1}{2}$.

Problem 90

Suppose all the integers have been colored with the three colors red, green and blue such that each integer has exactly one of those colors. Also suppose that the sum of any two green integers is blue, the sum of any two blue integers is green, the opposite of any green integer is blue, and the opposite of any blue integer is green. Finally, suppose that 1492 is red and that 2011 is green. Describe precisely which integers are red, which integers are green, and which integers are blue.

Answer. The green integers are those of the form $(3m+1)2011$ with integer m; the blue integers are those of the form $(3m+2)2011$ with integer m. All other integers are red.

Solution. First note that 0 must be red, for if 0 were green (blue) then $0 + 0 = 0$ would be blue (green). Also, if n is red, then $-n$ is also red (for if $-n$ were green then $n = -(-n)$ would be blue, and conversely). So it is enough to determine which *positive* integers are red, green, and blue.

Let x be the smallest positive integer that is not red, and, for now, let's suppose it is green. Then $2x = x + x$ is blue, $4x = 2x + 2x$ is green, $5x = 4x + x$ is blue, $7x = 5x + 2x$ is green, $8x = 4x + 4x$ is blue. In fact, we can see by induction on k that for any integer $k \geq 0$, $(3k+1)x$ is green and $(3k+2)x$ is blue: We've checked that this is true for $k = 0, 1, 2$, and if it's true for k, then $(3(k+1)+1)x = (3k+2)x + 2x$ (the sum of two blue integers) is green, and $(3(k+1)+2)x = (3k+1)x + 4x$ (the sum of two green integers) is blue. (If x were blue instead of green, we could switch the colors "blue" and "green" in this discussion without making any other changes.)

We claim that the remaining positive multiples of x: $3x, 6x, 9x, \ldots$ are red. For if $3kx$ is green, then $(3k+1)x = 3kx + x$ (sum of two green integers) is blue, contradiction. If $3kx$ is blue, then $(3k+2)x = 3kx + 2x$ (sum of two blue integers) is green, contradiction. The only remaining possibility is that $3kx$ is red.

We now claim that any positive integer that is not a multiple of x must be red. For suppose $y > 0$ is not a multiple of x, and consider the sequence

$$x, \ 2x, \ 4x, \ 5x, \ 7x, \ 8x, \ 10x, \ 11x, \ 13x, \ 14x, \ \ldots$$

which consists of alternating green and blue integers. If $y < x$ then y is red because x is the smallest non-red positive integer. Suppose y is between two integers of the sequence that differ by x: $(3k+1)x < y < (3k+2)x$. If y is blue, then because $-(3k+1)x$ is blue, $y - (3k+1)x$ is green. But the previous inequality implies that $0 < y - (3k+1)x < x$, contradicting that x is the smallest non-red positive integer. If y is green then $-y$ is blue, so $-y + (3k+2)x$ is green. But $0 < -y + (3k+2)x < x$, and again we have a contradiction. So y must be red.

The remaining case is that y is between two integers of the sequence that differ by $2x$: $(3k+2)x < y < (3k+4)x$. In this case, if y is green, then $y - (3k+2)x$ is blue, and if y is blue, then $(3k+4)x - y$ is blue. Now $0 < y - (3k+2)x < 2x$ and $0 < (3k+4)x - y < 2x$, so in either case that y is not red, we have a blue integer between 0 and $2x$. However, by the first case there can be no such blue integer between x and $2x$, and since x is the smallest positive non-red integer, there can be none between 0 and x. Also, neither $y - (3k+2)x$ nor $(3k+4)x - y$ can be x because y is not a multiple of x. This contradiction shows that y must be red.

We now know that the *only* positive non-red integers are the multiples of x of the form $(3k+1)x$ and $(3k+2)x$. We are given that 2011 is green, so 2011 must be such a multiple. But 2011 is prime, so the only possibilities for x are $x = 1$ and $x = 2011$. We are also given that 1492 is red; note that $1492 = 3 \cdot 497 + 1$. Therefore, $x = 1$ is impossible, leaving

$x = 2011$ as the only possibility. Thus, the green positive integers are of the form $(3k + 1)2011$, $k \geq 0$, while the blue positive integers are of the form $(3k + 2)2011$, $k \geq 0$. The green negative integers are therefore of the form $-(3k + 2)2011 = (3(-k - 1) + 1)2011$, $k \geq 0$, and the blue negative integers are $-(3k + 1)2011 = (3(-k - 1) + 2)2011$, $k \geq 0$. Because any integer m can be written as either k or $-k - 1$ with $k \geq 0$, our answer is justified.

Problem 91

Let A be an $m \times n$ matrix with every entry either 0 or 1. How many such matrices A are there for which the number of 1's in each row and each column is even?

Solution. There are $2^{(m-1)(n-1)}$ such matrices.

To see this, we consider only "0-1" matrices (matrices whose entries are either 0 or 1), and call a row or column of such a matrix *odd* or *even* according to whether it contains an odd or even number of 1's. We then want to find the number of $m \times n$ matrices for which all rows and columns are even.

Let A be such a matrix. The $(m - 1) \times (n - 1)$ submatrix in the upper left corner of A is also a 0-1 matrix. We shall show that any $(m-1) \times (n-1)$ 0-1 matrix B uniquely determines an $m \times n$ 0-1 matrix A whose rows and columns are all even.

So let B be an $(m - 1) \times (n - 1)$ matrix; we will construct a matrix of the form

$$A = \begin{pmatrix} & & & a_{1n} \\ & B & & \vdots \\ & & & a_{m-1,n} \\ a_{m1} & \cdots & a_{m,n-1} & a_{mn} \end{pmatrix}.$$

To make the first row of A even, a_{1n} must be 0 or 1 depending on whether the first row of B is even or odd. A similar argument for the other rows and columns of B shows that $a_{1n}, \ldots a_{m-1,n}$ and $a_{m1}, \ldots, a_{m,n-1}$ are all determined by the entries of B. Now A will have the desired property provided that the last row and column of A come out to be even. This can be arranged by making the right choice (0 or 1) for a_{mn}, provided the parity (even or odd) of the "partial row" $(a_{m1}, \ldots, a_{m,n-1})$ is the same as the parity of the "partial column"

$$\begin{pmatrix} a_{1n} \\ \vdots \\ a_{m-1,n} \end{pmatrix}.$$

Given how we got the entries in this partial row and in this partial column, this means that the parity of the number of odd rows of B must be the same as the parity of the number of odd columns of B. However, this is always true, because each of these parities is odd or even according to whether the whole matrix B contains an odd or even number of 1's. Thus there is exactly one way to complete an $(m-1) \times (n-1)$ matrix B to a matrix A with the desired property.

Clearly, different $(m-1) \times (n-1)$ matrices B lead to different matrices A of the desired type. Thus, we see that there is a 1-1 correspondence between $m \times n$ matrices A for which all rows and columns are even and arbitrary $(m-1) \times (n-1)$ matrices B. Since we can choose each of the $(m-1)(n-1)$ entries of B independently, there are $2^{(m-1)(n-1)}$ choices for B, and we are done.

Comments. In terms of modular arithmetic, the construction of A from B can be phrased as follows. Given $B = (b_{ij})$, set

$$a_{nj} \equiv \sum_{i=1}^{n-1} b_{ij} \pmod{2} \quad \text{and} \quad a_{im} \equiv \sum_{j=1}^{m-1} b_{ij} \pmod{2}.$$

As for a_{nm}, the row condition requires

$$a_{nm} \equiv \sum_{j=1}^{m-1} a_{nj} \equiv \sum_{j=1}^{m-1} \sum_{i=1}^{n-1} b_{ij} \pmod{2},$$

whereas the column condition requires

$$a_{nm} \equiv \sum_{i=1}^{n-1} a_{im} \equiv \sum_{i=1}^{n-1} \sum_{j=1}^{m-1} b_{ij} \pmod{2}.$$

Since these sums are identical, a_{nm} is well defined.

More generally, a similar argument shows that there are $N^{(m-1)(n-1)}$ $m \times n$ matrices, with entries from $\{0, 1, 2, \ldots, N-1\}$, whose rows and columns each add to a multiple of N.

Problem 92

For $0 \le x \le 1$, let $T(x) = \begin{cases} x & \text{if } x \le 1/2 \\ 1-x & \text{if } x \ge 1/2 \end{cases}$, and define $f(x) = \sum_{n=1}^{\infty} T(x^n)$.

a. Evaluate $f\left(\dfrac{1}{\sqrt[3]{2}}\right)$.

b. Find all x for which $f(x) = 2012$.

Solution. a. For $x = 1/\sqrt[3]{2}$, $T(x) = 1-x$ because $x > 1/2$, $T(x^2) = 1-x^2$ because $x^2 = 1/\sqrt[3]{4} > 1/2$; on the other hand, for $n \ge 3$, $T(x^n) = x^n$ because $x^n \le x^3 = 1/2$. So

$$f(x) = (1-x) + (1-x^2) + \sum_{n=3}^{\infty} x^n$$

$$= 2 - x - x^2 + \frac{x^3}{1-x} = \frac{2 - 2x - x + x^2 - x^2 + x^3 + x^3}{1-x}$$

$$= \frac{3 - 3x}{1-x} \quad \text{(because } x^3 = 1/2\text{)}$$

$$= 3.$$

b. Because $f(0) = \sum_{n=1}^{\infty} T(0) = 0$ and $f(1) = \sum_{n=1}^{\infty} T(1) = 0$, we may assume that $0 < x < 1$. Then there is a positive integer k such that $x^k \le 1/2$ but $x^j > 1/2$ for $j < k$. For this choice of k, $T(x^j) = 1 - x^j$ for $j < k$ and $T(x^j) = x^j$ for $j \ge k$, so

$$f(x) = \sum_{j=1}^{k-1}(1 - x^j) + \sum_{j=k}^{\infty} x^j$$

$$= k-1 - \sum_{j=1}^{k-1} x^j + \frac{x^k}{1-x}$$

$$= k-1 - \frac{x-x^k}{1-x} + \frac{x^k}{1-x}$$

$$= k-1 + \frac{2x^k - x}{1-x}.$$

Now $x^k \le 1/2$ implies $2x^k - x \le 1-x$, and $1-x > 0$ so $\dfrac{2x^k - x}{1-x} \le 1$. On the other hand, $x^{k-1} > 1/2$, so $x^k > x/2$, and $\dfrac{2x^k - x}{1-x} > 0$. So we have

$f(x) = k-1+\varepsilon$, where $\varepsilon = \dfrac{2x^k - x}{1-x}$ satisfies $0 < \varepsilon \leq 1$. Because $k-1$ is an integer, the only way that $k-1+\varepsilon$ can equal the integer 2012 is to have $k-1 = 2011$, $\varepsilon = 1$. That is, $k = 2012$, $2x^k = 1$, so $x = \dfrac{1}{\sqrt[2012]{2}}$. This is the unique x for which $f(x) = 2012$.

Problem 93

Suppose A and B are convex subsets of $\mathbb{R}^3$. Let C be the set of all points R for which there are points P in A and Q in B such that R lies between P and Q. Does C have to be convex?

Solution. The set C is convex.

We identify points in $\mathbb{R}^3$ with vectors in order to be able to add points and to multiply points by scalars. Let R_1 and R_2 be points of C. We show that every point between R_1 and R_2 is in C by showing that for every t, $0 < t < 1$,

$$R = t\,R_1 + (1-t)\,R_2$$

is in C.

Let R_i, $i = 1, 2$, be between P_i in A and Q_i in B. Then there must be $u_i, 0 \leq u_i \leq 1$, for which

$$R_i = u_i\,P_i + (1-u_i)\,Q_i.$$

If $u_1 = u_2 = 0$, then $R = tQ_1 + (1-t)Q_2$ is in B, hence in C. Similarly, if $u_1 = u_2 = 1$, then R is in A, hence in C. We now assume neither holds. Combining the previous equations, we have

$$R = \bigl(t\,u_1\,P_1 + (1-t)u_2\,P_2\bigr) + \bigl(t(1-u_1)\,Q_1 + (1-t)(1-u_2)\,Q_2\bigr).$$

To use the convexity of A, the coefficients of P_1 and P_2 must be nonnegative (which they are) and sum to 1; therefore, we will factor $d = t\,u_1 + (1-t)u_2$ from the first bracketed term above. Now $1-d = t(1-u_1) + (1-t)(1-u_2)$, so we factor this from the second term, and we have

$$R = d\left(\dfrac{tu_1}{d}P_1 + \dfrac{(1-t)u_2}{d}P_2\right) + (1-d)\left(\dfrac{t(1-u_1)}{1-d}Q_1 + \dfrac{(1-t)(1-u_2)}{1-d}Q_2\right).$$

(We will see below that $0 < d < 1$, so $d \neq 0$ and $(1-d) \neq 0$.) Now since

$$\dfrac{tu_1}{d} + \dfrac{(1-t)u_2}{d} = 1,$$

the convexity of A allows us to conclude that

$$\frac{tu_1}{d}P_1 + \frac{(1-t)u_2}{d}P_2 = P$$

for some P in A. Similarly,

$$\frac{t(1-u_1)}{1-d}Q_1 + \frac{(1-t)(1-u_2)}{1-d}Q_2 = Q$$

for some Q in B. Since $0 < d = tu_1 + (1-t)u_2 < t + (1-t) = 1$, $R = dP + (1-d)Q$ is in C, and we are done.

Problem 94

Start with a circle and inscribe a regular n-gon in it, then inscribe a circle in that regular n-gon, then inscribe a regular n-gon in the new circle, then a third circle in the second n-gon, and so forth. Continuing in this way, the disk inside the original circle will be divided into infinitely many regions, some of which are bounded by a circle on the outside and one side of a regular n-gon on the inside (call these "type I" regions) while others are bounded by two sides of a regular n-gon on the outside and a circle on the inside ("type II" regions).

Let $f(n)$ be the fraction of the area of the original disk that is occupied by type I regions. What is the limit of $f(n)$ as n tends to infinity?

Answer. The limit is 2/3.

Solution. Suppose we have a circle of radius r and we inscribe a regular n-gon in it. Then we can divide the region inside the n-gon into n congruent triangles with a vertex at the center, such as $\triangle OAB$ in the diagram. If OC is the bisector of $\angle AOB$,

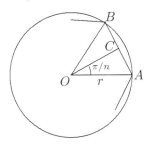

then the area of this triangle is

$$\tfrac{1}{2}\cdot OC \cdot AB = \tfrac{1}{2}\cdot r\cos\tfrac{\pi}{n}\cdot 2r\sin\tfrac{\pi}{n} = \tfrac{1}{2}r^2 \sin\tfrac{2\pi}{n},$$

so the area inside the n-gon is $n\cdot \tfrac{1}{2}r^2 \sin\tfrac{2\pi}{n}$. Because the area inside the circle is πr^2, the n type I regions between the circle on the outside and our inscribed n-gon have a total area of $\pi r^2 - n\tfrac{1}{2}r^2\sin\tfrac{2\pi}{n} = r^2\left(\pi - \tfrac{n}{2}\sin\tfrac{2\pi}{n}\right)$.

The next step is to inscribe a circle in our regular n-gon. The radius of that circle is $OC = r\cos\tfrac{\pi}{n}$, so the area inside that circle is $\cos^2\tfrac{\pi}{n}$ times the area of the original circle. The whole figure then "scales down" by that factor, so the type I regions between the new circle and *its* inscribed n-gon will have a total area of $\left(r^2\cos^2\tfrac{\pi}{n}\right)\left(\pi - \tfrac{n}{2}\sin\tfrac{2\pi}{n}\right)$. The next round of n type I regions will have a total area of $r^2(\cos^2\tfrac{\pi}{n})^2\left(\pi - \tfrac{n}{2}\sin\tfrac{2\pi}{n}\right)$, and so forth, so the total area of *all* the type I regions will be the sum of the geometric series

$$r^2\left(\pi - \tfrac{n}{2}\sin\tfrac{2\pi}{n}\right) + r^2(\cos^2\tfrac{\pi}{n})\left(\pi - \tfrac{n}{2}\sin\tfrac{2\pi}{n}\right) + r^2(\cos^2\tfrac{\pi}{n})^2\left(\pi - \tfrac{n}{2}\sin\tfrac{2\pi}{n}\right) + \cdots$$
$$= r^2\left(\pi - \tfrac{n}{2}\sin\tfrac{2\pi}{n}\right)\left(1 + \cos^2\tfrac{\pi}{n} + \cos^4\tfrac{\pi}{n} + \cos^6\tfrac{\pi}{n} + \cdots\right)$$
$$= r^2 \frac{\pi - \tfrac{n}{2}\sin\tfrac{2\pi}{n}}{1 - \cos^2\tfrac{\pi}{n}} = r^2 \frac{\pi - \tfrac{n}{2}\sin\tfrac{2\pi}{n}}{\sin^2\tfrac{\pi}{n}}.$$

Because the area of the original disk is πr^2, the fraction of this area occupied by type I regions is

$$f(n) = \frac{\pi - \tfrac{n}{2}\sin\tfrac{2\pi}{n}}{\pi \sin^2 \tfrac{\pi}{n}},$$

so to finish the problem we have to find the limit of this as $n \to \infty$. To do so, let's make the substitution $\theta = \tfrac{\pi}{n}$. Because $\theta \to 0$ as $n \to \infty$, we get

$$\lim_{n\to\infty} \frac{\pi - \tfrac{n}{2}\sin\tfrac{2\pi}{n}}{\pi \sin^2\tfrac{\pi}{n}} = \lim_{\theta\to 0} \frac{\pi - \tfrac{\pi}{2\theta}\sin 2\theta}{\pi \sin^2 \theta} = \lim_{\theta\to 0} \frac{\theta - \tfrac{1}{2}\sin 2\theta}{\theta \sin^2 \theta}.$$

This limit can be found in a variety of ways, for instance by using the Taylor expansion for the sine. A more elementary method starts with l'Hôpital's rule:

$$\lim_{\theta\to 0}\frac{\theta-\frac{1}{2}\sin 2\theta}{\theta\sin^2\theta}=\lim_{\theta\to 0}\frac{1-\cos 2\theta}{\sin^2\theta+\theta\cdot 2\sin\theta\cos\theta}=\lim_{\theta\to 0}\frac{2\sin^2\theta}{\sin^2\theta+2\theta\sin\theta\cos\theta}$$
$$=\lim_{\theta\to 0}\frac{2}{1+2(\theta/\sin\theta)\cos\theta}=\frac{2}{1+2}=\frac{2}{3},$$

where we have used $\lim_{\theta\to 0}\dfrac{\sin\theta}{\theta}=1$.

Problem 95

Suppose we have a configuration (set) of finitely many points in the plane which are not all on the same line. We call a point in the plane a *center* for the configuration if for every line through that point, there is an equal number of points of the configuration on either side of the line.

a. Give a necessary and sufficient condition for a configuration of four points to have a center.

b. Is it possible for a finite configuration of points (not all on the same line) to have more than one center?

Solution. a. For a configuration of four noncollinear points to have a center, it is necessary and sufficient for the four points to be the vertices of a convex quadrilateral (by which we mean a quadrilateral whose interior angles all measure less than 180°).

Given a convex quadrilateral, it is clear that the intersection of the two diagonals is a center for the configuration of the four vertices. Conversely, let A, B, C, and D be four noncollinear points with center O. The line AO must contain a second point, say C. The points B and D must be on opposite sides of the line AO. We claim $ABCD$ is a convex quadrilateral. The interior angle at B is an angle of the triangle ABC, hence measures less than 180°. Similarly, the interior angle at D measures less than 180°. Since no line contains A, B, C, and D, the line BO must also contain D. Arguing as above, the interior angles of $ABCD$ at A and C must measure less than 180°. We have proved the quadrilateral is convex.

b. No configuration of noncollinear points can have more than one center. Suppose a configuration has center O. Let O' be a second point in the plane. Choose a point A of the configuration not on the line OO'. Since O is a center for the configuration, the line AO must have an equal number of points (of the configuration) on either side of it. Now consider the line L

through O' parallel to AO. The side of L containing A has at least one more point (in fact, at least two more points) than the side of AO not containing O'. The other side of L has at most the same number of points as the side of AO containing O'. Thus, the side of L containing A has more points of the configuration than the other side of L, and O' cannot be a center for the configuration.

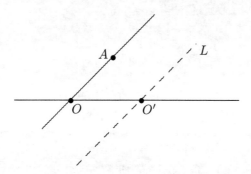

Comment. For O to be a center for a configuration of $2n$ points, no three on a line, it is still necessary and sufficient that every line through O and one point contain a second point, with $n-1$ points on either side of the line. Even with six points, the configuration may not form a convex hexagon, as shown below.

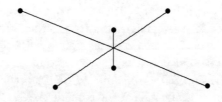

Problem 96

Define a sequence of matrices by $A_1 = \begin{pmatrix} 0 & 1/2 \\ -1/2 & 1 \end{pmatrix}$, $A_2 = \begin{pmatrix} 3 & -2 \\ 1 & 0 \end{pmatrix}$, and for $n \geq 1$, $A_{n+2} = A_{n+1} A_n A_{n+1}^{-1}$. Find approximations to the matrices A_{2010} and A_{2011}, with each entry correct to within 10^{-300}.

Answer. $A_{2010} \approx \begin{pmatrix} 2 & -1 \\ 0 & 1 \end{pmatrix}$ and $A_{2011} \approx \begin{pmatrix} 0.5 & 0 \\ 0 & 0.5 \end{pmatrix}$.

Solution. From $A_{n+2} = A_{n+1} A_n A_{n+1}^{-1}$, we see that $A_{n+2} A_{n+1} = A_{n+1} A_n$ and so the product $A_{n+2} A_{n+1}$ does not depend on n. In other words, if we set $B = A_2 A_1$, then $A_{n+2} A_{n+1} = B$, or equivalently, $A_{n+2} = B A_{n+1}^{-1}$, for all n. Hence, $A_{n+2} = B A_{n+1}^{-1} = B (B A_n^{-1})^{-1} = B A_n B^{-1}$. Using this repeatedly, we can rewrite each of the matrices A_n in terms of B and either A_1 (for n odd) or A_2 (for n even). Specifically, for $n = 2k+1$,

$$A_n = A_{2k+1} = B A_{2k-1} B^{-1} = B \cdot (B A_{2k-3} B^{-1}) \cdot B^{-1} = \cdots = B^k A_1 B^{-k},$$

and for $n = 2k+2$,

$$A_n = A_{2k+2} = B A_{2k} B^{-1} = \cdots = B^k A_2 B^{-k}.$$

In particular, $A_{2010} = B^{1004} A_2 B^{-1004}$, $A_{2011} = B^{1005} A_1 B^{-1005}$.

Now

$$B = \begin{pmatrix} 3 & -2 \\ 1 & 0 \end{pmatrix} \begin{pmatrix} 0 & 1/2 \\ -1/2 & 1 \end{pmatrix} = \begin{pmatrix} 1 & -1/2 \\ 0 & 1/2 \end{pmatrix}, \text{ so}$$

$$B^{-1} = \frac{1}{1/2} \begin{pmatrix} 1/2 & 1/2 \\ 0 & 1 \end{pmatrix} = \begin{pmatrix} 1 & 1 \\ 0 & 2 \end{pmatrix}.$$

An easy induction shows that

$$B^{-k} = \begin{pmatrix} 1 & 2^k - 1 \\ 0 & 2^k \end{pmatrix}, \text{ so } B^k = \frac{1}{2^k} \begin{pmatrix} 2^k & 1 - 2^k \\ 0 & 1 \end{pmatrix} = \begin{pmatrix} 1 & 2^{-k} - 1 \\ 0 & 2^{-k} \end{pmatrix}.$$

Alternatively, we can use the eigenvalues and eigenvectors of B: We find $\begin{pmatrix} 1 \\ 0 \end{pmatrix}$ as an eigenvector for $\lambda = 1$ and $\begin{pmatrix} 1 \\ 1 \end{pmatrix}$ as an eigenvector for $\lambda = 1/2$, so

$$B = \begin{pmatrix} 1 & 1 \\ 0 & 1 \end{pmatrix} \begin{pmatrix} 1 & 0 \\ 0 & 1/2 \end{pmatrix} \begin{pmatrix} 1 & 1 \\ 0 & 1 \end{pmatrix}^{-1}.$$

Thus

$$B^k = \begin{pmatrix} 1 & 1 \\ 0 & 1 \end{pmatrix} \begin{pmatrix} 1 & 0 \\ 0 & 1/2 \end{pmatrix}^k \begin{pmatrix} 1 & 1 \\ 0 & 1 \end{pmatrix}^{-1}$$

$$= \begin{pmatrix} 1 & 1 \\ 0 & 1 \end{pmatrix} \begin{pmatrix} 1 & 0 \\ 0 & 2^{-k} \end{pmatrix} \begin{pmatrix} 1 & -1 \\ 0 & 1 \end{pmatrix} = \begin{pmatrix} 1 & 2^{-k} - 1 \\ 0 & 2^{-k} \end{pmatrix},$$

and $B^{-k} = \begin{pmatrix} 1 & 2^k - 1 \\ 0 & 2^k \end{pmatrix}$. Therefore, from earlier work,

$$A_{2010} = B^{1004} A_2 B^{-1004} = \begin{pmatrix} 1 & 2^{-1004} - 1 \\ 0 & 2^{-1004} \end{pmatrix} \begin{pmatrix} 3 & -2 \\ 1 & 0 \end{pmatrix} \begin{pmatrix} 1 & 2^{1004} - 1 \\ 0 & 2^{1004} \end{pmatrix}$$
$$= \begin{pmatrix} 2 + 2^{-1004} & -1 - 2^{-1004} \\ 2^{-1004} & 1 - 2^{-1004} \end{pmatrix},$$

$$A_{2011} = B^{1005} A_1 B^{-1005} = \begin{pmatrix} 1 & 2^{-1005} - 1 \\ 0 & 2^{-1005} \end{pmatrix} \begin{pmatrix} 0 & 1/2 \\ -1/2 & 1 \end{pmatrix} \begin{pmatrix} 1 & 2^{1005} - 1 \\ 0 & 2^{1005} \end{pmatrix}$$
$$= \begin{pmatrix} 1/2 - 2^{-1006} & 2^{-1006} \\ -2^{-1006} & 1/2 + 2^{-1006} \end{pmatrix}.$$

Note that $2^{10} > 10^3$, so $2^{1004} = 16 \cdot (2^{10})^{100} > 16 \cdot (10^3)^{100} = 16 \cdot 10^{300}$, so

$$A_{2010} \approx \begin{pmatrix} 2 & -1 \\ 0 & 1 \end{pmatrix} \quad \text{and} \quad A_{2011} \approx \begin{pmatrix} 0.5 & 0 \\ 0 & 0.5 \end{pmatrix},$$

with each entry accurate to within 10^{-300}.

Problem 97

Find all real solutions x of the equation

$$x^{10} - x^8 + 8x^6 - 24x^4 + 32x^2 - 48 = 0.$$

Answer. The only real solutions are $\sqrt{2}$ and $-\sqrt{2}$.

Solution 1. Since the powers of x in $x^{10} - x^8 + 8x^6 - 24x^4 + 32x^2 - 48$ are all even, it is sufficient to find all nonnegative solutions of

$$y^5 - y^4 + 8y^3 - 24y^2 + 32y - 48 = 0.$$

We make use of the rational root theorem to search for roots. This theorem states that the only possible rational roots of a polynomial with integer coefficients are quotients of a factor of the constant term (possibly negative) by a factor of the leading coefficient. In our case, it implies any rational root must be an integral factor of 48. The moderate size of the coefficients (as well as ease of computation) leads us to begin with the small factors. We soon see that $y = 2$ is a solution. Long division yields

$$y^5 - y^4 + 8y^3 - 24y^2 + 32y - 48 = (y - 2)(y^4 + y^3 + 10y^2 - 4y + 24).$$

The small size of the lone negative coefficient of the quartic factor above leads us to suspect there are no further nonnegative roots of the polynomial. One of many ways to show this is to rewrite the quartic as

$$y^4 + y^3 + 9y^2 + (y-2)^2 + 20.$$

This latter expression is clearly at least 20 for $y \geq 0$. We conclude that the only real solutions to the original tenth-degree equation are $\sqrt{2}$ and $-\sqrt{2}$.

Solution 2. As in the first solution, we are led to consider the polynomial

$$y^5 - y^4 + 8y^3 - 24y^2 + 32y - 48.$$

It may be rewritten as

$$y^5 - y^4 + 4 \cdot 2y^3 - 6 \cdot 2^2 y^2 + 4 \cdot 2^3 y - 2^4 - 32, \quad \text{or} \quad y^5 - (y-2)^4 - 32.$$

We find the root $y = 2$ by inspection. The derivative of the above polynomial is $5y^4 - 4(y-2)^3$. It is then easy to see that $y^5 - y^4 + 8y^3 - 24y^2 + 32y - 48$ is strictly increasing for all real y (consider the cases $y \leq 2$ and $y \geq 2$ separately), hence there are no other real zeros. Again, this shows the only real solutions to the original tenth-degree equation are $\sqrt{2}$ and $-\sqrt{2}$.

Problem 98

An ordinary die is cubical, with each face showing one of the numbers 1, 2, 3, 4, 5, 6. Each face borders on four other faces; each number is "surrounded" by four of the other numbers. Is it possible to make a die in the shape of a regular dodecahedron, where each of the numbers 1, 2, 3, 4, 5, 6 occurs on two different faces and is "surrounded" both times by all five other numbers? If so, in how many essentially different ways can it be done?

Solution. Yes, it is possible to have each of the numbers 1, 2, 3, 4, 5, 6 occurring on two of the faces of the dodecahedron, surrounded each time by the other five numbers. The figure shows one way to do this; there are twelve essentially different ways in which it can be done.

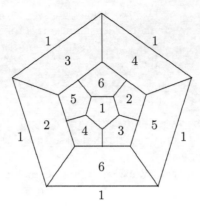

For the conditions to be satisfied, opposite faces must always have the same number. Thus, a solution is completely determined by the numbers placed on one face and its five neighbors. To count the number of essentially different solutions, we can assume (by rotating the dodecahedron) that the "central" face has a 1; by rotating the dodecahedron around this "central" face, we can further assume that the face represented by the pentagon "northwest" of the central pentagon has a 2 (as shown below). At first sight, every placement of the numbers 3, 4, 5, 6 in the other four faces adjacent to the central one will give rise to an essentially different solution. There are $4! = 24$ such placements. But wait! Turning the dodecahedron around so that the front face is turned to the back will replace the front face, labeled 1, and the clockwise pattern surrounding it, labeled $(2, a, b, c, d)$, by the back face, labeled 1, and the mirror image surrounding it, namely $(2, d, c, b, a)$. Thus, each labeled dodecahedron yields two placements of 3, 4, 5, 6 as described above, and therefore the number of essentially different labelings is $24/2 = 12$.

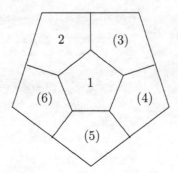

Problem 99

Arrange the positive integers in an array with three columns, as follows. The first row is [1, 2, 3]; for $n > 1$, row n is $[a, b, a+b]$, where a and b, with $a < b$, are the two smallest positive integers that have not yet appeared as entries in rows $1, 2, \ldots, n-1$. For each non-zero digit d and each positive integer m, in which column will the m-digit number $\underbrace{ddd\ldots d}_{m}$ end up, and why?

Answer. $\underbrace{ddd\ldots d}_{m}$ will be in column 1 if $d = 1, 4, 6$, or 8, or if $d = 9$ and m is even; in column 2 if $d = 2, 5$, or 7; in column 3 if $d = 3$ or if $d = 9$ and m is odd.

Solution. To get an idea, here are the first twenty rows of the array:

	Column 1	Column 2	Column 3
Row 1	1	2	3
Row 2	4	5	9
Row 3	6	7	13
Row 4	8	10	18
Row 5	11	12	23
Row 6	14	15	29
Row 7	16	17	33
Row 8	19	20	39
Row 9	21	22	43
Row 10	24	25	49
Row 11	26	27	53
Row 12	28	30	58
Row 13	31	32	63
Row 14	34	35	69
Row 15	36	37	73
Row 16	38	40	78
Row 17	41	42	83
Row 18	44	45	89
Row 19	46	47	93
Row 20	48	50	98

From this beginning we can make the following conjecture: for each integer $k \geq 0$, we have

	Column 1	Column 2	Column 3
Row $4k+1$	$10k+1$	$10k+2$	$10(2k)+3$
Row $4k+2$	$10k+4$	$10k+5$	$10(2k)+9$
Row $4k+3$	$10k+6$	$10k+7$	$10(2k+1)+3$

Exactly one of $10k+8$, $10k+9$ has already been placed in column 3.
If $10k+9$ has already been placed in column 3, then

| Row $4k+4$ | $10k+8$ | $10k+10$ | $10(2k+1)+8$ |

If $10k+8$ has already been placed in column 3, then

| Row $4k+4$ | $10k+9$ | $10k+10$ | $10(2k+1)+9$ |

We'll use induction to prove that these patterns persist for all k. They hold for $k = 0, 1, 2, 3, 4$ as we've seen, so suppose they hold for all non-negative integers smaller than n, where $n \geq 5$. The induction hypothesis implies that the first $4n$ rows contain all numbers less than or equal to $10n$, as well as, in column 3, $10i + 3$ and exactly one of $10i+8$, $10i+9$, for each i ($0 \leq i \leq 2n-1$). The two smallest positive integers that have not been placed yet are $10n+1, 10n+2$, so they must go into columns 1 and 2 of row $4n+1$, and their sum, $10(2n) + 3$, is in column 3. The next four smallest positive integers that have not yet been placed are $10n+4, 10n+5, 10n+6, 10n+7$ and these will go in the next two rows as prescribed. Finally, since exactly one of $10n+8$ or $10n+9$ is already in column 3 (because $n < 2n-1$), the next two smallest numbers are either $10n+8$ and $10n+10$ (if $10n+9$ has already been placed) or $10n+9$ and $10n+10$ (if $10n+8$ has already been placed). So the next four rows of the array are

	Column 1	Column 2	Column 3
Row $4n+1$	$10n+1$	$10n+2$	$10(2n)+3$
Row $4n+2$	$10n+4$	$10n+5$	$10(2n)+9$
Row $4n+3$	$10n+6$	$10n+7$	$10(2n+1)+3$

Exactly one of $10n+8$, $10n+9$ has already been placed in column 3.
If $10n+9$ has already been placed in column 3, then

| Row $4n+4$ | $10n+8$ | $10n+10$ | $10(2n+1)+8$ |

If $10n+8$ has already been placed in column 3, then

| Row $4n+4$ | $10n+9$ | $10n+10$ | $10(2n+1)+9$ |

This has the desired form, so by induction, the proof is complete.

From what we've shown, $\underbrace{ddd\ldots d}_{m}$ is in column 1 if $d = 1, 4$, or 6; in column 2 if $d = 2, 5$, or 7; in column 3 if $d = 3$. Because the number $\underbrace{888\ldots 8}_{m-1}9 = 10(2 \cdot \underbrace{444\ldots 4}_{m-1}) + 9$ has the form $10n + 9$ with n even, it is in

THE SOLUTIONS

column 3 (and row $4 \cdot \underbrace{444\ldots4}_{m-1}+2$). Thus, $\underbrace{888\ldots8}_{m} = 10 \cdot \underbrace{888\ldots8}_{m-1}+8$ is in column 1 (and row $4 \cdot \underbrace{888\ldots8}_{m-1}+4$).

Now suppose that $10n+9$ is in column 3. Then $10n+8$ is in column 1 and $10(2n+1)+8$ is in column 3, which in turn implies that $10(2n+1)+9$ is in column 1. On the other hand, if $10n+9$ is in column 1, then $10(2n+1)+9$ is in column 3. That is, of the two numbers $10n+9$, $10(2n+1)+9$, one is in column 1 and the other is in column 3.

As an application, suppose that n, in binary notation, has exactly m ($m > 0$) trailing 1's, that is, it has the form $abc\ldots d0\underbrace{111\ldots1}_{m}$. Then n is odd, so it has the form $2k+1$ for some integer k, so exactly one of $10n+9, 10k+9$ is in column 3. Also, note that in binary, k has the form $abc\ldots d0\underbrace{111\ldots1}_{m-1}$ (compare with n). By repeating this argument, we generate a sequence of numbers of the form

$$10n+9 = 10N_m + 9, \ 10N_{m-1}+9, \ \ldots, \ 10N_2+9, \ 10N_1+9, \ 10N_0+9,$$

where the binary expansion of N_s is $abc\ldots d0\underbrace{111\ldots1}_{s}$. The terms of this sequence alternate between columns 1 and 3 (or 3 and 1). But $10N_0+9$ is in column 3 because it has the form $10(2k)+9$, and therefore, $10n+9 = 10N_m+9$ is in column 3 if m is even and in column 1 if m is odd.

Thus, to find the column that lists $\underbrace{999\ldots9}_{m} = 10(\underbrace{999\ldots9}_{m-1})+9$ it is enough to know the number of trailing 1's in the binary representation of $\underbrace{999\ldots9}_{m-1}$.
Now $\underbrace{999\ldots9}_{m-1} = 10^{m-1} - 1$, and for some integer C we have

$$(10^{m-1} - 1) - (2^{m-1} - 1) = 10^{m-1} - 2^{m-1} = 2^{m-1}(5^{m-1} - 1) = 2^m C,$$

so $\underbrace{999\ldots9}_{m-1} = C \cdot 2^m + (2^{m-1} - 1) = C \cdot 2^m + (2^{m-2} + \cdots + 2^2 + 2 + 1)$.
This shows that $\underbrace{999\ldots9}_{m-1}$ has exactly $m-1$ trailing 1's in its binary representation. Therefore, from the previous paragraph, $\underbrace{999\ldots9}_{m} = 10(\underbrace{999\ldots9}_{m-1})+9$ is in column 3 if m is odd ($m-1$ is even) and in column 1 if m is even. This completes the analysis for $d = 9$.

Problem 100

For what positive integers n is $17^n - 1$ divisible by 2^n ?

Answer. The only positive values of n for which $17^n - 1$ is divisible by 2^n are $n = 1, 2, 3, 4$.

Solution. First note that for any positive integer n,

$$17^n - 1 = (17 - 1)(17^{n-1} + 17^{n-2} + \cdots + 1),$$

so $17^n - 1$ is always divisible by $17 - 1 = 16 = 2^4$; in particular, for $n = 1, 2, 3, 4$, $17^n - 1$ is divisible by 2^n, as claimed.

Now let n be any positive integer for which $17^n - 1$ is divisible by 2^n, and write n in the form $n = qm$, $q = 2^k$ ($k \geq 0$), m odd. We then have

$$17^n - 1 = 17^{qm} - 1 = (17^q - 1)(17^{q(m-1)} + 17^{q(m-2)} + \cdots + 1).$$

The second factor on the right is the sum of m odd numbers, so it is odd. Thus, because $17^n - 1$ is divisible by 2^n, the first factor $17^q - 1$ must be divisible by 2^n. In particular, $17^q - 1$ must be divisible by 2^q, and we now investigate for what powers q of 2 this could occur. We already know it happens for $q = 1, 2, 4$.

Note that $17^{2q} - 1 = (17^q - 1)(17^q + 1)$. Since 17 is 1 plus a multiple of 4, so is 17^q, so $17^q + 1$ is 2 plus a multiple of 4, and therefore $17^q + 1$ has exactly one factor 2, so $17^{2q} - 1$ has exactly one more factor 2 than $17^q - 1$. Starting with $17 - 1 = 2^4$ for $q = 1$, this shows that $17^2 - 1$ has 5 factors of 2 (exactly), $17^4 - 1$ has 6 factors 2, $17^8 - 1$ has 7 factors 2, and so on, and we see that for $q \geq 8$ a power of 2, $17^q - 1$ can never have as many as q factors 2.

So, for $17^n - 1$ to be divisible by 2^n, we must have $n = qm$ with m odd and $q = 1, 2,$ or 4. Also, as we saw earlier, $17^q - 1$ must be divisible by 2^n. If $q = 1$ this means $n \leq 4$, and we're done. If $q = 2$ it means $n \leq 5$ (because $17^2 - 1$ has 5 factors 2), but then $n = qm = 2m$ implies $n \leq 4$. And if $q = 4$ it means $n \leq 6$, but then $n = qm = 4m$, so again $n \leq 4$, and we're done.

Problem 101

Let f be a continuous function on the real numbers. Define a sequence of functions $f_0 = f, f_1, f_2, \ldots$ as follows:

$$f_0(x) = f(x) \quad \text{and} \quad f_{i+1}(x) = \int_0^x f_i(t)\,dt, \quad \text{for } i = 0, 1, 2, \ldots.$$

Show that for *any* continuous function f and any real number x,

$$\lim_{n \to \infty} f_n(x) = 0.$$

Solution. Because f is continuous on the interval between 0 and x, there is a constant M such that $|f(t)| \leq M$ for all t in that interval.

Claim: For any integer $n \geq 0$, $|f_n(t)| \leq M \dfrac{|t|^n}{n!}$ for all t between 0 and x.

Once we show this we'll be done, because we can apply the result to $t = x$ and we know that $\dfrac{|x|^n}{n!} \to 0$ as $n \to \infty$ (for instance, because $\sum \dfrac{|x|^n}{n!}$ converges by the ratio test), so $\lim_{n \to \infty} f_n(x) = 0$ by the squeeze principle.

Proof of claim: We'll prove the claim by induction on n. The case $n = 0$ is immediate from the definition of M. Suppose the claim is true for n, and first suppose that $x \geq 0$. Then for $t \in [0, x]$,

$$|f_{n+1}(t)| = \left| \int_0^t f_n(u)\,du \right| \leq \int_0^t |f_n(u)|\,du$$

$$\leq \int_0^t M \frac{|u|^n}{n!}\,du = \int_0^t M \frac{u^n}{n!}\,du$$

$$= M \frac{u^{n+1}}{(n+1)!} \bigg|_0^t = M \frac{t^{n+1}}{(n+1)!} = M \frac{|t|^{n+1}}{(n+1)!}.$$

The proof for $x < 0$ is similar: This time, for $t \in [x, 0]$,

$$|f_{n+1}(t)| = \left| \int_0^t f_n(u)\,du \right| = \left| \int_t^0 f_n(u)\,du \right| \leq \int_t^0 |f_n(u)|\,du$$

$$\leq \int_t^0 M \frac{|u|^n}{n!}\,du = \int_t^0 M \frac{(-1)^n u^n}{n!}\,du$$

$$= M \frac{(-1)^n u^{n+1}}{(n+1)!} \bigg|_t^0 = M \frac{(-1)^{n+1} t^{n+1}}{(n+1)!} = M \frac{|t|^{n+1}}{(n+1)!}.$$

Problem 102

Consider an arbitrary circle of radius 2 in the coordinate plane. Let n be the number of lattice points inside, but not on, the circle.
a. What is the smallest possible value for n?
b. What is the largest possible value for n?

Answer. The smallest possible value of n is 9; the largest is 14.

Solution. We begin by making several reductions, in order to simplify the calculations. First, by shifting the origin if necessary, we may assume that no lattice point is closer to the center of the circle than $(0,0)$. By rotating the plane through a multiple of $90°$, we may assume the center lies in the first quadrant or on its boundary. Finally, by reflecting the plane in the line $y = x$ if necessary, we may assume the center of the circle lies in the region R (shaded in the figure) defined by $0 \le x \le 1/2$, $0 \le y \le x$.

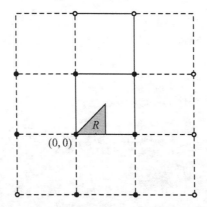

Any lattice point outside the square $-1 \le x \le 2$, $-1 \le y \le 2$ has distance at least two from the unit square ($0 \le x \le 1$, $0 \le y \le 1$), hence cannot be inside the circle. Simple calculations show that the eight lattice points indicated in the figure by "•" have distance less than 2 from *every* point in R. Further computation shows that only the six lattice points indicated in the figure by "∘" have distance less than 2 to some, but not all, of R. We see that $(-1,-1)$ is the only such point whose distance from the origin is less than 2. On the other hand, each point of R *except* the origin has distance less than 2 to $(2,0)$. Therefore, the smallest possible n is 9, obtained, for instance, when the circle has center at $(0,0)$.

The largest n is clearly no more than 14, the number of lattice points under consideration. If there is a point in R whose distance from all 14 of the lattice points is less than 2, the symmetry of the problem implies that there is such a point on the line $x = 1/2$. Any point $(1/2, y), 0 \le y \le 1/2$, has distance less than 2 to $(2, 0)$ and to $(2, 1)$. The distance to the other four questionable lattice points will be less than 2 if and only if

$$\left(\frac{3}{2}\right)^2 + (y+1)^2 < 4 \quad \text{and} \quad \left(\frac{1}{2}\right)^2 + (2-y)^2 < 4.$$

These inequalities simplify to

$$y < \frac{\sqrt{7}}{2} - 1 \quad \text{and} \quad 2 - \frac{\sqrt{15}}{2} < y.$$

Since

$$0 < 2 - \frac{\sqrt{15}}{2} < \frac{\sqrt{7}}{2} - 1 < \frac{1}{2},$$

there is a y for which the circle of radius 2 and center $(1/2, y)$ encloses 14 lattice points. As one might suspect, one such circle is centered at $(1/2, 1/4)$.

Problem 103

Let a and b be nonzero real numbers and (x_n) and (y_n) be sequences of real numbers. Given that

$$\lim_{n \to \infty} \frac{ax_n + by_n}{\sqrt{x_n^2 + y_n^2}} = 0$$

and that x_n is never 0, show that

$$\lim_{n \to \infty} \frac{y_n}{x_n}$$

exists and find its value.

Idea. Let $z_n = y_n/x_n$. Note that

$$\frac{ax_n + by_n}{\sqrt{x_n^2 + y_n^2}} = \pm \frac{a + bz_n}{\sqrt{1 + z_n^2}}.$$

If we assume for the moment that $L = \lim_{n \to \infty} z_n$ exists, then we have

$$0 = \lim_{n \to \infty} \frac{a + bz_n}{\sqrt{1 + z_n^2}} = \frac{a + bL}{\sqrt{1 + L^2}},$$

and so $L = -a/b$.

Solution 1. To show that L exists, we first show that the sequence (z_n) is bounded. We know that

$$\lim_{n\to\infty} \frac{a+bz_n}{\sqrt{1+z_n^2}} = 0;$$

on the other hand,

$$\lim_{z\to\infty} \frac{a+bz}{\sqrt{1+z^2}} = b \quad \text{and} \quad \lim_{z\to-\infty} \frac{a+bz}{\sqrt{1+z^2}} = -b.$$

Now, if (z_n) were unbounded, it would have some subsequence with limit either ∞ or $-\infty$, contradicting the assumption that $b \neq 0$.

Now that we know (z_n) is bounded, say $|z_n| \leq M$ for all n, we have

$$\left|\frac{a+bz_n}{\sqrt{1+z_n^2}}\right| \geq \left|\frac{a+bz_n}{\sqrt{1+M^2}}\right|$$

and hence by the squeeze principle,

$$\lim_{n\to\infty} \frac{a+bz_n}{\sqrt{1+M^2}} = 0,$$

which implies $\lim_{n\to\infty}(a+bz_n) = 0$, and finally $\lim_{n\to\infty} z_n = -a/b$.

Solution 2. (The late Meyer Jerison, Purdue University) With no loss of generality, assume $x_n^2 + y_n^2 = 1$. This amounts to a change of notation:

$$x_n \text{ to } \frac{x_n}{\sqrt{x_n^2+y_n^2}}, \quad y_n \text{ to } \frac{y_n}{\sqrt{x_n^2+y_n^2}}.$$

Now, $ax_n + by_n$ is the dot product of the unit vector (x_n, y_n) with (a, b). The hypothesis is that the dot product tends to 0, so that y_n/x_n should tend to the slope orthogonal to (a, b), namely, $-a/b$. Note that the sequence of unit vectors (x_n, y_n) need not converge; they may flip back and forth.

For an algebraic version of this argument, express the points (x_n, y_n) and (b, a) in polar coordinates: $(1, \theta_n)$ and (A, α). Then

$$ax_n + by_n = A\sin\alpha\cos\theta_n + A\cos\alpha\sin\theta_n = A\sin(\theta_n + \alpha).$$

The hypothesis becomes $\lim_{n\to\infty} \sin(\theta_n + \alpha) = 0$; hence, $\lim_{n\to\infty} \tan(\theta_n + \alpha) = 0$. This implies

$$\lim_{n\to\infty} \tan\theta_n = \lim_{n\to\infty} \frac{\tan(\theta_n + \alpha) - \tan\alpha}{1 + \tan(\theta_n + \alpha)\tan\alpha} = -\tan\alpha.$$

But $\tan\theta_n = y_n/x_n$ and $\tan\alpha = a/b$.

Problem 104

a. Show that there is a color pattern of black and white "grid" cubes in 3-space for which every cube has exactly 13 neighbors of each color.

b. For $n > 3$, is it still always possible to find a color pattern for a regular grid of "hypercubes" so that every hypercube, whether black or white, has an equal number of black and white neighbors? If so, show why; if not, give an example of a specific n for which it is impossible.

Answer. Yes, for all n there is a coloring of the n-dimensional grid in which each hypercube has an equal number of white and black neighbors.

Solution 1. a. One way the grid cubes can be colored so that every cube has exactly 13 neighbors of each color is to partition the cubes into $2 \times 2 \times 2$ subsets and use alternating coloring (in each direction) for these subsets, giving each $2 \times 2 \times 2$ cube a solid color. In such a coloring all $1 \times 1 \times 1$ grid cubes occupy symmetrical positions, since each of them is at a corner of a unique solid-colored $2 \times 2 \times 2$ cube. For example, if we look at a white $1 \times 1 \times 1$ cube (denoted in the figure by a dot) at the top left front corner of its $2 \times 2 \times 2$ cube, the colors of its neighbors are:

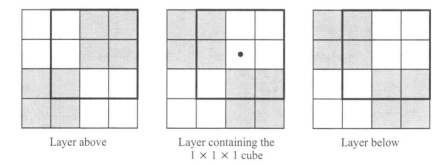

Layer above   Layer containing the   Layer below
$1 \times 1 \times 1$ cube

As shown by the heavy lines, our $1 \times 1 \times 1$ cube with a dot has 4 white and 5 black neighbors in the layer above, 4 white and 4 black neighbors in its own layer, and 5 white and 4 black neighbors in the layer below, for a total of 13 neighbors of each color.

b. We'll use induction on the dimension to show that such a coloring is possible. The construction is similar to that of part (a), if you think of the latter construction as starting from the coloring of the *plane* with solid 2×2 squares in a checkerboard pattern:

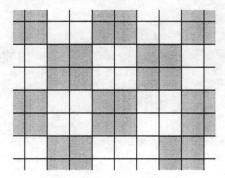

To show the inductive step, suppose we have a coloring pattern for a regular grid of hypercubes in k-dimensional space in which each hypercube has an equal number (which must be $(3^k - 1)/2$) of black and of white neighbors. Consider a regular grid of hypercubes in $(k{+}1)$-dimensional space, and divide it into k-dimensional layers. Color two successive layers using the k-dimensional coloring pattern that exists (by the induction hypothesis), then the next two layers using the *opposite* of that pattern (interchanging black and white), then the next two layers using the pattern itself, then two layers using the opposite pattern, and so forth. Then any of the hypercubes in $(k{+}1)$-dimensional space will have an equal number of black and of white neighbors in its own layer. Because the layers "above" and "below" it have opposite patterns, in those two layers taken together the hypercube will also have an equal number of black and of white neighbors. So we have constructed a coloring of the desired type in $(k{+}1)$-dimensional space, and our proof by induction is complete.

Solution 2. We can orient the tiling so that the centers of the hypercubes are the lattice points $(k_1, \ldots, k_n)$ with integer coordinates. Color a cube white if $k_1 + \cdots + k_n \equiv 0, 1 \pmod 4$ and black if not. We claim that every hypercube has the same number of black and white neighbors.

Consider a hypercube whose coordinates sum to k. Its neighbors have coordinates summing to between $k - n$ and $k + n$, inclusive. Suppose it has a_m neighbors whose coordinates sum to $k + m$. Then the a_m are the coefficients of x^m in the expansion of $(x + 1 + 1/x)^n - 1$. Let

$$S_r = \sum_{\substack{-n \leq m \leq n \\ m \equiv r \pmod 4}} a_m.$$

The number of neighbors that are the same color as our hypercube will be either $S_0 + S_1$ or $S_0 + S_3$ and the number of the opposite color will be either $S_2 + S_3$ or $S_1 + S_2$, respectively. By symmetry, $S_1 = S_3$, so it suffices to prove $S_0 = S_2$.

Substituting $x = \pm i$ into $(x + 1 + 1/x)^n - 1$ and its expansion in terms of the a_m yields

$$0 = [(i + 1 + 1/i)^n - 1] + [(-i + 1 - 1/i)^n - 1]$$
$$= (S_0 + S_1 i - S_2 - S_3 i) + (S_0 - S_1 i - S_2 + S_3 i) = 2S_0 - 2S_2,$$

so $S_0 = S_2$ and we are done.

Problem 105

What is the angle between adjacent faces of the regular icosahedron?

Answer. The angle is $\cos^{-1}(-\sqrt{5}/3) \approx 138.2°$.

Solution. More generally, suppose that AU, AW, and AV are three concurrent lines in $\mathbb{R}^3$, and consider the problem of finding the angle θ between the planes containing triangles UAW and WAV. Let $\alpha = \angle UAW$, $\beta = \angle WAV$, and $\gamma = \angle UAV$, and let $\mathbf{u}$, $\mathbf{v}$, and $\mathbf{w}$ be unit vectors in the directions of AU, AV and AW respectively (see figure).

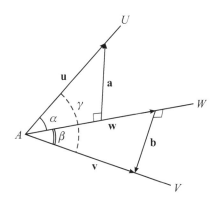

Comments. The preceding graph suggests several questions for further exploration. For example, if σ is a permutation of $1, 2, \ldots, n$, let $H(\sigma)$, the *height* of σ, denote the maximum number of steps that may be required to unscramble σ into increasing order.

a. Find an efficient way of evaluating $H(\sigma)$.
b. Characterize those permutations σ for which $H(\sigma) = 2^{n-1} - 1$, or more generally, those permutations σ for which $H(\sigma) = k$, for $k = 1, 2, \ldots, 2^{n-1} - 1$.
c. Characterize those permutations σ that have no predecessors of height $H(\sigma) + 1$.

This problem came about because of a misinterpretation of a related problem posed by Barry Cipra. A complete account of this history is given in the extended abstract "Sorting by Placement and Shift" by S. Elizalde and P. Winkler, Proceedings of the 20th ACM-SIAM Symposium on Discrete Algorithms (SODA'09), New York, 2009. The article contains Noam Elkies' proof that the directed graph is acyclic (for every n), and a proof that the maximum value of $H(\sigma)$ is $2^{n-1} - 1$. Also, see a subsequent paper by Villő Csiszár, "Jump Home and Shift: An Acyclic Operation on Permutations", *American Mathematical Monthly*, October, 2009, pp. 735-738.

Problem 107

Given a constant C, find all functions f such that

$$f(x) + C f(2 - x) = (x - 1)^3$$

for all x.

Solution. For $C \neq 1, -1$, the unique solution is given by

$$f(x) = \frac{(x - 1)^3}{1 - C};$$

for $C = 1$, there is no solution; for $C = -1$, the solutions are the functions of the form $f(x) = \frac{1}{2}(x - 1)^3 + E(x - 1)$, where E is any even function.

Replace x by $2 - x$ in the original equation, to get

$$f(2 - x) + Cf(x) = -(x - 1)^3.$$

We now have a system of two linear equations in the two unknowns $f(x)$ and $f(2-x)$:

$$\begin{pmatrix} 1 & C \\ C & 1 \end{pmatrix} \begin{pmatrix} f(x) \\ f(2-x) \end{pmatrix} = \begin{pmatrix} (x-1)^3 \\ -(x-1)^3 \end{pmatrix}.$$

The determinant of the matrix of coefficients is $1 - C^2$. Thus, if $C^2 \neq 1$, there is a unique solution, namely

$$f(x) = \frac{(x-1)^3}{1-C}.$$

If $C = 1$, then there is no solution, because then for any $x \neq 1$, the equations are inconsistent.

Finally, if $C = -1$, it is straightforward to show that any function of the form $f(x) = \frac{1}{2}(x-1)^3 + E(x-1)$, where E is any even function, satisfies the equation. Conversely, suppose that f satisfies the equation. Then we can write f in the form

$$\begin{aligned} f(x) &= \tfrac{1}{2}\bigl(f(x) - f(2-x)\bigr) + \tfrac{1}{2}\bigl(f(x) + f(2-x)\bigr) \\ &= \tfrac{1}{2}(x-1)^3 + \tfrac{1}{2}\bigl(f(x) + f(2-x)\bigr) \\ &= \tfrac{1}{2}(x-1)^3 + \tfrac{1}{2}\Bigl(f\bigl(1 + (x-1)\bigr) + f\bigl(1 - (x-1)\bigr)\Bigr) \\ &= \tfrac{1}{2}(x-1)^3 + E(x-1), \end{aligned}$$

where the even function E is defined by $E(x) = \frac{1}{2}\bigl(f(1+x) + f(1-x)\bigr)$.

Comment. Another way to solve $f(x) - f(2-x) = (x-1)^3$ is to observe that it is an inhomogeneous linear equation in f, which we could try to solve by first solving the corresponding homogeneous equation $f(x) - f(2-x) = 0$ and then finding a particular solution to the inhomogeneous equation. It is not hard to guess a particular solution, $\frac{1}{2}(x-1)^3$, and it is easy to see that the homogeneous equation is satisfied exactly when f is an even function in $x - 1$. Thus, the general solution is $f(x) = \frac{1}{2}(x-1)^3 + E(x-1)$, where E is an even function.

Problem 108

A digital time/temperature display flashes back and forth between time, temperature in degrees Fahrenheit, and temperature in degrees Centigrade. Suppose that over the course of a week in summer, the temperatures measured are between 15°C and 25°C and that they are randomly and uniformly distributed over that interval. What is the probability that, at any given time, the rounded value in degrees F of the converted temperature (from degrees C) is not the same as the value obtained by first rounding the temperature in degrees C, then converting to degrees F and rounding once more?

Idea. A good way to begin this problem is to consider the step functions which represent the two methods of converting.

Solution 1. The probability that the display will seem to be in error as explained in the problem is 4/9.

To see this, first note that since 5°C exactly equals 9°F, we can restrict ourselves to the interval from 15°C to 20°C. (Whatever happens there will exactly repeat from 20°C to 25°C, except that all readings—rounded or unrounded—will be 5°C, or 9°F, higher.) If we round a temperature in this interval to the nearest degree C and then convert to degrees F, we will get one of the following: 59°F (= 15°C), 60.8°F (= 16°C), 62.6°F, 64.4°F, 66.2°F, 68°F (= 20°C). These round to 59, 61, 63, 64, 66, 68 degrees F respectively. Thus we have the following:

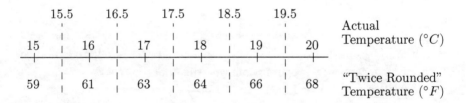

On the other hand, if we first convert the temperature in degrees C to degrees F and then round off, we can get any value from 59 to 68, as shown in the next figure. Comparing the two "thermometers", we see that the "once rounded" and "twice rounded" readings in degrees F are different when the "once rounded" temperature is 60, 62, 65, or 67°F, but the same in all other cases. So out of a total temperature interval of 5°C, there are four intervals, each of length $10/18 = 5/9$°C, in which the display will appear to be in error; thus the probability of this occurring is $(4 \cdot 5/9)/5 = 4/9$, as stated.

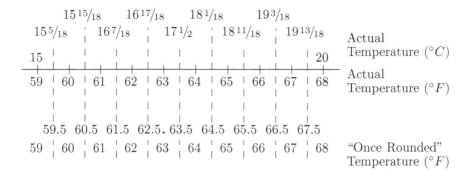

Solution 2. Let x be the exact Centigrade temperature. If $f(x)$ denotes the result of converting to Fahrenheit and then rounding, we have

$$f(x) = \lfloor 1.8x + 32.5 \rfloor.$$

This function is a step function with discontinuities 5/9 apart. On the other hand, if we round, and then convert to Fahrenheit, and round again, we obtain the function

$$F(x) = \lfloor 1.8 \lfloor x + .5 \rfloor + 32.5 \rfloor,$$

which has discontinuities at the half-integers.

If n is an integer, we observe that

$$f(n - 1/18) = F(n - 1/18) = \lfloor 1.8n + 32.5 \rfloor,$$

and that

$$\lim_{x \to (n+1/18)^-} f(x) = \lim_{x \to (n+1/18)^-} F(x) = \lfloor 1.8n + 32.5 \rfloor.$$

Thus, f and F agree on the interval $[n-1/18, n+1/18)$, which is the middle ninth of a step for $F(x)$; therefore this step must include exactly one entire step of f.

Since $f(x+5) = f(x) + 9$ and $F(x+5) = F(x) + 9$, the probability of agreement is the same on any interval of length 5°C. Since the interval from 15°C to 25°C is made up of two intervals of length 5°C, we may, equivalently, compute the probability on the interval from $-.5$°C to 4.5°C. This interval is comprised of exactly five full steps of $F(x)$, hence the probability of agreement is 5/9 and the probability of discrepancy is 4/9.

Comment. The probability of disagreement if one converts from Fahrenheit to Centigrade over an interval of 9°F is drastically different; similar arguments to those above show that the probability of disagreement is only 2/15. This is due to the fact that the Fahrenheit scale provides a finer measurement.

Problem 109

Do there exist sequences of positive real numbers in which each term is the product of the two previous terms and for which *all* odd-numbered terms are greater than 1, while all even-numbered terms are less than 1? If so, find all such sequences. If not, prove that no such sequence is possible.

Answer. Yes, there are such sequences; in fact, for each $a > 1$ there is a unique such sequence starting with a. The first two terms of that sequence are a, $a^{(1-\sqrt{5})/2}$.

Solution 1. Let the first two terms of the sequence be a, b. Then the sequence continues

$$a, \quad b, \quad ab, \quad ab^2, \quad a^2b^3, \quad a^3b^5, \quad a^5b^8, \quad a^8b^{13}, \ldots$$

We note, and an easy induction shows, that after a, b, the nth subsequent term of the sequence is $a^{F_n} b^{F_{n+1}}$, where the F_n are the Fibonacci numbers, defined by $F_1 = F_2 = 1$, $F_n = F_{n-1} + F_{n-2}$ ($n \geq 2$). The conditions in the problem imply that $a > 1 > b$, and after that, $a^{F_n} b^{F_{n+1}} > 1$ for n odd, $a^{F_n} b^{F_{n+1}} < 1$ for n even. It is easier to deal with these inequalities if we introduce the quantities $c = \ln a$ and $d = -\ln b$. In terms of c and d the conditions are $c > 0$ (from $a > 1$), $d > 0$ (from $b < 1$), $F_n c - F_{n+1} d > 0$ for n odd, and $F_n c - F_{n+1} d < 0$ for n even. That is, we are looking for positive numbers c and d such that

$$\frac{c}{d} > \frac{F_{n+1}}{F_n} \text{ for all odd } n \quad \text{and} \quad \frac{c}{d} < \frac{F_{n+1}}{F_n} \text{ for all even } n.$$

Since c/d is a single number, the real question is: For what positive numbers x is $F_{n+1}/F_n < x$ for all odd n and $F_{n+1}/F_n > x$ for all even n?

Now recall that the ratio of successive terms of the Fibonacci sequence converges to $\varphi = (1 + \sqrt{5})/2$, the golden ratio. A proof of this, which also shows that $F_{n+1}/F_n < \varphi$ for odd n and $F_{n+1}/F_n > \varphi$ for even n, is

THE SOLUTIONS

outlined in the note below. Therefore, the one possible value for $x = c/d$ is $(1 + \sqrt{5})/2$. So the possible starting values for a and b are those positive numbers for which

$$x = \frac{c}{d} = \frac{\ln a}{-\ln b} = \frac{1 + \sqrt{5}}{2} \iff \ln a = -\frac{1 + \sqrt{5}}{2} \ln b$$
$$\iff a = b^{-(1+\sqrt{5})/2}$$
$$\iff b = a^{-2/(1+\sqrt{5})} = a^{(1-\sqrt{5})/2}$$

with $a > 1$.

Note. The general term of the Fibonacci sequence is given by Binet's formula $F_n = \frac{1}{\sqrt{5}}\left(\varphi^n - (-1/\varphi)^n\right)$. This formula can be found by solving the linear recurrence $F_{n+2} - F_{n+1} - F_n = 0$ with initial conditions $F_0 = 0$, $F_1 = 1$ (for example, by the method of characteristic roots or by the method of generating functions). It follows that for odd n, $F_n > \frac{1}{\sqrt{5}}\varphi^n$, and for even n, $F_n < \frac{1}{\sqrt{5}}\varphi^n$. Thus, $\frac{F_{n+1}}{F_n} < \varphi^{n+1}/\varphi^n = \varphi$ for n odd, and $\frac{F_{n+1}}{F_n} > \varphi$ for n even.

Finally,

$$\lim_{n \to \infty} \frac{F_{n+1}}{F_n} = \lim_{n \to \infty} \frac{\frac{1}{\sqrt{5}}\left[\varphi^{n+1} - (-1/\varphi)^{n+1}\right]}{\frac{1}{\sqrt{5}}\left[\varphi^n - (-1/\varphi)^n\right]}$$
$$= \lim_{n \to \infty} \frac{\left[1 - (-1/\varphi^2)^{n+1}\right]\varphi^{n+1}}{\left[1 - (-1/\varphi^2)^n\right]\varphi^n} = \varphi.$$

Solution 2. (Eugene Luks, University of Oregon) Let $(a_n)_{n \geq 0}$ be such a sequence of positive numbers, and let $b_n = \ln a_n$ for each n. Then we have $b_{n+2} = b_{n+1} + b_n$ for $n \geq 0$, and the terms alternate in sign, with $b_0 > 0$. By the standard method for solving linear recurrence relations with constant coefficients, we find

$$b_n = c_1 \left(\frac{1 + \sqrt{5}}{2}\right)^n + c_2 \left(\frac{1 - \sqrt{5}}{2}\right)^n$$

for some constants c_1 and c_2. Since $\frac{1+\sqrt{5}}{2} > 1$ and $-1 < \frac{1-\sqrt{5}}{2} < 0$, the terms will alternate in the prescribed manner if and only if $c_1 = 0$ and $c_2 > 0$. This means that $a = a_0 = e^{b_0} = e^{c_2} > 1$, and $a_1 = e^{b_1} = e^{c_2(1-\sqrt{5})/2} = a^{(1-\sqrt{5})/2}$.

Problem 110

Sketch the set of points (x, y) in the plane which satisfy

$$(x^2 - y^2)^{2/3} + (2xy)^{2/3} = (x^2 + y^2)^{1/3}.$$

Idea. The presence of sums and differences of squares suggests changing the equation into polar form.

Answer.

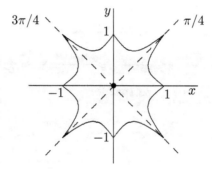

Solution. To find this graph (which includes an isolated point at the origin), start by transforming the given equation to polar coordinates (using $x = r \cos \theta$, $y = r \sin \theta$, $x^2 + y^2 = r^2$). We get

$$\left(r^2(\cos^2 \theta - \sin^2 \theta)\right)^{2/3} + \left(2r^2 \cos \theta \sin \theta\right)^{2/3} = (r^2)^{1/3}.$$

This is certainly true at the origin (where $r = 0$). For $r \neq 0$, we can divide through by $r^{2/3}$ to obtain

$$\left(r(\cos^2 \theta - \sin^2 \theta)\right)^{2/3} + \left(2r \cos \theta \sin \theta\right)^{2/3} = 1,$$
$$(r \cos 2\theta)^{2/3} + (r \sin 2\theta)^{2/3} = 1.$$

If we introduce new "rectangular" coordinates (X, Y) by $X = r \cos 2\theta$, $Y = r \sin 2\theta$, then the last equation reads $X^{2/3} + Y^{2/3} = 1$, which is the equation of an astroid in the XY-plane. Its graph may be found using standard calculus techniques and is shown below.

THE SOLUTIONS

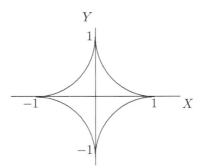

Now note that if (x, y) has polar coordinates (r, θ), then the corresponding (X, Y) has polar coordinates $(r, 2\theta)$: the same distance to the origin, but twice the polar angle. Thus we can find the actual graph (in the xy-plane) by replacing each point of the astroid by one with *half* its polar angle, and including the origin. The resulting curve will have cusps at $\theta = 0, \pi/4, \pi/2, 3\pi/4, \ldots$ to correspond to $2\theta = 0, \pi/2, \pi, 3\pi/2, \ldots$, and we get the eight-pointed "star" shown in the answer.

Problem 111

Let $f_1, f_2, \ldots, f_{2012}$ be functions such that the derivative of each of them is the sum of the others. Let $F = f_1 f_2 \cdots f_{2012}$ be the *product* of all these functions. Find all possible values of r, given that $\lim_{x \to -\infty} F(x) e^{rx}$ is a finite nonzero number.

Answer. The possible values of r are

$$2012, \ 0, \ -2012, \ -2 \cdot 2012, \ -3 \cdot 2012, \ \ldots, \ -2009 \cdot 2012, \ -2011 \cdot 2012,$$

but *not* $-2010 \cdot 2012$.

Solution. Let $S = f_1 + f_2 + \cdots + f_{2012}$ be the sum of all the functions. Then the derivative of S is

$$\begin{aligned} S' &= f'_1 + f'_2 + \cdots + f'_{2012} \\ &= \left(f_2 + f_3 + \cdots + f_{2012}\right) + \left(f_1 + f_3 + \cdots + f_{2012}\right) + \cdots \\ &\quad + \left(f_1 + f_2 + \cdots + f_{2011}\right) \\ &= 2011\, S, \end{aligned}$$

so $S(x) = Ce^{2011x}$ for some constant C. Thus, for any i ($1 \le i \le 2012$) we have

$$f'_i(x) = S(x) - f_i(x) = Ce^{2011x} - f_i(x).$$

So our functions are all solutions of the first-order inhomogeneous equation $y' = Ce^{2011x} - y$, and it follows that there are constants $A_1, A_2, \ldots, A_{2012}$ such that

$$f_i(x) = \frac{C}{2012} e^{2011x} + A_i e^{-x}.$$

Note that since $\sum f_i = S$, we have $\sum A_i = 0$. Conversely, given constants C and A_i ($1 \le i \le 2012$) with $\sum A_i = 0$, the functions f_i we just found satisfy the conditions of the problem.

As $x \to -\infty$, the term $A_i e^{-x}$ in $f_i(x)$ will be dominant *unless* $A_i = 0$. To keep track of this more easily, let's renumber the functions, if necessary, so that for some k, $A_i \ne 0$ for $i \le k$ and $A_i = 0$ for $i > k$. The extreme possibilities for k are $k = 0$ (when all the A_i are 0) and $k = 2012$ (when none of the A_i are 0), and k can take on all values in between *except $k = 1$*, which is impossible because $\sum A_i = 0$.

Taking out a factor e^{-x} from each $f_i(x)$ with $A_i \ne 0$ and a factor e^{2011x} from each $f_i(x)$ with $A_i = 0$, and setting $B = C/2012$, we see that

$$F(x)e^{rx} = (Be^{2012x} + A_1) \cdots (Be^{2012x} + A_k) B^{2012-k} e^{(rx - kx + (2012-k)2011x)}.$$

Note that for $1 \le i \le k$, $\lim_{x \to -\infty} (Be^{2012x} + A_i) = A_i \ne 0$. We can take $B \ne 0$, so $\lim_{x \to -\infty} F(x)e^{rx}$ can be a finite nonzero number if and only if $r - k + (2012 - k)2011 = 0$, that is, $r = (k - 2011)2012$. Remembering that k can take on the values $0, 2, 3, 4, \ldots, 2012$, we see that r can be

$$-2011 \cdot 2012,\ -2009 \cdot 2012,\ -2008 \cdot 2012,\ -2007 \cdot 2012,\ \ldots,\ -2012,\ 0,\ 2012.$$

Comment. An alternative method for solving the system of linear differential equations

$$\begin{pmatrix} f_1 \\ f_2 \\ f_3 \\ \vdots \\ f_n \end{pmatrix}' = \begin{pmatrix} 0 & 1 & 1 & \cdots & 1 \\ 1 & 0 & 1 & \cdots & 1 \\ 1 & 1 & 0 & \cdots & 1 \\ \vdots & \vdots & \vdots & \ddots & \vdots \\ 1 & 1 & 1 & \cdots & 0 \end{pmatrix} \begin{pmatrix} f_1 \\ f_2 \\ f_3 \\ \vdots \\ f_n \end{pmatrix}$$

THE SOLUTIONS   201

is to find the eigenvalues of the coefficient matrix, C, first. Note that

$$C + I = \begin{pmatrix} 1 & 1 & 1 & \cdots & 1 \\ 1 & 1 & 1 & \cdots & 1 \\ 1 & 1 & 1 & \cdots & 1 \\ \vdots & \vdots & \vdots & \ddots & \vdots \\ 1 & 1 & 1 & \cdots & 1 \end{pmatrix}$$

is a matrix of rank 1, so -1 is an eigenvalue of multiplicity $n-1$. Because the trace of C is zero and equals the sum of the eigenvalues, the remaining eigenvalue must be $n-1$. A basis for the eigenspace for $\lambda = -1$ can be taken to be

$$\begin{pmatrix} 1 \\ 0 \\ 0 \\ \vdots \\ 0 \\ -1 \end{pmatrix}, \begin{pmatrix} 0 \\ 1 \\ 0 \\ \vdots \\ 0 \\ -1 \end{pmatrix}, \cdots, \begin{pmatrix} 0 \\ 0 \\ 0 \\ \vdots \\ 1 \\ -1 \end{pmatrix}.$$

An eigenvector for $\lambda = n-1$ is $(1, 1, 1, \ldots, 1)^T$. So the general solution is

$$\begin{pmatrix} f_1(x) \\ f_2(x) \\ f_3(x) \\ \vdots \\ f_{n-1}(x) \\ f_n(x) \end{pmatrix} = \begin{pmatrix} 1 & 0 & 0 & \cdots & 0 & 1 \\ 0 & 1 & 0 & \cdots & 0 & 1 \\ 0 & 0 & 1 & \cdots & 0 & 1 \\ \vdots & \vdots & \vdots & \ddots & \vdots & \vdots \\ 0 & 0 & 0 & \cdots & 1 & 1 \\ -1 & -1 & -1 & \cdots & -1 & 1 \end{pmatrix} \begin{pmatrix} A_1 e^{-x} \\ A_2 e^{-x} \\ A_3 e^{-x} \\ \vdots \\ A_{n-1} e^{-x} \\ B e^{(n-1)x} \end{pmatrix}.$$

That is, $f_i(x) = Be^{(n-1)x} + A_i e^{-x}$, where B and A_i are constants such that $A_n = -A_1 - A_2 - \cdots - A_{n-1}$, as in the preceding solution.

Problem 112

On a table in a dark room there are n hats, each numbered clearly with a different number from the set $\{1, 2, \ldots, n\}$. A group of k intelligent students, with $k < n$, comes into the room, and each student takes a hat at random and puts it on his or her head. The students go back outside, where they can see the numbers on each other's hats. Each student now looks carefully at all the other students and announces

(i) the largest hat number that (s)he can see, as well as

(ii) the smallest hat number that (s)he can see.

After all these announcements are made, what is the probability that *all* students should be able to deduce their own hat numbers:

a. if $n = 6, k = 5$;

b. in general, as a function of n and k?

Answer. For $k = 1$ the probability is 0. For $k = 2, 3, 4$, the probability is 1. For $5 \leq k < n$, the probability is $\dfrac{\binom{n-k+3}{3}}{\binom{n}{k}}$. In particular, for $n = 6, k = 5$, the probability is $\dfrac{\binom{4}{3}}{\binom{6}{5}} = 2/3$.

Solution. a. The "largest hat numbers" that will be announced are the true largest hat number (by the four students who don't have that hat number) and the next largest hat number (by the student whose own hat has the largest number) among the five numbers chosen. Similarly, the "smallest hat numbers" announced will be the smallest (four times) and the next smallest (once) among the five numbers actually chosen. In particular, the only number that was chosen but *isn't* announced is the middle number of the five. In other words, if the numbers chosen are $a_1 < a_2 < a_3 < a_4 < a_5$, then after all the announcements are made everyone will know a_1, a_2, a_4, a_5. Then everyone who has one of those four numbers will be able to deduce her/his number (because it's not visible to her/him and yet it's announced). How about the person with a_3? That student can see the numbers a_1, a_2, a_4, a_5 which have been announced, so he/she knows his/her number is a_3. However, he/she will only be able to deduce the *value* of a_3 from this information if there is only one possibility for a_3; this happens if and only if $a_4 - a_2 = 2$ (so that there is only one number between a_2 and a_4). Now, five numbers will be chosen from the six, and all six numbers are equally likely not to be chosen. If 1, 2, 5, or 6 is not chosen, the numbers $a_1, \ldots, a_5$ will be 2, 3, 4, 5, 6; 1, 3, 4, 5, 6; 1, 2, 3, 4, 6; 1, 2, 3, 4, 5 respectively, and in each case $a_4 - a_2 = 2$. But if 3 or 4 is not chosen, we get 1, 2, 4, 5, 6 or 1, 2, 3, 5, 6, and in these cases $a_4 - a_2 = 3$ and the "middle" student has no way of deducing where (s)he has a 3 or a 4. Therefore, the desired probability is $4/6 = 2/3$.

b. Clearly, if $k = 1$ the probability is 0 (the one student doesn't have anything to get a clue from). If $k = 2$ the probability is 1 (each student's hat number will be announced by the other one), and if $k = 3$ or $k = 4$

the probability will still be 1 (all hat numbers will be announced, so the announced number that you can't see is yours). So assume $k > 4$. As in (a), if the chosen numbers are $a_1 < a_2 < \ldots < a_{k-1} < a_k$, then the announced numbers will be a_1 ($(k-1)$ times), a_2 (once), a_{k-1} (once), and a_k ($(k-1)$ times). The students with those hat numbers will be able to deduce them, because those numbers are announced and they can't see their own number.

If $a_{k-1} - a_2 = k-3$, so that there are only $k-4$ numbers between a_2 and a_{k-1}, then everyone will know that those numbers are $a_3 = a_2+1, \ldots, a_{k-2} = a_{k-1}-1$, and now that everyone knows all the numbers chosen, everyone can deduce that his/her own number is the "invisible" one (to them). On the other hand, if $a_{k-1} - a_2 > k - 3$, then there is a "gap" in the sequence $a_2 < a_3 < \cdots < a_{k-2} < a_{k-1}$; that is, there is an i ($2 \le i \le k-2$) for which $a_{i+1} - a_i > 1$. But then the student with a_i will have no way of deciding that that number really is a_i and not $a_i + 1$ — unless $i = 2$ (so that a_i was announced), in which case the student with a_3 will have no way of deciding that his/her number is really a_3 and not $a_3 - 1$.

We have just seen that the desired probability is the probability that $a_{k-1} - a_2 = k-3$. Note that $2 \le a_2 \le n-k+2$. For a particular choice of a_2, say $a_2 = i$, there are $(i-1)$ possible choices for a_1 and since $a_{k-1} = k+i-3$, there are $(n-k-i+3)$ possible choices for $a_k > a_{k-1}$. So of the $\binom{n}{k}$ total choices of $a_1, a_2, \ldots, a_k$, the number of choices with $a_{k-1} - a_2 = k-3$ is

$$\sum_{i=2}^{n-k+2} (i-1)(n-k-i+3) = \sum_{i=2}^{n-k+2} \left(-i^2 + (n-k+4)i - (n-k+3)\right)$$

$$= \sum_{i=1}^{n-k+2} \left(-i^2 + (n-k+4)i - (n-k+3)\right)$$

$$= -\frac{(n-k+2)(n-k+3)(2n-2k+5)}{6}$$

$$+ \frac{(n-k+4)(n-k+2)(n-k+3)}{2} - (n-k+2)(n-k+3)$$

$$= (n-k+2)(n-k+3)\left(-\tfrac{1}{3}n + \tfrac{1}{3}k - \tfrac{5}{6} + \tfrac{1}{2}n - \tfrac{1}{2}k + 2 - 1\right)$$

$$= \frac{(n-k+2)(n-k+3)(n-k+1)}{6} = \binom{n-k+3}{3}.$$

Thus the probability is $\dfrac{\binom{n-k+3}{3}}{\binom{n}{k}}$, and we are done.

Solution 2. We can simplify the calculation in part (b) of Solution 1 by finding a one-to-one correspondence between the choices with $a_{k-1}-a_2 = k-3$ and the number of ways to choose 3 numbers from $n-(k-3)$ given ones. To do so, for any choice with $a_{k-1}-a_2 = k-3$, collapse the consecutive numbers $a_2, \ldots, a_{k-1}$ down to just the number a_2 and replace a_k by $a_k - (k-3)$. Then $a_1, a_2, a_k - (k-3)$ can be chosen arbitrarily (in increasing order) from $1, 2, \ldots, n-(k-3)$, and to each such choice corresponds a unique choice of $a_1, \ldots, a_k$ from $1, 2, \ldots, n$ with $a_{k-1}-a_2 = k-3$. Thus the numerator of the probability is $\binom{n-k+3}{3}$, as computed earlier. (This combinatorial argument is due to John Perry '98, student, Carleton College.)

Problem 113

Find all integer solutions to $x^2 + 615 = 2^n$.

Idea. If n were even, then $615 = 2^n - x^2$ would be a difference of squares and could be factored accordingly. Is there any reason for n to be even?

Solution. The only solutions are $(x, n) = (\pm 59, 12)$.

We first factor 615 as $3 \cdot 5 \cdot 41$. Because 2 is not a square modulo 3 (or mod 5), n must be even, say $n = 2m$. Then $615 = 2^{2m} - x^2 = (2^m - x)(2^m + x)$. Now 615 can be written as a product of two positive integers in four ways: $1 \cdot 615$, $3 \cdot 205$, $5 \cdot 123$, and $15 \cdot 41$. Since $(2^m - x) + (2^m + x) = 2^{m+1}$, we want the sum of the two factors to be a power of 2. This occurs only for the factorization $615 = 5 \cdot 123$, which yields the solutions $x = \pm 59$, $n = 12$.

Problem 114

Sum the infinite series

$$\sum_{n=1}^{\infty} \sin \frac{2\alpha}{3^n} \sin \frac{\alpha}{3^n}.$$

Idea. The best hope of computing the partial sums is to show that they telescope.

THE SOLUTIONS

Solution. The sum is $\dfrac{1-\cos\alpha}{2}$.

Using the identity $\sin\theta\sin\phi = \tfrac{1}{2}\cos(\theta-\phi) - \tfrac{1}{2}\cos(\theta+\phi)$, we find that the kth partial sum is

$$S_k = \sum_{n=1}^{k} \sin\left(\frac{2\alpha}{3^n}\right)\sin\left(\frac{\alpha}{3^n}\right)$$

$$= \sum_{n=1}^{k}\left(\frac{1}{2}\cos\left(\frac{\alpha}{3^n}\right) - \frac{1}{2}\cos\left(\frac{\alpha}{3^{n-1}}\right)\right)$$

$$= \frac{1}{2}\cos\left(\frac{\alpha}{3^k}\right) - \frac{1}{2}\cos\alpha.$$

It follows that

$$\sum_{n=1}^{\infty}\sin\left(\frac{2\alpha}{3^n}\right)\sin\left(\frac{\alpha}{3^n}\right) = \lim_{k\to\infty}\left(\frac{1}{2}\cos\left(\frac{\alpha}{3^k}\right) - \frac{1}{2}\cos\alpha\right) = \frac{1-\cos\alpha}{2}.$$

Problem 115

A fair coin is flipped repeatedly. Starting from $x = 0$, each time the coin comes up "heads," 1 is added to x, and each time the coin comes up "tails," 1 is subtracted from x. Let a_n be the expected value of $|x|$ after n flips of the coin. Does $a_n \to \infty$ as $n \to \infty$?

Answer. Yes, $a_n \to \infty$ as $n \to \infty$.

Solution. Note that $|x|$ is the distance from x to the origin. If $x \neq 0$ at some stage, then because x is equally likely to change to $x-1$ as to $x+1$, the expected distance will be unaffected by the next flip. On the other hand, if $x = 0$, the distance will increase from 0 to 1. Now the probability that $x = 0$ after n flips is 0 for n odd. For n even, there are $\binom{n}{n/2}$ possible sequences of outcomes of n flips that include exactly $n/2$ heads, so the probability that $x = 0$ after n flips is $\binom{n}{n/2}/2^n$. We can conclude that

$$a_{n+1} = \begin{cases} a_n & \text{if } n \text{ is odd,} \\ a_n + \binom{n}{n/2}/2^n & \text{if } n \text{ is even.} \end{cases}$$

Therefore,

$$a_{2n+2} = a_{2n+1} = \sum_{k=0}^{n} \frac{\binom{2k}{k}}{2^{2k}},$$

and so

$$\lim_{n \to \infty} a_n = \sum_{k=0}^{\infty} \frac{\binom{2k}{k}}{2^{2k}};$$

the question is whether this series diverges.

Now $\binom{2k}{k}$ is the largest among the $2k+1$ binomial coefficients $\binom{2k}{j}$ with $j = 0, 1, 2, \ldots, 2k$. The sum of these coefficients is 2^{2k}, and so $\frac{\binom{2k}{k}}{2^{2k}} > \frac{1}{2k+1}$. Therefore, the series diverges by comparison with $\sum_{k=0}^{\infty} \frac{1}{2k+1}$, and $a_n \to \infty$ as $n \to \infty$.

Comment. The estimate $\frac{\binom{2k}{k}}{2^{2k}} > \frac{1}{2k+1}$ is very rough. In fact, for $k > 1$,

$$\frac{\binom{2k}{k}}{2^{2k}} = \frac{(2k)!}{\left(2 \cdot 4 \cdots (2k)\right)^2} = \frac{1 \cdot 3 \cdot 5 \cdots (2k-1)}{2 \cdot 4 \cdot 6 \cdots (2k)}$$

$$= \frac{1}{\sqrt{2}} \cdot \frac{3}{\sqrt{2 \cdot 4}} \cdot \frac{5}{\sqrt{4 \cdot 6}} \cdots \frac{2k-1}{\sqrt{(2k-2) \cdot 2k}} \cdot \frac{1}{\sqrt{2k}}$$

$$> \frac{1}{\sqrt{2}} \cdot \frac{1}{\sqrt{2k}} = \frac{1}{2\sqrt{k}}$$

and

$$\frac{\binom{2k}{k}}{2^{2k}} = \frac{\sqrt{1 \cdot 3}}{2} \cdot \frac{\sqrt{3 \cdot 5}}{4} \cdots \frac{\sqrt{(2k-1) \cdot (2k+1)}}{2k} \cdot \frac{1}{\sqrt{2k+1}} < \frac{1}{\sqrt{2k+1}}.$$

Problem 116

a. Show that there is no cubic polynomial whose graph passes through the points $(0,0)$, $(1,1)$, and $(2,16)$ and which is increasing for *all x*.
b. Show that there is a polynomial whose graph passes through $(0,0)$, $(1,1)$, and $(2,16)$ and which is increasing for all x.

c. Show that if $x_1 < x_2 < x_3$ and $y_1 < y_2 < y_3$, there is always an increasing polynomial (for all x) whose graph passes through (x_1, y_1), (x_2, y_2), and (x_3, y_3).

Solution. a. Any cubic polynomial $p(x)$ with $p(0) = 0$, $p(1) = 1$, and $p(2) = 16$ will be of the form $p(x) = ax^3 + bx^2 + cx$ with $a + b + c = 1$, $8a + 4b + 2c = 16$, from which we find that $b = 7 - 3a$ and $c = 2a - 6$.

In order for $p(x)$ to be an increasing polynomial, we need, for all x, $p'(x) = 3ax^2 + 2bx + c \geq 0$. This requires $a > 0$ and $D \leq 0$, where D is the discriminant. But

$$D = (2b)^2 - 4(3a)c = 4(b^2 - 3ac)$$
$$= 4\Big((7-3a)^2 - 3a(2a-6)\Big) = 4(3a^2 - 24a + 49)$$
$$= 12(a-4)^2 + 4,$$

showing that $D \leq 0$ is impossible.

b. Note that x and x^5 are both increasing polynomials whose graphs pass through $(0, 0)$ and $(1, 1)$. The graph of x passes below $(2, 16)$ while the graph of x^5 passes above, so we can look for some weighted average of x and x^5 to do the trick. Specifically, if we set $p_w(x) = wx + (1-w)x^5$, then $p_w(x)$ will be increasing (for all x) for any w with $0 \leq w \leq 1$, and the graph of such a polynomial always passes through $(0, 0)$ and $(1, 1)$. An easy calculation shows that for $w = \frac{8}{15}$, it passes through $(2, 16)$ as well.

c. Consider the line L through (x_1, y_1) and (x_2, y_2). If (x_3, y_3) lies on L, then we can use the increasing linear polynomial whose graph is L. If (x_3, y_3) lies below L, then $(-x_1, -y_1)$ lies above the line through $(-x_3, -y_3)$ and $(-x_2, -y_2)$; note that $-x_3 < -x_2 < -x_1$ and $-y_3 < -y_2 < -y_1$. If $p(x)$ is an increasing polynomial whose graph passes through these last three points, then $-p(-x)$ is an increasing polynomial whose graph passes through the original three points.

So we may assume, without loss of generality, that (x_3, y_3) lies above L. By translating (shifting) (x_1, y_1) to the origin and adjusting the scales along both axes, we may further assume that the points are $(0, 0)$, $(1, 1)$, and (r, s) with $1 < r < s$. Now we can use a construction similar to the one we used for part (b): Choose a positive integer n such that $r^{2n+1} > s$ (this is possible because $r > 1$). Then let $p_w(x) = wx + (1-w)x^{2n+1}$. We then have $p_w(0) = 0$, $p_w(1) = 1$, and $p_w(x)$ is increasing for any w with $0 \leq w \leq 1$.

Because $p_0(r) = r^{2n+1} > s$ and $p_1(r) = r < s$, by the Intermediate Value Theorem there is a value of w between 0 and 1 for which $p_w(r) = s$, giving us the desired polynomial.

Comments. In general, if $x_1 < x_2 < \cdots < x_n$ and $y_1 < y_2 < \cdots < y_n$, then it can be shown that there is an increasing polynomial passing through $(x_1, y_1), (x_2, y_2), \ldots, (x_n, y_n)$. Also, even for $n = 3$, there is no upper bound on the minimal degree required.

Problem 117

Do there exist five rays emanating from the origin in $\mathbb{R}^3$ such that the angle between any two of these rays is obtuse?

Answer. No, not all the angles formed by pairs of the five rays can be obtuse.

Solution. Let $\mathbf{v}_i = (x_i, y_i, z_i)$, $1 \leq i \leq 5$, denote a unit vector along the ith ray, and suppose that all the angles are obtuse. By proper choice of axes (corresponding to a rotation), we may assume $\mathbf{v}_5 = (0, 0, -1)$. We have

$$\mathbf{v}_i \cdot \mathbf{v}_j = |\mathbf{v}_i||\mathbf{v}_j| \cos \theta_{ij} < 0,$$

where θ_{ij} is the (obtuse) angle between $\mathbf{v}_i$ and $\mathbf{v}_j$. In particular, taking $j = 5$ we see that $z_i > 0$ for $i = 1, 2, 3, 4$.

Now let $\mathbf{w}_i$ be the projection of $\mathbf{v}_i$ onto the xy-plane, i.e., $\mathbf{w}_i = (x_i, y_i, 0)$. Some pair of $\mathbf{w}_1, \mathbf{w}_2, \mathbf{w}_3, \mathbf{w}_4$ form a non-obtuse angle. We may assume that $\mathbf{w}_1, \mathbf{w}_2$ is such a pair, so that $0 \leq \mathbf{w}_1 \cdot \mathbf{w}_2 = x_1 x_2 + y_1 y_2$. But then

$$\mathbf{v}_1 \cdot \mathbf{v}_2 = x_1 x_2 + y_1 y_2 + z_1 z_2 > 0,$$

and the angle between $\mathbf{v}_1$ and $\mathbf{v}_2$ is acute, completing the proof.

Comment. No two of the *six* unit vectors $(\pm 1, 0, 0)$, $(0, \pm 1, 0)$, $(0, 0, \pm 1)$ along the coordinate axes form an *acute* angle.

Problem 118

Are there infinitely many triples (a, b, c) such that the integers a, b, c have no common factors and the square of b is the average of the squares of a and c?

THE SOLUTIONS

Answer. Yes, there are infinitely many such triples.

Solution 1. Rather than just writing them down, let's think about how one might come up with a set of solutions. Start by considering the odd squares

$$1, \quad 9, \quad 25, \quad 49, \quad 81, \quad 121, \quad 169, \quad 225, \quad 289, \quad 361, \quad \ldots$$

and note that the differences between successive odd squares are multiples of 8; in fact, we have $(2k+1)^2 - (2k-1)^2 = 8k$. From this point of view, the fact that 25 is the average of 1 and 49, or $49 - 25 = 25 - 1$, is the result of

$$7^2 - 5^2 = 8 \cdot 3 = 8 \cdot 2 + 8 \cdot 1 = (5^2 - 3^2) + (3^2 - 1^2),$$

or, cancelling the common factor 8, of

$$3 = 2 + 1. \qquad (*)$$

Another example: $169 = 13^2$ is the average of $49 = 7^2$ and $289 = 17^2$; this comes down to $289 - 169 = 169 - 49$, or

$$17^2 - 13^2 = (17^2 - 15^2) + (15^2 - 13^2) = (13^2 - 11^2) + (11^2 - 9^2) + (9^2 - 7^2),$$

so

$$8 \cdot 8 + 8 \cdot 7 = 8 \cdot 6 + 8 \cdot 5 + 8 \cdot 4,$$

$$8 + 7 = 6 + 5 + 4. \qquad (**)$$

Note that in both $(*)$ and $(**)$ both sides are sums of consecutive integers with the left side starting where the right side left off. Conversely, each such equation will lead to a triple of squares in which the middle square is the average of the other two (we won't worry about common factors yet).

Now $(*)$ and $(**)$ are the first two in an infinite list of such equations. The next two are

$$15 + 14 + 13 = 12 + 11 + 10 + 9,$$

$$24 + 23 + 22 + 21 = 20 + 19 + 18 + 17 + 16.$$

The general form of these equations is

$$\underbrace{(k^2+2k) + \cdots + (k^2+k+1)}_{k \text{ terms}} = \underbrace{(k^2+k) + (k^2+k-1) + \cdots + k^2}_{(k+1) \text{ terms}},$$

which is easily checked using the fact that the sum of a finite arithmetic series is the sum of the number of terms multiplied by the average of the first and last term: The left side sums to

$$k \cdot \tfrac{1}{2}\left[(k^2+2k) + (k^2+k+1)\right] = \tfrac{1}{2}k(2k^2+3k+1) = \tfrac{1}{2}k(k+1)(2k+1)$$

and the right side sums to

$$(k+1) \cdot \tfrac{1}{2}\left[(k^2+k) + k^2\right] = \tfrac{1}{2}(k+1)(2k^2+k) = \tfrac{1}{2}k(k+1)(2k+1),$$

so they're equal.

Multiplying the general equation by 8 and then converting both sides to differences of odd squares, we get

$$8(k^2+2k) + \cdots + 8(k^2+k+1) = 8(k^2+k) + 8(k^2+k-1) + \cdots + 8k^2$$
$$[2(k^2+2k)+1]^2 - [2(k^2+2k)-1]^2 + \cdots + [2(k^2+k+1)+1]^2 - [2(k^2+k+1)-1]^2 =$$
$$[2(k^2+k)+1]^2 - [2(k^2+k)-1]^2 + \cdots + [2k^2+1]^2 - [2k^2-1]^2$$

and each side telescopes to give

$$(2k^2+4k+1)^2 - (2k^2+2k+1)^2 = (2k^2+2k+1)^2 - (2k^2-1)^2.$$

In other words, the square of $2k^2+2k+1$ is always the average of the square of $2k^2-1$ and the square of $2k^2+4k+1$.

How about common factors? Suppose that $2k^2+2k+1$ and $2k^2-1$ have a common factor d. Then d is also a factor of their difference $2k+2$, hence also of $k(2k+2) = 2k^2+2k$, hence also of $2k^2+2k+1 - (2k^2+2k) = 1$, so $d = 1$. Thus we have infinitely many triples $(a, b, c) = (2k^2-1,\ 2k^2+2k+1,\ 2k^2+4k+1)$ of the desired type.

Solution 2. We begin with a standard parametrization of the primitive Pythagorean triples (found in many introductory books on number theory, e.g., Kenneth Rosen, *Elementary Number Theory and its Applications*, Addison-Wesley, 6th Edition, 2011, Theorem 13.1)

$$x = m^2 - n^2, \qquad y = 2mn, \qquad z = m^2 + n^2,$$

where $m > n$ are relatively prime positive integers of opposite parity. We note that $z^2 = x^2 + y^2$ implies that $2z^2 = (x-y)^2 + (x+y)^2$, so z^2 is the average of $(x-y)^2$ and $(x+y)^2$. Any common factor of two of $x-y$, $x+y$, and z must divide x, y, and z because all are odd. Thus $x-y$, $x+y$, and z are relatively prime, so we get infinitely many triples as desired by taking

$$a = |m^2 - 2mn - n^2|, \qquad b = m^2 + n^2, \qquad c = m^2 + 2mn - n^2.$$

For example, if we take $m = 2k$ and $n = 1$, then we get $a = |4k^2 - 4k - 1|$, $b = 4k^2 + 1$, $c = 4k^2 + 4k - 1$.

Solution 3. Set $b_1 = 5$ and $c_1 = 7$ and for each $n \geq 1$, set $b_{n+1} = 3b_n + 2c_n$ and $c_{n+1} = 4b_n + 3c_n$. Then an easy induction shows that $(1, b_n, c_n)$ satisfies the conditions of the problem.

Problem 119

Find all twice continuously differentiable functions f for which there exists a constant c such that, for all real numbers a and b,

$$\left| \int_a^b f(x)\,dx - \frac{b-a}{2}(f(b) + f(a)) \right| \leq c(b-a)^4.$$

Idea. Consider

$$\lim_{b \to a} \frac{\int_a^b f(x)\,dx - \frac{b-a}{2}(f(b) + f(a))}{(b-a)^3}.$$

Solution. The functions f are the linear polynomials.

First of all, if f is a linear polynomial, then

$$\int_a^b f(x)\,dx = \frac{b-a}{2}(f(b) + f(a)).$$

(This can be shown either by direct computation, or by interpreting the right-hand side as the signed area of the trapezoid bounded by $x = a$, $x = b$, the x-axis, and $y = f(x)$.) Thus if f is a linear polynomial, then the inequality in the problem is satisfied for $c = 0$.

Conversely, suppose that f is a twice continuously differentiable function which satisfies the given inequality for some c. For fixed a, we can consider

$$L = \lim_{b \to a} \frac{\int_a^b f(x)\,dx - \frac{b-a}{2}(f(b) + f(a))}{(b-a)^3}.$$

On the one hand, the inequality shows that $L = 0$. On the other hand, L is the limit of an indeterminate form, and l'Hôpital's rule can be applied. By the Fundamental Theorem of Calculus and two applications of l'Hôpital's rule, we have

$$L = \lim_{b \to a} \frac{f(b) - \frac{1}{2}(f(b) + f(a)) - \frac{b-a}{2}f'(b)}{3(b-a)^2} = \lim_{b \to a} \frac{f(b) - f(a) - (b-a)f'(b)}{6(b-a)^2}$$

$$= \lim_{b \to a} \frac{f'(b) - f'(b) - (b-a)f''(b)}{12(b-a)} = \lim_{b \to a} -\frac{f''(b)}{12} = -\frac{f''(a)}{12},$$

because f'' is continuous. Therefore, $f''(a) = 0$ for all a, that is, f'' is identically zero. So we have $f(x) = C_1 x + C_2$ for some constants C_1 and C_2, and we are done.

Comment. By the same argument, our conclusion still holds if we replace the exponent 4 on the right-hand side of the inequality by $3+\varepsilon$ for any $\varepsilon > 0$. On the other hand, the error in the trapezoidal estimate $\frac{b-a}{2}(f(b) + f(a))$ for the integral $\int_a^b f(x)\,dx$ is known to have the form $-f''(\xi)\frac{(b-a)^3}{12}$ for some ξ between a and b. Thus if the exponent 4 were replaced by 3 instead of $3 + \varepsilon$, any function f whose second derivative is bounded and continuous would qualify.

Problem 120

Let A be a set of n real numbers. Because A has 2^n subsets, we can get 2^n sums by choosing a subset B of A and taking the sum of the numbers in B. What is the least number of *different* sums we must get by taking the 2^n possible sums of subsets of a set with n numbers?

Answer. The least number of different sums of subsets of a set A with n numbers is

$$\begin{cases} \frac{1}{4}n^2 + 1 & \text{if } n \text{ is even,} \\ \frac{1}{4}(n^2 + 3) & \text{if } n \text{ is odd.} \end{cases}$$

Solution. To see that we can achieve that number of different sums, take the set

$$A = \left\{ -n/2 + 1, -n/2 + 2, \ldots, -1, 0, 1, \ldots, n/2 \right\}$$

if n is even, and the set

$$A = \Big\{ -(n-1)/2,\ -(n-3)/2,\ \ldots,\ -1,\ 0,\ 1,\ \ldots,\ (n-3)/2,\ (n-1)/2 \Big\}$$

if n is odd. Whether n is even or odd, it's then not hard to see that we can get any desired integer between the sums of all the negative numbers in A and the sum of all the positive numbers in A (inclusive), but no other result, as the sum of some subset B of A. If n is odd, the sum of the positive numbers in A is

$$1 + 2 + \cdots + (n-1)/2 = \frac{n^2 - 1}{8},$$

and the sum of the negative numbers in A is $-(n^2 - 1)/8$, so the number of different sums of subsets for A is

$$1 + 2\frac{n^2 - 1}{8} = \frac{1}{4}(n^2 + 3).$$

If n is even, the sum of the positive numbers in A is

$$1 + 2 + \cdots + n/2 = \frac{n(n+2)}{8},$$

while the sum of the negative numbers in A is

$$-\Big((n/2 - 1) + (n/2 - 2) + \cdots + 1\Big) = -\frac{n(n-2)}{8},$$

so we get

$$\frac{n(n-2)}{8} + 1 + \frac{n(n+2)}{8} = \frac{1}{4}n^2 + 1$$

different sums of subsets. However, we still need to show that we can't somehow (by a different choice of A) get fewer.

So now let A be any set of n real numbers. Suppose that A contains k positive numbers, arranged in increasing order as $a_1 < a_2 < \cdots < a_k$, and ℓ negative numbers. (Note that because A may contain 0, $k+\ell$ could be either n or $n-1$.) Then we can construct at least $k(k+1)/2$ different positive sums of subsets of A, namely,

$$a_1 < a_2 < \cdots < a_k < \qquad (k \text{ in this row})$$
$$a_k + a_1 < a_k + a_2 < \cdots < a_k + a_{k-1} < \qquad (k-1 \text{ in this row})$$
$$(a_k + a_{k-1}) + a_1 < \cdots < (a_k + a_{k-1}) + a_{k-2} < \qquad (k-2 \text{ in this row})$$
$$\vdots$$
$$(a_k + a_{k-1} + \cdots + a_3) + a_1 < (a_k + a_{k-1} + \cdots + a_3) + a_2 < \qquad (2 \text{ in this row})$$
$$(a_k + a_{k-1} + \cdots + a_2) + a_1. \qquad (1 \text{ in this row})$$

(Note that this list has $k + (k-1) + (k-2) + \cdots + 2 + 1 = k(k+1)/2$ sums.) Similarly, there are at least $\ell(\ell+1)/2$ negative sums of subsets, so counting 0 (from the empty subset), there are at least $k(k+1)/2 + \ell(\ell+1)/2 + 1$ sums in all. Now $k + \ell \geq n-1$, so $\ell \geq n-1-k$ and

$$\frac{\ell(\ell+1)}{2} \geq \frac{(n-1-k)(n-k)}{2},$$

so we have at least

$$\frac{k(k+1)}{2} + \frac{(n-1-k)(n-k)}{2} + 1 = k^2 - k(n-1) + \frac{n(n-1)}{2} + 1 \qquad (*)$$

different sums of subsets. By standard techniques (completing the square, or calculus) the minimum value of $(*)$, as k ranges through the positive integers from 1 to n, occurs when $k = (n-1)/2$ if n is odd and when $k = n/2$ or $n/2 - 1$ (same value) if n is even. The minimum value is

$$\left(\frac{n-1}{2}\right)^2 - \frac{n-1}{2}(n-1) + \frac{n(n-1)}{2} + 1 = \frac{1}{4}(n^2 + 3)$$

if n is odd and

$$\left(\frac{n}{2}\right)^2 - \frac{n}{2}(n-1) + \frac{n(n-1)}{2} + 1 = \frac{1}{4}n^2 + 1$$

if n is even, so we're done.

Problem 121

Consider the following two-person game, in which players take turns coloring edges of a cube. Three colors (red, green, and yellow) are available. The cube starts off with all edges uncolored; once an edge is colored, it cannot be colored again. Two edges with a common vertex are not allowed to have the same color. The last player to be able to color an edge wins the game.

a. Given best play on both sides, should the first or the second player win? What is the winning strategy?

b. Since there are twelve edges in all, a game can last at most twelve turns. How many twelve-turn end positions are essentially different?

Solution. a. The second player should win. A winning strategy for the second player is to duplicate each of the first player's moves on the opposite side of the cube. That is, whenever the first player colors an edge, the second player counters by coloring the diametrically opposite edge the same color. After each of the second player's moves, the partial coloring of the cube will be symmetric with respect to the center of the cube. Therefore, any new move the first player makes which is legal (i.e., which does not cause two edges of the same color to meet) can be duplicated on the diametrically opposite side by the second player, and so the second player will be the last to be able to move and thus wins the game.

b. There are four essentially different end positions for a twelve-turn game, as shown below.

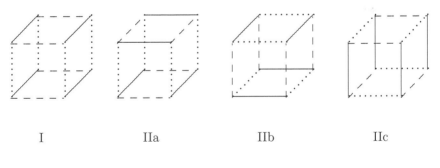

    I   IIa   IIb   IIc

(······ = red, —— = yellow, – – – – = green)

To see why, first note that in an end position for a twelve-turn game, each edge is colored and the three edges that meet at any vertex have the three different colors red, green, yellow, in some order. Now consider the four edges around a face of the cube. Since only three colors are available, two of these edges (at least) must have the same color; these edges must be parallel, or they would meet at a corner of the face. Thus for each face, either one or both pairs of parallel edges of that face must have the same color. Now we can distinguish two cases.

Case 1. For each face, *both* pairs of parallel edges of that face have the same color. It follows that for each of the three possible directions for edges of a cube, all four edges in that direction have the same color. By rotating the cube, we can assume that the red edges are vertical and the green and yellow edges are as shown in I above. Thus there is only one (essentially different) end position possible in this case.

Case 2. On at least one face, only one pair of parallel edges has the same color. By rotating the cube, we can assume that this is the front face and that the edges of the same color are vertical. Suppose this color is red. Then by turning the cube upside down if necessary, we can also assume that the top edge of the front face is yellow and the bottom edge is green. But then the colors of the edges leading back are completely determined, and so are the colors of the back edges. Thus the cube is colored as shown in IIa.

Similarly, if the color of the vertical edges in the front face is green, the cube can be turned upside down, if necessary, to match IIb above, whereas if the color is yellow, the cube will match IIc. In each case, all vertical edges have the same color but the edges in each of the other directions are not the same color, so the cases shown in IIa, IIb, and IIc are essentially different.

Thus, there are four essentially different end positions in all.

Problem 122

The unit circle has the property that distances along the curve are numerically equal to the difference of the corresponding angles (in radians) at the origin. Are there other differentiable curves in the plane with this property? If so, what do they look like?

Answer. Yes, there are. Besides the unit circle, there are the circles of radius 1/2 that pass through the origin. Also, there are "hybrids" formed by following one of these circles of radius 1/2 and then "switching" to the unit circle (which is externally tangent to all the other circles), and then possibly switching back to a circle of radius 1/2, and so forth.

Solution. Suppose such a curve is given in polar coordinates by $r = f(\theta)$. Then it has parametric equations $x = r\cos\theta = f(\theta)\cos\theta$, $y = f(\theta)\sin\theta$, and therefore the distance along the curve corresponding to a change from $\theta = \theta_1$ to $\theta = \theta_2$ is

$$\int_{\theta_1}^{\theta_2} \sqrt{\left(\frac{dx}{d\theta}\right)^2 + \left(\frac{dy}{d\theta}\right)^2}\, d\theta$$

$$= \int_{\theta_1}^{\theta_2} \sqrt{\Big(f'(\theta)\cos\theta - f(\theta)\sin\theta\Big)^2 + \Big(f'(\theta)\sin\theta + f(\theta)\cos\theta\Big)^2}\, d\theta$$

$$= \int_{\theta_1}^{\theta_2} \sqrt{\Big(f'(\theta)\Big)^2 + \Big(f(\theta)\Big)^2}\, d\theta.$$

So we are looking for functions f such that

$$\int_{\theta_1}^{\theta_2} \sqrt{\Big(f'(\theta)\Big)^2 + \Big(f(\theta)\Big)^2}\, d\theta = \theta_2 - \theta_1$$

for all θ_1, θ_2. By taking the derivative of each side with respect to θ_2 and then replacing θ_2 by θ, we see that we need

$$\sqrt{\Big(f'(\theta)\Big)^2 + \Big(f(\theta)\Big)^2} = 1$$

for all θ. Conversely, if this happens, then the integral above will be $\int_{\theta_1}^{\theta_2} 1\, d\theta = \theta_2 - \theta_1$, as desired. Squaring each side now leads to a pair of separable differential equations for $r = f(\theta)$:

$$\Big(f'(\theta)\Big)^2 + \Big(f(\theta)\Big)^2 = 1$$

$$\left(\frac{dr}{d\theta}\right)^2 + r^2 = 1$$

$$\frac{dr}{d\theta} = \pm\sqrt{1 - r^2}$$

where we will take $r \geq 0$, so $\sqrt{1 - r^2} = 0$ only for $r = 1$. For $r \neq 1$,

$$\frac{1}{\sqrt{1-r^2}}\frac{dr}{d\theta} = \pm 1$$

$$\int \frac{dr}{\sqrt{1-r^2}} = \pm \int d\theta$$

$$\arcsin r = \pm(\theta + C)$$

$$r = \pm \sin(\theta + C).$$

So, for any constant C, we can take $r = \sin(\theta + C)$ on intervals where $\sin(\theta + C) \geq 0$, and $r = -\sin(\theta + C)$ on intervals where $\sin(\theta + C) \leq 0$.

If $r = \sin(\theta + C)$, $(0 \leq \theta + C \leq \pi)$, we get

$$x = \sin(\theta + C)\cos\theta = \tfrac{1}{2}\left(\sin(2\theta + C) + \sin C\right) \quad \text{and}$$
$$y = \sin(\theta + C)\sin\theta = \tfrac{1}{2}\left(\cos C - \cos(2\theta + C)\right).$$

Because $0 \leq \theta + C \leq \pi$ we get $0 \leq 2(\theta + C) \leq 2\pi$, which implies $-C \leq 2\theta + C \leq 2\pi - C$, so $2\theta + C$ runs through an interval of length 2π. This means that $\big(\cos(2\theta + C), \sin(2\theta + C)\big)$ runs through the entire unit circle. Thus, from

$$x - \tfrac{1}{2}\sin C = \tfrac{1}{2}\sin(2\theta + C) \quad \text{and} \quad y - \tfrac{1}{2}\cos C = -\tfrac{1}{2}\cos(2\theta + C),$$

we see that our curve is the circle

$$\left(x - \tfrac{1}{2}\sin C\right)^2 + \left(y - \tfrac{1}{2}\cos C\right)^2 = \tfrac{1}{4},$$

with center at $\left(\tfrac{1}{2}\sin C, \tfrac{1}{2}\cos C\right)$ and radius $\tfrac{1}{2}$. As C varies, the centers will trace a circle of radius $1/2$ with center at the origin. Combining these circles with the unit circle $r = 1$, we get the curves described in the answer above.

If $r = -\sin(\theta + C)$, then $r = \sin(\theta + C + \pi)$, so this is not really a different case; we just have $r = \sin(\theta + D)$, where $D = C + \pi$ is a new constant.

Problem 123

> Fifty-two is the sum of two squares;
> And three less is a square! So who cares?
> You may think it's curious,
> Perhaps it is spurious,
> Are there other such numbers somewheres?

Answer. There are infinitely many integer solutions.

Solution 1. The only number smaller than 52 with the stated property is 4 (a somewhat degenerate case); the first number larger than 52 with the property is 292. We will show that there are infinitely many integers with the stated property.

Rewrite the equation $x^2 + y^2 = z^2 + 3$ in the form
$$x^2 - 3 = z^2 - y^2 = (z+y)(z-y).$$
If we set $a = z + y$ and $b = z - y$, we have $x^2 - 3 = ab$; also $y = (a-b)/2$ and $z = (a+b)/2$, so y and z will be integers provided a and b have the same parity. This is impossible if x is odd, since $x^2 - 3 \equiv 0 \pmod{4}$ has no solution, so let x be even, say $x = 2k$. Then $x^2 - 3 = (4k^2 - 3) \cdot 1$, and therefore we can take $a = 4k^2 - 3$, $b = 1$ and have a and b both be odd. In this way we find the infinite family of solutions
$$\begin{cases} x = 2k \\ y = 2k^2 - 2 \\ z = 2k^2 - 1. \end{cases}$$

Solution 2. Here is an improvement on Solution 1: we will find a doubly infinite family of solutions.

We begin as in Solution 1, but we will now choose x to be a suitable polynomial in k and m, so that $x^2 - 3$ will have a polynomial factorization.

Keeping x even, we can make sure that $x^2 - 3$ has the same polynomial factor $4k^2 - 3$ as in Solution 1 by setting $x = 2m(4k^2 - 3) + 2k$, for then
$$x^2 - 3 = (4k^2 - 3)\bigl(4m^2(4k^2 - 3) + 8mk + 1\bigr).$$
Solving
$$\begin{cases} z + y = 4m^2(4k^2 - 3) + 8mk + 1 \\ z - y = 4k^2 - 3, \end{cases}$$
we find
$$\begin{cases} x = 2m(4k^2 - 3) + 2k \\ y = 2m^2(4k^2 - 3) + 4mk - 2k^2 + 2 \\ z = 2m^2(4k^2 - 3) + 4mk + 2k^2 - 1. \end{cases}$$

Problem 124

Let E be an ellipse in the plane. Describe the set S of all points P outside the ellipse such that the two tangent lines to the ellipse that pass through P form a right angle.

Answer. The set is always a circle with center at the center of the ellipse and with radius $\sqrt{c^2 + d^2}$, where $2c$ and $2d$ are the lengths of the major and minor axes of the ellipse.

Solution 1. We can choose the x- and y-axes along the major and minor axes of the ellipse, respectively, and we can choose the unit of length so the y-intercepts of the ellipse are ± 1 (if $2c$ and $2d$ are the original lengths of the major and minor axes of the ellipse, rescale by dividing all coordinates by d). Then the ellipse will have equation $\frac{x^2}{a^2} + y^2 = 1$, or $x^2 + a^2 y^2 = a^2$, where the x-intercepts are $\pm a$, $a = c/d$ and $a \geq 1$. Now let $P = (x_0, y_0)$ be a point outside the ellipse, and let $y - y_0 = m(x - x_0)$ be a nonvertical line through P. That line will be tangent to the ellipse exactly if the equation

$$x^2 + a^2 \Big(m(x - x_0) + y_0\Big)^2 = a^2,$$

obtained by intersecting the line and the ellipse, has a double root x. The equation works out to

$$x^2 + a^2 \Big(mx + (y_0 - mx_0)\Big)^2 = a^2,$$
$$(1 + a^2 m^2)x^2 + 2a^2 m(y_0 - mx_0)x + a^2 \Big[(y_0 - mx_0)^2 - 1\Big] = 0,$$

which has a double root exactly if its discriminant is zero. That is, the line $y - y_0 = m(x - x_0)$ is tangent to the ellipse exactly if

$$\Big(2a^2 m(y_0 - mx_0)\Big)^2 - 4(1 + a^2 m^2)\Big(a^2(y_0 - mx_0)^2 - a^2\Big) = 0,$$

which simplifies to

$$m^2(a^2 - x_0^2) + 2x_0 y_0 m + (1 - y_0^2) = 0.$$

We want the two roots m to this quadratic equation (which are the slopes of the tangent lines to the ellipse that pass through P) to multiply to -1 (so the tangent lines will be at right angles to each other). That is, we want

the equation to factor as
$$(a^2 - x_0^2)(m - m_1)(m - m_2) = 0 \quad \text{with} \quad m_1 m_2 = -1.$$
Then the constant term should be
$$(a^2 - x_0^2)(-1) = 1 - y_0^2, \quad \text{so} \quad x_0^2 + y_0^2 = a^2 + 1.$$
Note that any point P satisfying this condition does lie outside the ellipse. So the two non-vertical, non-horizontal tangent lines to the ellipse through P are perpendicular exactly if P lies on the circle with radius $\sqrt{a^2 + 1}$ and center at the center of the ellipse. But the case of a vertical or horizontal tangent line to the ellipse also leads to points on that same circle, for the possible vertical tangent lines to the ellipse are $x = \pm a$ while the possible horizontal tangent lines are $y = \pm 1$, and these intersect at the four points $(\pm a, \pm 1)$ on the circle. For these points, the quadratic equation $m^2(a^2 - x_0^2) + 2x_0 y_0 m + (1 - y_0^2) = 0$ degenerates to a linear equation whose solution is $m = 0$.

Solution 2. In a way similar to Solution 1, we may choose coordinates so that the ellipse has the form $a^2 x^2 + y^2 = 1$. As we rotate E and S through 360 degrees about the origin, every point on S will arise (four times) as the intersection of horizontal and vertical tangents to the rotated ellipse, and conversely.

So, take an arbitrary angle θ, $0 \leq \theta < 2\pi$, and make a change of variables corresponding to a clockwise rotation of the plane. Specifically, let
$$\begin{pmatrix} u \\ v \end{pmatrix} = \begin{pmatrix} \cos\theta & \sin\theta \\ -\sin\theta & \cos\theta \end{pmatrix} \begin{pmatrix} x \\ y \end{pmatrix},$$
or equivalently
$$\begin{pmatrix} x \\ y \end{pmatrix} = \begin{pmatrix} \cos\theta & -\sin\theta \\ \sin\theta & \cos\theta \end{pmatrix} \begin{pmatrix} u \\ v \end{pmatrix}.$$
Then in the rotated plane, the equation of the ellipse is
$$a^2(u\cos\theta - v\sin\theta)^2 + (u\sin\theta + v\cos\theta)^2 = 1,$$
$$(a^2\cos^2\theta + \sin^2\theta)u^2 + 2(1 - a^2)\sin\theta\cos\theta\, uv + (a^2\sin^2\theta + \cos^2\theta)v^2 = 1.$$

To find the horizontal and vertical tangents we use implicit differentiation. From $Au^2 + 2Buv + Cv^2 = 1$ (where $A = a^2\cos^2\theta + \sin^2\theta > 0$, $B = (1 - a^2)\sin\theta\cos\theta$, $C = a^2\sin^2\theta + \cos^2\theta > 0$) we get
$$2(Bu + Cv)\frac{dv}{du} = -2(Au + Bv).$$

So the horizontal tangent lines, obtained when $Au = -Bv$, are given by

$$A(-Bv/A)^2 + 2B(-Bv/A)v + Cv^2 = 1,$$
$$v^2(-B^2 + AC) = A,$$
$$v = \pm\sqrt{\frac{A}{AC - B^2}},$$

and the vertical tangent lines, obtained when $Bu = -Cv$, are given by

$$Au^2 + 2Bu(-Bu/C) + C(-Bu/C)^2 = 1,$$
$$u^2(AC - B^2) = C,$$
$$u = \pm\sqrt{\frac{C}{AC - B^2}}.$$

Now the discriminant $AC - B^2$ of the conic is invariant under rotation, so $AC - B^2 = a^2$ (which can be checked by direct substitution). Thus the points in S are of the form

$$(u, v) = \left(\frac{\pm\sqrt{a^2 \sin^2\theta + \cos^2\theta}}{a}, \frac{\pm\sqrt{a^2 \cos^2\theta + \sin^2\theta}}{a} \right).$$

For these points we see that $u^2 + v^2 = (a^2 + 1)/a^2$. Therefore, the points in S lie on a circle about the origin having radius $\sqrt{a^2 + 1}/a$.

Now consider any ray from the origin. The tangents from a point on this ray just outside the ellipse will form an angle that is just under 180°. The tangents from a point on this ray far from the origin will form an angle that is just over 0°. By the Intermediate Value Theorem, the angle formed by the tangents will be 90° somewhere on this ray. The ray must intersect the set S. It follows that S is the entire circle.

Problem 125

Starting with a positive number $x_0 = a$, let $(x_n)_{n \geq 0}$ be the sequence of numbers such that

$$x_{n+1} = \begin{cases} x_n^2 + 1 & \text{if } n \text{ is even,} \\ \sqrt{x_n} - 1 & \text{if } n \text{ is odd.} \end{cases}$$

For what positive numbers a will there be terms of the sequence arbitrarily close to 0?

Solution. The sequence has terms arbitrarily close to 0 for all positive numbers a. In fact, the subsequence $(x_{2n})_{n \geq 0}$ has limit 0.

To show this, let $y_n = x_{2n}$. We have

$$y_{n+1} = \sqrt{x_{2n+1}} - 1 = \sqrt{x_{2n}^2 + 1} - 1 = \sqrt{y_n^2 + 1} - 1, \quad (*)$$

from which we conclude $y_n > 0$ for all n. Furthermore,

$$y_{n+1} < \sqrt{y_n^2 + 2y_n + 1} - 1 = y_n.$$

We have just seen that (y_n) is a decreasing sequence of positive numbers, and as such it must have a limit L, $L \geq 0$. We finish the solution by proving that $L = 0$. Taking the limit of each side of $(*)$ yields

$$L = \sqrt{L^2 + 1} - 1, \quad \text{and so} \quad (L+1)^2 = L^2 + 1.$$

The only solution to this equation is $L = 0$, and we are done.

Problem 126

Suppose you pick the one million entries of a 1000×1000 matrix independently and at random from the set of digits. Is the determinant of the resulting matrix more likely to be even or odd?

Answer. The determinant is more likely to be even than odd.

Solution. First note that there are equally many possible even digits as odd digits, so we may as well replace the even digits with 0 and the odd digits with 1 and consider the determinant modulo 2. In other words, the question is whether for a random 1000×1000 matrix with entries in the field $\mathbb{Z}/2\mathbb{Z} = \{0, 1\}$ of integers modulo 2, the determinant is more likely to be 0 or 1. Another way of phrasing this is: if we pick 1000 vectors (the columns of the matrix) at random in $(\mathbb{Z}/2\mathbb{Z})^{1000}$, are those vectors more likely to be linearly dependent (determinant 0) over $\mathbb{Z}/2\mathbb{Z}$ or linearly independent (determinant 1)?

Now if we pick the first 999 vectors $\mathbf{v}_1, \ldots, \mathbf{v}_{999}$ to be linearly independent, then the span of those vectors (consisting of all linear combinations $c_1 \mathbf{v}_1 + c_2 \mathbf{v}_2 + \cdots + c_{999} \mathbf{v}_{999}$, with each $c_i = 0$ or 1) will have exactly 2^{999} vectors in it, which is exactly half the vectors in $(\mathbb{Z}/2\mathbb{Z})^{1000}$. So if the first 999 vectors are linearly independent over $\mathbb{Z}/2\mathbb{Z}$, then the last vector $\mathbf{v}_{1000}$

is equally likely to be in the span of the others (which will result in determinant 0) or not to be in that span (in which case $\mathbf{v}_1, \ldots, \mathbf{v}_{1000}$ will be linearly independent, resulting in determinant 1). However, the first 999 vectors may be linearly dependent, and if so it is certain that $\mathbf{v}_1, \ldots, \mathbf{v}_{1000}$ will also be linearly dependent and the determinant will be 0. Thus the probability that the determinant will be 1 modulo 2 is $(1/2) \cdot$ (the probability that $\mathbf{v}_1, \ldots, \mathbf{v}_{999}$ are linearly independent) $< (1/2) \cdot 1 = 1/2$, showing that the determinant is less likely to be odd, and more likely to be even.

Comment. With slightly more effort we can find the exact probability that $\mathbf{v}_1, \ldots, \mathbf{v}_{1000}$ are linearly independent. Let P_k be the probability that k randomly chosen vectors in $(\mathbb{Z}/2\mathbb{Z})^{1000}$ are linearly independent. Then,

$P_{k+1} =$ probability that randomly chosen vectors $\mathbf{v}_1, \ldots \mathbf{v}_{k+1}$ are linearly independent
$=$ probability that randomly chosen vectors $\mathbf{v}_1, \ldots, \mathbf{v}_k$ are linearly independent *and* $\mathbf{v}_{k+1}$ is not in the span of $\mathbf{v}_1, \ldots, \mathbf{v}_k$
$=$ (probability that randomly chosen vectors $\mathbf{v}_1, \ldots, \mathbf{v}_k$ are linearly independent) $\cdot$ (probability that $\mathbf{v}_{k+1}$ is not in the span of $\mathbf{v}_1, \ldots, \mathbf{v}_k$)
$= P_k \cdot \dfrac{2^{1000} - 2^k}{2^{1000}}$ (because there are 2^k vectors in the span of $\mathbf{v}_1, \ldots, \mathbf{v}_k$).

Clearly P_1 is equal to the probability that a randomly chosen vector is not the zero vector, and this probability is $\frac{2^{1000}-1}{2^{1000}}$. So, using the recurrence relation above,

$$P_2 = \left(\frac{2^{1000}-1}{2^{1000}}\right)\left(\frac{2^{1000}-2}{2^{1000}}\right), \; P_3 = \left(\frac{2^{1000}-1}{2^{1000}}\right)\left(\frac{2^{1000}-2}{2^{1000}}\right)\left(\frac{2^{1000}-2^2}{2^{1000}}\right),$$

and continuing in this way,

$$P_{1000} = \left(\frac{2^{1000}-1}{2^{1000}}\right)\left(\frac{2^{1000}-2}{2^{1000}}\right)\left(\frac{2^{1000}-2^2}{2^{1000}}\right) \cdots \left(\frac{2^{1000}-2^{999}}{2^{1000}}\right)$$
$$= \left(1 - \frac{1}{2^{1000}}\right)\left(1 - \frac{1}{2^{999}}\right)\left(1 - \frac{1}{2^{998}}\right) \cdots \left(1 - \frac{1}{2^2}\right)\left(1 - \frac{1}{2}\right)$$
$$= \frac{1}{2} \cdot \frac{3}{4} \cdot \frac{7}{8} \cdot \frac{15}{16} \cdot \frac{31}{32} \cdot \ldots \cdot \frac{2^{1000}-1}{2^{1000}} \approx 0.2887880951.$$

Therefore, the probability that the determinant is even is approximately 0.711211905.

Problem 127

a. Find all positive numbers T for which
$$\int_0^T x^{-\ln x}\,dx = \int_T^\infty x^{-\ln x}\,dx.$$

b. Given that $\int_0^\infty e^{-x^2}\,dx = \sqrt{\pi}/2$, evaluate the above integrals for all such T.

Solution. We will show that $T = e^{1/2}$ is the only such number and that
$$\int_0^T x^{-\ln x}\,dx = \int_T^\infty x^{-\ln x}\,dx = \frac{e^{1/4}\sqrt{\pi}}{2}.$$

Substituting $y = \ln x$ converts the integrals
$$\int_0^T x^{-\ln x}\,dx \quad \text{and} \quad \int_T^\infty x^{-\ln x}\,dx$$

to
$$\int_{-\infty}^{\ln T} e^{-y^2+y}\,dy \quad \text{and} \quad \int_{\ln T}^\infty e^{-y^2+y}\,dy.$$

In order to make these integrals more symmetric, we complete the square in the exponent and substitute $z = y - 1/2$, obtaining
$$e^{1/4}\int_{-\infty}^{\ln T - 1/2} e^{-z^2}\,dz \quad \text{and} \quad e^{1/4}\int_{\ln T - 1/2}^\infty e^{-z^2}\,dz.$$

These improper integrals both converge.

a. Because e^{-z^2} is a positive, even function, the two integrals are equal if and only if $\ln T - 1/2 = 0$, or equivalently, $T = e^{1/2}$.

b. Since $\int_0^\infty e^{-x^2}\,dx = \sqrt{\pi}/2$, we find
$$\int_0^{e^{1/2}} x^{-\ln x}\,dx = \int_{e^{1/2}}^\infty x^{-\ln x}\,dx = \frac{e^{1/4}\sqrt{\pi}}{2}.$$

Problem 128

Let a_n be the number of different strings of length n that can be formed from the symbols X and O with the restriction that a string may not consist of identical smaller strings. Prove or disprove each of the following conjectures.

a. For any $n > 2$, a_n is divisible by 6.

b. $\lim\limits_{n \to \infty} \dfrac{a_{n+1}}{a_n} = 2$.

Answer. Both conjectures are correct.

Solution. There are 2^n possible strings of length n. A string s of length n that consists of identical smaller strings of length d, where d is the smallest integer for which this is true, can be identified with the string of length d consisting of its first d symbols. Thus, each string of length n is among the strings counted by a_d for a unique divisor d of n. As a result, we have

$$\sum_{d|n} a_d = 2^n;$$

we'll use this recurrence relation to prove our conjectures.

a. We use induction on n. We get a_3 from

$$\sum_{d|3} a_d = 2^3 \implies a_1 + a_3 = 8 \implies a_3 = 8 - a_1 = 8 - 2 = 6;$$

in particular, a_3 is divisible by 6. Now suppose that a_k is divisible by 6 for all k with $2 < k < n$.

If n is odd, then, by the induction hypothesis, all terms in the summation $\sum_{d|n} a_d$ except a_1 and a_n are divisible by 6, so that $a_1 + a_n \equiv 2^n \pmod{6}$. Because $a_1 = 2$, this means that $a_n \equiv 2^n - 2 \pmod{6}$.

If n is even, the induction hypothesis implies that in $\sum_{d|n} a_d$, all terms except a_1, a_2, and a_n are divisible by 6, so that $a_1 + a_2 + a_n \equiv 2^n \pmod{6}$, and because $a_1 = a_2 = 2$, $a_n \equiv 2^n - 4 \pmod{6}$.

So our induction proof will be complete if we can show that for n odd, $2^n - 2$ is divisible by 6 while for n even, $2^n - 4$ is divisible by 6. But $2^n - 2$ and $2^n - 4$ are both even, so it is enough to show that $2^n - 2$ is divisible by 3 when n is odd and $2^n - 4$ is divisible by 3 when n is even. Modulo 3, the powers of 2 $(2, 4, 8, 16, \ldots)$ are congruent to $2, 1, 2, 1, \ldots$, and the result follows.

b. The idea is that for large n, a_n is "reasonably close" to 2^n. More precisely, because we have $\sum_{d|n} a_d = 2^n$ and any divisor d of n with $d < n$ is at most $n/2$, we have

$$a_n \geq 2^n - a_{\lfloor n/2 \rfloor} - a_{\lfloor n/2 \rfloor - 1} - \cdots - a_1$$
$$\geq 2^n - 2^{\lfloor n/2 \rfloor} - 2^{\lfloor n/2 \rfloor - 1} - \cdots - 2 > 2^n - 2^{\lfloor n/2 \rfloor + 1}$$

THE SOLUTIONS

because a sum of consecutive powers of 2, say $2 + 2^2 + \cdots + 2^k$, is two less than the next power, 2^{k+1}. So we have

$$2^n - 2^{\lfloor n/2 \rfloor + 1} < a_n \le 2^n.$$

Therefore, we have both

$$\frac{a_{n+1}}{a_n} > \frac{2^{n+1} - 2^{\lfloor (n+1)/2 \rfloor + 1}}{2^n} = 2 - \frac{1}{2^{n-1-\lfloor (n+1)/2 \rfloor}}$$

and

$$\frac{a_{n+1}}{a_n} < \frac{2^{n+1}}{2^n - 2^{\lfloor n/2 \rfloor}} = \frac{2}{1 - \frac{1}{2^{n-\lfloor n/2 \rfloor}}}$$

As $n \to \infty$,

$$\frac{2}{1 - \frac{1}{2^{n-\lfloor n/2 \rfloor}}} \longrightarrow 2 \quad \text{and} \quad 2 - \frac{1}{2^{n-1-\lfloor (n+1)/2 \rfloor}} \longrightarrow 2,$$

so because $\dfrac{a_{n+1}}{a_n}$ is between these two expressions, $\lim_{n \to \infty} \dfrac{a_{n+1}}{a_n} = 2$ by the squeeze principle, and we are done.

Problem 129

Let $f(x, y) = x^2 + y^2$ and $g(x, y) = x^2 - y^2$. Are there differentiable functions $F(z)$, $G(z)$, and $z = h(x, y)$ such that $f(x, y) = F(z)$ and $g(x, y) = G(z)$?

Solution. No. If there were such functions, taking partial derivatives would yield

$$2x = \frac{\partial f}{\partial x} = F'(z)\frac{\partial h}{\partial x}, \qquad 2x = \frac{\partial g}{\partial x} = G'(z)\frac{\partial h}{\partial x},$$

$$2y = \frac{\partial f}{\partial y} = F'(z)\frac{\partial h}{\partial y}, \qquad -2y = \frac{\partial g}{\partial y} = G'(z)\frac{\partial h}{\partial y}.$$

We would then have

$$2x\frac{\partial h}{\partial y} = F'(z)\frac{\partial h}{\partial x}\frac{\partial h}{\partial y} = 2y\frac{\partial h}{\partial x}, \qquad 2x\frac{\partial h}{\partial y} = G'(z)\frac{\partial h}{\partial x}\frac{\partial h}{\partial y} = -2y\frac{\partial h}{\partial x},$$

which combine to yield $2x\frac{\partial h}{\partial y} = 2y\frac{\partial h}{\partial x} = 0$. Thus we would have $\frac{\partial h}{\partial y} = \frac{\partial h}{\partial x} = 0$, at least off the x- and y-axes, so h, and therefore f and g, would be constant in the interior of each quadrant, a contradiction.

Problem 130

For $x \geq 0$, let $y = f(x)$ be continuously differentiable, with positive, increasing derivative. Consider the ratio between the distance from $(0, f(0))$ to $(x, f(x))$ along the curve $y = f(x)$ and the straight-line distance from $(0, f(0))$ to $(x, f(x))$. Must this ratio have a limit as $x \to \infty$? If so, what is the limit?

Answer. Yes, the limit of the ratio is always 1.

Solution 1. Because replacing $f(x)$ by $f(x) - f(0)$ will not affect the conditions on the derivative, we may assume without loss of generality that $f(0) = 0$. The ratio of the arc length to the straight-line distance is then given by

$$\frac{\int_0^x \sqrt{1 + [f'(u)]^2}\, du}{\sqrt{x^2 + [f(x)]^2}}.$$

Note that as $x \to \infty$, the numerator and denominator both tend to ∞. Therefore, by l'Hôpital's rule and the Fundamental Theorem of Calculus, the limit of the ratio will equal

$$\lim_{x \to \infty} \frac{\sqrt{1 + [f'(x)]^2}}{\frac{x + f(x)f'(x)}{\sqrt{x^2 + [f(x)]^2}}},$$

provided this last limit exists. We know that $f'(x) > 0$, and we'll soon see that for large enough x, $f(x) > 0$. Using this, we can rewrite our (putative) limit as

$$L = \lim_{x \to \infty} \frac{\sqrt{[1/f'(x)]^2 + 1} \cdot \sqrt{[x/f(x)]^2 + 1}}{x/[f(x)f'(x)] + 1};$$

we still need to show that L exists and that $L = 1$.

Because $f'(x)$ is positive and increasing, $\lim_{x \to \infty} f'(x)$ is either ∞ or finite and positive (specifically, the least upper bound B of all the values of $f'(x)$). We now consider these cases separately.

Case 1. $\lim_{x \to \infty} f'(x) = \infty$. Then for every B, there exists $a > 0$ such that $f'(x) \geq B$ for $x \geq a$. For such x,

$$f(x) = f(a) + \int_a^x f'(u)\, du \geq f(a) + B(x - a).$$

In particular, $f(x) \geq (B/2)x$ for x sufficiently large. Since this is true for all B, we have $\lim_{x \to \infty} \dfrac{x}{f(x)} = 0$; it follows from this and $\lim_{x \to \infty} f'(x) = \infty$ that

$$L = \lim_{x \to \infty} \frac{\sqrt{[1/f'(x)]^2 + 1} \cdot \sqrt{[x/f(x)]^2 + 1}}{x/[f(x)f'(x)] + 1} = 1,$$

as desired.

Case 2. $\lim_{x \to \infty} f'(x) = B$. This time, for any $\varepsilon > 0$ we have $B - \varepsilon \le f'(x) \le B$ for x large enough, and as in the first case we conclude that

$$f(a) + (B - \varepsilon)(x - a) \le f(x) \le f(a) + B(x - a)$$

for some $a > 0$ (which depends on ε) and all $x \ge a$. Dividing by x, we see that by taking x large enough, we can get $\dfrac{f(x)}{x}$ within an arbitrarily small distance of the interval $[B - \varepsilon, B]$. This is true for every ε, so $\lim_{x \to \infty} \dfrac{f(x)}{x} = B$. Along with $\lim_{x \to \infty} f'(x) = B$, this yields

$$L = \frac{\sqrt{1/B^2 + 1} \cdot \sqrt{1/B^2 + 1}}{1/B^2 + 1} = 1,$$

and we are done.

Solution 2. (Bjorn Poonen, MIT) We will parametrize the curve by arc length s, so that the position vector of a point on the curve is given by $\mathbf{r}(s) = (x(s), y(s))$. Then $\mathbf{r}'(s)$ is a unit vector whose argument (angle to the horizontal) is increasing and bounded by $\pi/2$, so this unit vector has a limit $\mathbf{v} = \lim_{s \to \infty} \mathbf{r}'(s)$.

We now show more generally that for any smooth path $\mathbf{r}(t)$ such that $\lim_{t \to \infty} \mathbf{r}'(t) = \mathbf{v} \ne \mathbf{0}$, the ratio of the arc length to the distance to the origin (or to any other fixed point) tends to 1. The arc length along the path from $t = 0$ until $t = T$ is $\int_0^T |\mathbf{r}'(t)| \, dt$, and by our assumption this can be written as $T|\mathbf{v}| + o(T)$, where $o(T)$ denotes an unspecified function $g(T)$ for which $\lim_{T \to \infty} \dfrac{g(T)}{T} = 0$. On the other hand, we have

$$\mathbf{r}(T) = \mathbf{r}(0) + \int_0^T \mathbf{r}'(t) \, dt = T\mathbf{v} + o(T).$$

Therefore, the distance to the origin, $|\mathbf{r}(T)|$, is of the form $T|\mathbf{v}| + o(T)$, just as the arc length is. This shows that the ratio tends to 1 as $T \to \infty$, and we are done.

Problem 131

Do there exist four positive integers such that the sum of any two of them is a perfect square and such that the six squares found in this way are all different? If so, exhibit four such positive integers; if not, show why this cannot be done.

Answer. Yes, there exist four such positive integers; one example is given by $a = 4082$, $b = 5522$, $c = 3314$, $d = 407$, for which the six squares are $a+b = 98^2$, $a+c = 86^2$, $a+d = 67^2$, $b+c = 94^2$, $b+d = 77^2$, $c+d = 61^2$.

Solution. One way of coming up with such examples without a computer search is to notice first that if $a+b, a+c, a+d, b+c, b+d, c+d$ are all distinct squares, then $a+b+c+d$ must be the sum of two squares in three essentially distinct ways: $(a+b)+(c+d)$, $(a+c)+(b+d)$, $(a+d)+(b+c)$. Conversely, if an integer N can be written as $N = k_1^2 + k_2^2 = k_3^2 + k_4^2 = k_5^2 + k_6^2$, then we can find constants a, b, c, d such that $a+b = k_1^2$, $c+d = k_2^2$, $a+c = k_3^2$, $a+d = k_5^2$ (and $b+d = k_4^2$, $b+c = k_6^2$ will be automatic from those four equations). Our task will be to find k_1, k_2, k_3, k_5 so that a, b, c, d will be distinct positive integers.

Note that if m and n are integers, then the complex number $z = m + ni$ has $|z|^2 = (m+ni)(m-ni) = m^2+n^2$ and $w = n+mi$ also has $|w|^2 = m^2+n^2$; in addition, if z_1, z_2 are complex numbers, then $|z_1 z_2|^2 = |z_1|^2 |z_2|^2$. As a result, if we pick three complex numbers $m_1 + n_1 i$, $m_2 + n_2 i$, $m_3 + n_3 i$ and multiply them together and we also look at products obtained from that one by replacing $m_1 + n_1 i$ by $n_1 + m_1 i$ and/or $m_2 + n_2 i$ by $n_2 + m_2 i$ and/or $m_3 + n_3 i$ by $n_3 + m_3 i$, the different products $m + ni$ we get will have the same values of $m^2 + n^2$. For example:

$$(2 + 3i)(3 + 4i)(5 + 4i) = -98 + 61i,$$

$$(3 + 2i)(3 + 4i)(4 + 5i) = -86 + 77i,$$

$$(3 + 2i)(3 + 4i)(5 + 4i) = -67 + 94i,$$

and

$$98^2 + 61^2 = 86^2 + 77^2 = 67^2 + 94^2 = (2^2 + 3^2)(3^2 + 4^2)(4^2 + 5^2) = 13325.$$

So $N = 13325$ is written as the sum of two squares in three different ways. If we solve for a, b, c, d such that $a+b = 98^2$, $c+d = 61^2$, $a+c = 86^2$, $a+d = 67^2$ as suggested above, we find the example given in the answer.

Alternatively, one can take any three distinct primitive Pythagorean triples $(x_1, y_1, z_1), (x_2, y_2, z_2), (x_3, y_3, z_3)$ (that is, $x_i^2 + y_i^2 = z_i^2$) and scale them up so that $z_1 = z_2 = z_3$ to obtain three distinct representations as a sum of two squares

$$z_1 = z_2 = z_3 = x_1^2 + y_1^2 = x_2^2 + y_2^2 = x_3^2 + y_3^2.$$

Now take $k_1 = x_1$, $k_2 = y_1$, $k_3 = x_2$, $k_4 = y_2$, $k_5 = x_3$, $k_6 = y_3$ and continue as above. For example, the triples $(3, 4, 5), (5, 12, 13), (15, 8, 17)$ yield $a = 174897, b = 264672, c = 5728, d = 775728$. (Note that if we use the triple $(8, 15, 17)$ instead of $(15, 8, 17)$, we get neither positive nor integral a, b, c, d.) (This approach is due to Eugene Luks, University of Oregon.)

Problem 132

A point $P = (a, b)$ in the plane is *rational* if both a and b are rational numbers. Find all rational points P such that the distance between P and every rational point on the line $y = 13x$ is a rational number.

Answer. There are no such points.

Solution 1. Let the rational point $P = (a, b)$ have rational distance to every rational point on the line $y = 13x$. If P is on the line, then the distance from P to the origin is $\sqrt{170}\,|a|$. Since 170 is not a perfect square, this distance is irrational, unless $a = 0$. However, for $a = 0$, the distance from P to $(1, 13)$ is irrational. Thus, P cannot be on the line.

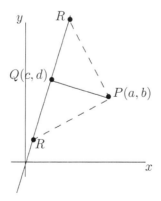

Now suppose that P is not on the line. Let $Q = (c, d)$ be the foot of the perpendicular from P to the line.

Then Q is the intersection of the lines
$$y = 13x \quad \text{and} \quad y = -\frac{1}{13}(x-a) + b,$$
so Q is a rational point. Therefore, the distance PQ is rational. Starting at Q, we move a distance PQ along the line $y = 13x$. The two points R we could end up at are
$$(b+c-d, -a+c+d) \quad \text{and} \quad (-b+c+d, a-c+d),$$
which are both rational. However, the distance PR is $\sqrt{2}$ times the distance PQ, hence it cannot be rational, a contradiction.

Solution 2. We show that, more generally, no *real* point P can satisfy the condition. Suppose $P = (a, b)$ does satisfy the condition. Then for any rational number x, the quantity
$$(x-a)^2 + (13x-b)^2 = 170x^2 - (2a+26b)x + (a^2+b^2)$$
is the square of a rational number. In particular, setting $x = 0$, we see that $a^2 + b^2$ is the square of a rational number. Setting $x = 1$ (or any nonzero rational) shows that $2a + 26b$ is rational. If $2a + 26b \neq 0$, then we can set $x = (a^2 + b^2)/(2a + 26b)$ and conclude that $170x^2$ is the square of a rational number, so 170 is a perfect square, a contradiction. On the other hand, if $2a + 26b = 0$, substituting $x = \sqrt{a^2 + b^2}$ implies $171(a^2 + b^2)$ is the square of a rational number. Since $a^2 + b^2$ is also a rational square, this implies $a = b = 0$. But then the distance from $P = (0,0)$ to $(1, 13)$ is the irrational number $\sqrt{170}$, a contradiction.

Solution 3. Let $P = (a, b)$ be a point with the given property. First we show that if k is any positive rational number, then $kP = (ka, kb)$ also has the property. To see this, note that for any rational point $Q = (t, u)$ on the line $y = 13x$, $Q/k = (t/k, u/k)$ is also a rational point on this line. So the distance $d(kP, Q)$ is rational, since $d(kP, Q) = k\, d(P, Q/k)$; thus, kP has the given property. Therefore, we may assume that a and b are *integers*; all the points $P_n = (na, nb)$ will then be lattice points with the given property. As in the first solution, we know that $P (= P_1)$ is not on the line $y = 13x$.

Let $Q = (1, 13)$, $O = (0, 0)$, and $d = d(O, P_1)$. For each positive integer n, let $L_n = d(P_n, Q)$. Because the coordinates of Q and all P_n are integers, the rational numbers d and L_n are in fact *integers*.

Now consider triangle QP_nP_{n+1} (see figure). As n increases, $\angle QP_nP_{n+1}$ approaches $180°$, and $L_{n+1} - L_n$ approaches d. (For a formal proof of this, let $c = d(Q, O)$, and let $\theta = \angle QOP_n$. Then, by the law of cosines,
$$L_n = \sqrt{c^2 + (nd)^2 - 2cnd\cos\theta},$$

THE SOLUTIONS

and therefore

$$\lim_{n\to\infty} (L_{n+1} - L_n) = \lim_{n\to\infty} \frac{\frac{1}{n}(L_{n+1}^2 - L_n^2)}{\frac{1}{n}(L_{n+1} + L_n)} = \frac{2d^2}{\sqrt{d^2} + \sqrt{d^2}} = d.)$$

Since any convergent sequence of integers must ultimately be constant, it follows that for large n, $L_{n+1} = L_n + d$. On the other hand, the triangle inequality implies that $L_{n+1} < L_n + d$, a contradiction.

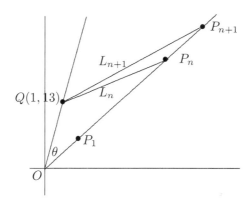

Solution 4. As in Solution 2, we will show that there cannot exist any *real* point $P = (a, b)$ whose distance to every rational point on $y = 13x$ is rational.

Suppose P is such a point. Arguing as in Solution 3, we see that for every nonzero rational k, the distance from (ka, kb) to $(1, 13)$ is rational. This clearly rules out $(a, b) = (0, 0)$. Also, the distance from (a, b) to $(0, 0)$ is rational, so the distance from (ka, kb) to $(0, 0)$ can be made to equal any rational $r > 0$ by suitable choice of rational $k > 0$.

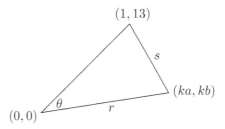

Now let r and s be the distances, and θ be the angle, shown in the previous figure. Then whenever $r > 0$ is rational, s is also, because k is. By

the law of cosines, r and s are related by
$$s^2 = r^2 - 2r\sqrt{170}\cos\theta + 170.$$
The right-hand side of this equation is a polynomial in r of the form
$$r^2 - \alpha r + 170,$$
where $\alpha = 2\sqrt{170}\cos\theta$. Since the value of this polynomial at $r = 1$ is a rational square, α is rational. Note that 170 is divisible by 2 but not by 4. By taking $r > 0$ to be an integer which is divisible by 4 times the denominator of α, we can arrange that $r^2 - \alpha r + 170$ is an integer which is divisible by 2 but not 4, and hence is not a square. But this contradicts the fact that $r^2 - \alpha r + 170$ must be the square of a rational number s.

Problem 133

For what positive numbers x does the series
$$\sum_{n=1}^{\infty}(1 - \sqrt[n]{x}) = (1 - x) + (1 - \sqrt{x}) + (1 - \sqrt[3]{x}) + \cdots$$
converge?

Answer. The series converges only for $x = 1$. Since the convergence for $x = 1$ is obvious, our solutions will assume $x \neq 1$.

Solution 1. We plan to use the limit comparison test, comparing the given series with $\sum_{n=1}^{\infty} \frac{1}{n}$. First we observe that, given x, $1 - \sqrt[n]{x}$ has the same sign for all n. Thus, it is enough to show that $\sum_{n=1}^{\infty}|1 - \sqrt[n]{x}|$ diverges. Writing $\sqrt[n]{x}$ as $e^{(\ln x)/n}$ and applying l'Hôpital's rule, we get
$$\lim_{n\to\infty}\frac{1 - \sqrt[n]{x}}{1/n} = \lim_{n\to\infty}\frac{1 - e^{(\ln x)/n}}{1/n}$$
$$= \lim_{n\to\infty}\frac{-e^{(\ln x)/n}\cdot((-\ln x)/n^2))}{-1/n^2} = -\ln x.$$
Therefore,
$$\lim_{n\to\infty}\frac{|1 - \sqrt[n]{x}|}{1/n} = |\ln x|,$$
which is a finite nonzero number (since $x \neq 1$). Because $\sum_{n=1}^{\infty}\frac{1}{n}$ diverges, so does $\sum_{n=1}^{\infty}|1 - \sqrt[n]{x}|$.

Solution 2. If the original series converges, then so does
$$\sum_{n=1}^{\infty} \frac{1-\sqrt[n]{x}}{1-x}.$$

Since $1-x = 1 - (\sqrt[n]{x})^n$, we have
$$\frac{1-\sqrt[n]{x}}{1-x} = \frac{1}{1+\sqrt[n]{x}+(\sqrt[n]{x})^2+\cdots+(\sqrt[n]{x})^{n-1}}.$$

Suppose first that $x < 1$. Then
$$\frac{1-\sqrt[n]{x}}{1-x} > \frac{1}{1+1+1+\cdots+1} = \frac{1}{n},$$

so since $\sum_{n=1}^{\infty} \frac{1}{n}$ diverges, we conclude that
$$\sum_{n=1}^{\infty}(1-\sqrt[n]{x})$$

also diverges. If, on the other hand, $x > 1$, then
$$\frac{1-\sqrt[n]{x}}{1-x} > \frac{1}{(\sqrt[n]{x})^{n-1}+(\sqrt[n]{x})^{n-1}+(\sqrt[n]{x})^{n-1}+\cdots+(\sqrt[n]{x})^{n-1}} > \frac{1}{nx},$$

and since $\sum_{n=1}^{\infty} \frac{1}{nx} = \frac{1}{x}\sum_{n=1}^{\infty} \frac{1}{n}$ diverges, so does $\sum_{n=1}^{\infty}(1-\sqrt[n]{x})$.

Comments. The comparison with $\sum_{n=1}^{\infty} \frac{1}{n}$ arises naturally in the second solution, but how could anyone have thought to use it in the first solution? One heuristic approach is to use the binomial theorem for noninteger exponents:
$$(1+z)^{1/n} = 1 + \frac{(\frac{1}{n})}{1!}z + \frac{(\frac{1}{n})(\frac{1}{n}-1)}{2!}z^2 + \cdots.$$

For $|z|$ small, we have $(1+z)^{1/n} \approx 1 + z/n$. Because convergence is most likely for $x \approx 1$, we might start by assuming that
$$1 - \sqrt[n]{x} \approx 1 - (1+(x-1)/n) = (1-x)/n.$$

The comparison with $\sum_{n=1}^{\infty} \frac{1}{n}$ is then quite natural.

In the first solution, we considered
$$\lim_{n\to\infty} \frac{1-\sqrt[n]{x}}{1/n}.$$

This limit can also be viewed as the derivative of $f(y) = -x^y = -e^{y\ln x}$ at $y = 0$, which again yields the value $-\ln x$.

Another approach to the problem is to apply the Mean Value Theorem to $f(y)$ to see that $1 - \sqrt[n]{x} = -\frac{1}{n}x^{c_n}\ln x$ for some c_n between 0 and $1/n$.

Since for any $x > 0$ and any n, $x^{c_n} \geq \min\{1, x\}$, we see that $|1 - \sqrt[n]{x}|$ is bounded below by a constant (depending on x) times $1/n$.

Problem 134

For each vertex of a triangle ABC, measure the distance along the opposite side from the midpoint to the "end" of the angle bisector, as a fraction of the total length of that opposite side. This yields a number between 0 and $1/2$; take the least of the three numbers found in this way. What are the possible values of this least number if ABC can be any triangle in the plane?

Answer. The possible values for the "index of scalenity" of a triangle in the plane are the numbers in the interval $[0, \frac{1}{2}\sqrt{5} - 1)$.

Solution. First a useful and standard preliminary result: In any triangle ABC, the angle bisector through C divides the opposite side in the ratio $AC : BC$ of the sides enclosing the angle; that is, in the figure, $\dfrac{AX}{XB} = \dfrac{AC}{BC}$.

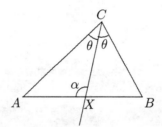

(Using the law of sines, $\dfrac{AX}{XB} = \dfrac{AX/\sin\theta}{XB/\sin\theta} = \dfrac{AC/\sin\alpha}{BC/\sin(\pi-\alpha)} = \dfrac{AC}{BC}$.)

Now consider a triangle ABC with side lengths $a = BC$, $b = AC$ and $c = AB$, where we may assume, by relabeling the vertices if necessary, that $a \leq b \leq c$. Let M be the midpoint of AC and L the end of the angle bisector at B (see figure).

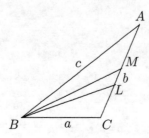

From our previous work,
$$\frac{CL}{LA} = \frac{a}{c}, \quad \text{because } BL \text{ is the angle bisector, so}$$

$$CL = \frac{a}{c} \cdot LA = \frac{a}{c} \cdot (CA - CL)$$
$$CL\left(1 + \frac{a}{c}\right) = \frac{a}{c} \cdot CA$$
$$\frac{CL}{CA} = \frac{a}{c(1 + a/c)} = \frac{a}{c+a} = \frac{1}{1 + c/a}.$$

In particular, because $c/a > 1$, the end of the angle bisector lies on the same side of the median as the shorter side BC. Thus, the quantity we need to compute to find the scalenity is the ratio

$$\frac{CM - CL}{CA} = \frac{\frac{b}{2} - \frac{b}{1 + c/a}}{b} = \frac{1}{2} - \frac{1}{1 + c/a}$$

We get similar expressions when we replace vertex B by A or C, and so we can rephrase our problem as follows:

What are the possible values of

$$\min\left\{\frac{1}{2} - \frac{1}{1 + c/a}, \; \frac{1}{2} - \frac{1}{1 + c/b}, \; \frac{1}{2} - \frac{1}{1 + b/a}\right\},$$

given that $a \leq b \leq c$ and that there exists a triangle with side lengths a, b, c?

Under these conditions, it is enough to find

$$\max\left\{\frac{1}{1 + c/a}, \; \frac{1}{1 + c/b}, \; \frac{1}{1 + b/a}\right\},$$

which in turn corresponds to finding

$$\min\left\{c/a, \; c/b, \; b/a\right\}.$$

Furthermore, we may assume, by suitable scaling, that $a = 1$, so we need to find, under these conditions, the possible values of

$$\min\left\{c, \; c/b, \; b\right\}.$$

Now $c \geq c/b$, so we ask: What is the maximum value of the minimum of c/b and b, as b and c range over all possible values for a triangle of lengths $1, b, c$, where $1 \leq b \leq c$? Because the length of each side of a (non-degenerate) triangle is less than the sum of the lengths of the other two sides, $c < 1 + b$, and therefore $c/b < (1+b)/b = 1 + 1/b$. A graph shows the situation:

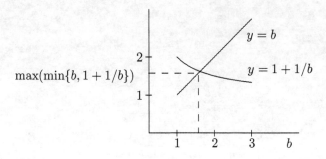

The maximum possible value for the minimum of $\{b, 1+1/b\}$ is at the point where $b = 1 + 1/b$, or $b^2 - b - 1 = 0$ with solution $b = (1+\sqrt{5})/2 = \varphi$ (the golden ratio); we reject the other root $(1-\sqrt{5})/2$ because $b > 0$. For any b, $1 \leq b < \varphi$, there is a c such that $b \leq c/b < 1 + 1/b$ (for $y \in [b, 1+1/b)$, let $c = yb$; then $b \leq b^2 \leq yb = c < (1+1/b)b = 1+b$), so there is a triangle for which $\min\{b, c/b\} = b$. Then, as b ranges from 1 to φ, the scalenity varies from $\frac{1}{2} - \frac{1}{1+1} = 0$ to

$$\frac{1}{2} - \frac{1}{1+\varphi} = \frac{1}{2} - \frac{\varphi-1}{\varphi^2-1} = \frac{1}{2} - \frac{\varphi-1}{\varphi} = \frac{1}{2} - 1 + \frac{1}{\varphi}$$
$$= -\frac{3}{2} + \varphi = -1 + \frac{1}{2}\sqrt{5} \approx 0.118.$$

The "limiting triangle" of maximal scalenity is the degenerate triangle with sides of lengths $a = 1$, $b = \varphi$, $c = a + b = 1 + \varphi = \varphi^2$.

Problem 135

Let R be a commutative ring with at least one, but only finitely many, (nonzero) zero divisors. Prove that R is finite.

Idea. The key observation is that the product of a zero divisor and an arbitrary element of a commutative ring is either 0 or a zero divisor.

Solution. Let $z_1, z_2, \ldots, z_n$ be the zero divisors in R. Let the set S consist of those nonzero elements of R which are not zero divisors. It is enough to show that S is finite. If $s \in S$, then $z_1 s$ is a zero divisor. Hence $z_1 s = z_j$ for some j. On the other hand, for a fixed j, there are at most $(n+1)$ elements $s \in S$ with $z_1 s = z_j$, because if $z_1 s' = z_1 s$, then $z_1(s' - s) = 0$, hence $s' - s$ is either 0 or a zero divisor. Thus, S has at most $n(n+1)$ elements, and the proof is complete.

Comments. We have actually proved that
$$|R| \leq 1 + n + n(n+1) = (n+1)^2.$$
This inequality is sharp for some n: the ring $\mathbb{Z}/p^2\mathbb{Z}$, p prime, has $p-1$ zero divisors.

An analogous result, proved similarly, holds for a noncommutative ring for which the number of left (or right) zero divisors is finite and nonzero.

Problem 136

Consider the transformation of the plane (except for the coordinate axes) defined by sending the point (x, y) to the point $(y + 1/x, x + 1/y)$. Suppose we apply this transformation repeatedly, starting with some specific point (x_0, y_0), to get a sequence of points (x_n, y_n).

a. Show that if (x_0, y_0) is in the first or the third quadrant, the sequence of points will tend to infinity.

b. Show that if (x_0, y_0) is in the second or the fourth quadrant, either the sequence will terminate because it lands at the origin, or the sequence will be eventually periodic with period 1 or 2, or there will be infinitely many n for which (x_n, y_n) is further from the origin than (x_{n-1}, y_{n-1}).

Solution. a. Let $u_n = x_n y_n$. Note that
$$u_{n+1} = \underbrace{(y_n + 1/x_n)}_{x_{n+1}}\underbrace{(x_n + 1/y_n)}_{y_{n+1}} = x_n y_n + 1/(x_n y_n) + 2 = u_n + 1/u_n + 2.$$

In particular, if (x_0, y_0) is in the first or third quadrant, we have $u_0 > 0$, $u_1 = u_0 + 1/u_0 + 2 > u_0 + 2$, $u_2 > u_1 + 2$, etc., so $u_n > 2n$ for all n. Therefore, for any n, at least one of $|x_n|, |y_n|$ is greater than $\sqrt{2n}$, so the distance from (x_n, y_n) to the origin is also greater than $\sqrt{2n}$. Thus that distance tends to ∞ as $n \to \infty$, and we are done.

b. First, a few examples. The point $(1, -1)$ lands at the origin right away. Because the point $(1/\sqrt{2}, -1/\sqrt{2})$ gets sent to itself, the sequence with $(x_0, y_0) = (1/\sqrt{2}, -1/\sqrt{2})$ is periodic of period 1. For $(x_0, y_0) = (1, -2)$, we get $(x_1, y_1) = (-1, 1/2)$, $(x_2, y_2) = (-1/2, 1)$, $(x_3, y_3) = (-1, 1/2)$, etc., so the sequence has become periodic with period 2. We'll see an example later for which (x_n, y_n) is infinitely often further from the origin than (x_{n-1}, y_{n-1}).

The points that land at the origin immediately are the points for which $y + 1/x = x + 1/y = 0$, so they are on the hyperbola $y = -1/x$. Assuming

(x_n, y_n) is never on that hyperbola, note that

$$\frac{y_{n+1}}{x_{n+1}} = \frac{x_n + 1/y_n}{y_n + 1/x_n} = \frac{(x_n y_n + 1)x_n}{(x_n y_n + 1)y_n} = \frac{x_n}{y_n}.$$

This suggests that it may be better to look at the sequences $x_0, y_1, x_2, y_3, \ldots$ and $y_0, x_1, y_1, x_3, \ldots$ rather than at (x_n) and (y_n), because the ratio of their corresponding terms is constant. So let

$$w_n = \begin{cases} x_n & \text{for } n \text{ even,} \\ y_n & \text{for } n \text{ odd,} \end{cases} \quad \text{and} \quad z_n = \begin{cases} y_n & \text{for } n \text{ even,} \\ x_n & \text{for } n \text{ odd.} \end{cases}$$

Then $\dfrac{z_{n+1}}{w_{n+1}} = \dfrac{z_n}{w_n}$ is a nonzero constant $r = y_0/x_0$, and we have

$$z_{n+1} = z_n + \frac{1}{w_n} = z_n + \frac{r}{z_n} \quad \text{for all } n.$$

Also, because $(x_0, y_0) = (w_0, z_0)$ is in the second or the fourth quadrant, r is negative, and all (x_n, y_n) will stay in the second or fourth quadrants. Note that the distance from (x_n, y_n) to the origin is given by

$$\sqrt{x_n^2 + y_n^2} = \sqrt{w_n^2 + z_n^2} = \sqrt{(z_n/r)^2 + z_n^2} = |z_n|\sqrt{1 + 1/r^2}.$$

We now study the behavior of the sequences (z_n) with $z_{n+1} = z_n + r/z_n$ for a fixed $r < 0$. Suppose that z_{n+1} and z_n have the same sign. Then $|z_{n+1}| = |z_n + r/z_n| < |z_n|$ and (x_{n+1}, y_{n+1}) is closer to the origin than (x_n, y_n). Thus (x_{n+1}, y_{n+1}) can be further from the origin than (x_n, y_n) only if z_{n+1} and z_n have different sign. So if $z_k > 0$, then the sequence $z_k, z_{k+1}, \ldots$ will decrease as long as it stays positive. However, it cannot have a limit L, because then we would have $L = L + r/L$, contradiction. Therefore, the sequence must eventually have a negative term. Similarly, if $z_k < 0$, the sequence $z_k, z_{k+1}, \ldots$ will increase as long as it stays negative, but it cannot have a limit, so it will eventually have a positive term.

Now z_{n+1} and z_n have different sign if and only if $|r/z_n| > |z_n|$, that is, when $|z_n| < \sqrt{|r|} = \sqrt{-r}$. (If $|z_n| = \sqrt{-r}$, then (x_n, y_n) is on the hyperbola $y = -1/x$, and (x_{n+1}, y_{n+1}) lands at the origin.) Also, $|z_{n+1}| > |z_n|$ if and only if $|r/z_n| > 2|z_n|$, that is, when $|z_n| < \sqrt{-r/2}$. If $|z_n| = \sqrt{-r/2}$, $z_{n+1} = -z_n$ and $z_{n+2} = z_n$, so from that point the sequence (z_n) is periodic with period 2, and the sequence of points (x_n, y_n) is periodic with period 1 or 2.

If it never happens that $|z_n| = \sqrt{-r}$ or $|z_n| = \sqrt{-r/2}$, then as the sequence (z_n) maintains the same sign (and $|z_n|$ decreases), either it reaches a stage with $|z_n| < \sqrt{-r/2}$, in which case $|z_{n+1}| > |z_n|$ and the distance to the origin increases at the next stage, or the sequence reaches a stage with $\sqrt{-r/2} < |z_n| < \sqrt{-r}$. But in the latter case,

$$|z_{n+1}| = |r/z_n| - |z_n| < -r/\sqrt{-r/2} - \sqrt{-r/2} = \sqrt{-r/2},$$

so $|z_{n+2}| > |z_{n+1}|$ and the distance to the origin increases one stage later. As we have seen that the sequence must change sign infinitely often (unless it terminates), we are done.

Finally, for a specific example in which (x_n, y_n) is infinitely often further from the origin than (x_{n-1}, y_{n-1}), we can take $x_0 = 1$, $y_0 = -1 + 1/\sqrt{2}$, leading to $r = -1 + 1/\sqrt{2}$ and the sequence

$$(z_n) = -1 + 1/\sqrt{2},\ 1/\sqrt{2},\ 1 - 1/\sqrt{2},\ -1/\sqrt{2},\ \ldots$$

of period 4, for which $|z_n|$ alternates between two different values.

Problem 137

Let $(a_n)_{n\geq 0}$ be a sequence of positive integers such that $a_{n+1} = 2a_n + 1$. Is there an a_0 such that the sequence consists only of prime numbers?

Idea. Experimenting with several such sequences, we observe that odd primes in the sequence show up as factors of later terms in the sequence, in fact, in a periodic way. This suggests a connection with Fermat's Little Theorem, which states that if p is a prime and a is not divisible by p, then $a^{p-1} - 1$ is divisible by p.

Solution. There is no such a_0. To see why, first note that $a_{n+1} = 2a_n + 1$ implies $a_{n+1} + 1 = 2(a_n + 1)$ and thus $a_{n+1} + 1 = 2^n(a_1 + 1)$, that is,

$$a_{n+1} = 2^n a_1 + 2^n - 1.$$

Now $a_1 = 2a_0 + 1$ is odd, so if a_1 is also prime, then $2^{a_1-1} - 1$ is divisible by a_1 by Fermat's Little Theorem. Thus, for $n = a_1 - 1$, $a_{n+1} = 2^n a_1 + (2^{a_1-1} - 1)$ is divisible by a_1, and since $a_{n+1} > a_1 > 1$, a_{n+1} cannot be prime.

Problem 138

Suppose $c > 0$ and $0 < x_1 < x_0 < 1/c$. Suppose also that $x_{n+1} = cx_n x_{n-1}$ for $n = 1, 2, \ldots$.

a. Prove that $\lim_{n\to\infty} x_n = 0$.
b. Let $\varphi = (1+\sqrt{5})/2$. Prove that
$$\lim_{n\to\infty} \frac{x_{n+1}}{x_n^\varphi}$$
exists, and find it.

Solution. (Joe Buhler, Reed College)

a. Define a new sequence by $y_n = cx_n$. Then $0 < y_1 < y_0 < 1$ and $y_{n+1} = y_n y_{n-1}$. An easy induction argument shows that $y_n < y_0^n$ for all $n > 0$, so $y_n \to 0$ and $x_n \to 0$ as $n \to \infty$.

b. Define $z_n = \dfrac{y_{n+1}}{y_n^\varphi} = c^{1-\varphi} \dfrac{x_{n+1}}{x_n^\varphi}$. We will find $\lim_{n\to\infty} z_n$. Since $y_{n+1} = y_n y_{n-1}$ and $\varphi(\varphi-1) = 1$, we have

$$z_n = \frac{y_{n+1}}{y_n^\varphi} = \frac{y_{n-1}}{y_n^{\varphi-1}} = \left(\frac{y_n}{y_{n-1}^\varphi}\right)^{1-\varphi} = z_{n-1}^{1-\varphi}.$$

By induction, $z_n = z_0^{(1-\varphi)^n}$, and since $|1-\varphi| < 1$, we have $\lim_{n\to\infty} z_n = 1$. Therefore,

$$\lim_{n\to\infty} \frac{x_{n+1}}{x_n^\varphi} = c^{\varphi-1}.$$

Comments. A straightforward induction shows that $y_{n+1} = y_1^{F_{n+1}} y_0^{F_n}$, where F_n denotes the nth Fibonacci number. ($F_0 = 0$, $F_1 = 1$, and for $n \geq 2$, $F_n = F_{n-1} + F_{n-2}$.)

This problem was sparked by the rate of convergence of the secant method in numerical analysis.

Problem 139

Consider the number of solutions to the equation $\sin x + \cos x = \alpha \tan x$ for $0 \leq x \leq 2\pi$, where α is an unspecified real number. What are the possibilities for this number of solutions, as α is allowed to vary?

Answer. The number of solutions can be $2, 3$, or 4.

Solution. For $\alpha = 0$ the equation is $\sin x + \cos x = 0$, or $\tan x = -1$, with two solutions, $x = 3\pi/4$ and $7\pi/4$. So assume $\alpha \neq 0$. Because

$$\sin x + \cos x = \sqrt{2}\left(\cos\left(\frac{\pi}{4}\right)\sin x + \sin\left(\frac{\pi}{4}\right)\cos x\right) = \sqrt{2}\sin\left(x + \frac{\pi}{4}\right),$$

we can write our equation as

$$\frac{\sqrt{2}}{\alpha}\sin\left(x+\frac{\pi}{4}\right)=\tan x.$$

Let $f(x) = \frac{\sqrt{2}}{\alpha}\sin\left(x+\frac{\pi}{4}\right)$ and $g(x) = \tan x$, and begin with the case $\alpha > 0$. Note that $f(0) > g(0)$ and $\lim_{x\to\pi/2^-} g(x) = \infty$, so, because f is concave down and g is concave up on the interval $[0, \pi/2)$, the graphs of f and g will intersect exactly once in this interval. Next, note that $\lim_{x\to\pi/2^+} g(x) = -\infty$ and $f(5\pi/4) < 0 < g(5\pi/4)$, so, because $f(x)$ is decreasing and $g(x)$ is increasing on the interval $(\pi/2, 5\pi/4)$, the graphs will intersect exactly once in this interval. There are no intersections in the interval $[5\pi/4, 3\pi/2)$, because $f(x)$ is negative on that interval whereas $g(x)$ is positive. We conclude that the two graphs intersect exactly twice in the interval $[0, 3\pi/2]$.

Finally, from $\lim_{x\to 3\pi/2^+} g(x) = -\infty$ and the fact that f is concave up and g is concave down on the interval $(3\pi/2, 7\pi/4]$ we conclude that the graphs of f and g might intersect 0, 1, or 2 times in the interval $(3\pi/2, 2\pi]$. (On the interval $(7\pi/4, 2\pi)$, f is positive and g is negative.) And each of these can happen. For example, for sufficiently small α the amplitude of f can be made arbitrarily large, and for large amplitudes, as illustrated in the following graph for $\alpha = \alpha_1$, the graphs will intersect twice. For large α the amplitude can be made arbitrarily small, and for small amplitudes, as illustrated in the graph for $\alpha = \alpha_2$, the graphs will not intersect at all. Because the amplitude of f is a continuous function of α, there will be an α, $\alpha_1 < \alpha < \alpha_2$, for which the graphs of f and g will be tangent and therefore intersect exactly once in the interval $(3\pi/2, 2\pi]$.

In summary, as $\alpha > 0$ varies from 0 to ∞, the number of intersections on $[0, 2\pi]$ will go from 4 to 3 to 2.

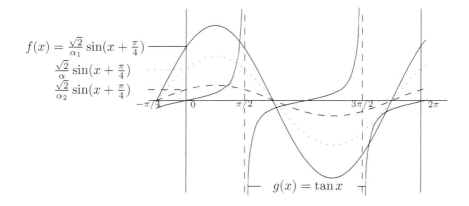

One could use a similar argument for $\alpha < 0$, or note that because both f and g have 2π as a period, we can shift the interval for x to *any* interval of length 2π as long as we check that there are no solutions at the endpoints. If we shift to the interval $[\pi, 3\pi]$, we are essentially replacing x by $x + \pi$, which will change $f(x)$ to its opposite while leaving $g(x)$ unchanged. That is, we essentially get the same equation with α replaced by $-\alpha$. So the number of solutions on our interval of length 2π will be the same for α as for $-\alpha$, and this completes the proof.

Solution 2. We've seen there are two solutions when $\alpha = 0$, so henceforth we assume that $\alpha \neq 0$.

We set $u = \tan(x/2)$ and substitute the half-angle formulas

$$\cos x = 2\cos^2(x/2) - 1 = \frac{2 - \sec^2(x/2)}{\sec^2(x/2)} = \frac{1 - \tan^2(x/2)}{1 + \tan^2(x/2)} = \frac{1 - u^2}{1 + u^2},$$

$$\sin x = 2\sin(x/2)\cos(x/2) = \frac{2\tan(x/2)}{\sec^2(x/2)} = \frac{2\tan(x/2)}{1 + \tan^2(x/2)} = \frac{2u}{1 + u^2},$$

$$\tan x = \frac{\sin x}{\cos x} = \frac{2u}{1 - u^2}$$

into the equation. Clearing denominators we get $p(u) = 0$, where the polynomial p is given by $p(u) = u^4 - (2 + 2\alpha)u^3 - 2u^2 + (2 - 2\alpha)u + 1$.

We observe that $x = 0, \pi/2, \pi, 3\pi/2, 2\pi$ are never solutions of the original equation. Also, $u = 0$ is never a root of p. Because $p(\pm 1) = \mp 4\alpha$, each of $u = 1$ and $u = -1$ is a root if and only if $\alpha = 0$, so (for $\alpha \neq 0$) we didn't introduce new roots by clearing the denominator $1 - u^2$. As x runs through $(0, 2\pi)$, u takes on every nonzero value exactly once, so the solutions of the original equation are in one-to-one correspondence with the roots of p.

Obviously, p has at most four roots. If $\alpha > 0$, then $p(1) < 0$ and because $p(0) = 1$ and $\lim_{u \to \infty} p(u) = \infty$, p has at least two positive roots. If $\alpha < 0$, then $p(-1) < 0$ and so p has at least two negative roots. Because the original equation has two solutions when $\alpha = 0$, we need only show that it is possible for p to have either three or four roots.

For $\alpha = 1/5$, $p(-3/5) = -4/125$, so p has two negative roots, for a total of four.

Because p changes sign at least twice and $\lim_{u \to -\infty} p(u) = \lim_{u \to \infty} p(u) = \infty$, if p has a double (but not triple) root u, it will have exactly three roots. We'll show explicitly that this can happen.

A double root u is a solution to $p(u) = p'(u) = 0$ for which $p''(u) \neq 0$. This means

$$u^4 - 2u^3 - 2u^2 + 2u + 1 - (2u^3 + 2u)\alpha = 0$$

and
$$4u^3 - 6u^2 - 4u + 2 - (6u^2 + 2)\alpha = 0$$
while
$$12u^2 - 12u - 4 - 12u\alpha \neq 0.$$
Solving for α, this is equivalent to
$$\alpha = \frac{u^4 - 2u^3 - 2u^2 + 2u + 1}{2u^3 + 2u} = \frac{4u^3 - 6u^2 - 4u + 2}{6u^2 + 2}$$
while $\alpha \neq u - 1 - 1/(3u)$. Clearing denominators and simplifying, we see that u must satisfy
$$q(u) \equiv u^6 + 5u^4 - 8u^3 - 5u^2 - 1 = 0.$$
This polynomial has real roots in each of the intervals $(-1, 0)$ and $(1, 2)$. It can be checked that each of these roots u gives rise to a value of α with $\alpha \neq u - 1 - 1/(3u)$, and so there are two values of α for which p has three roots.

Comment. Although $q(u)$ is irreducible over the rationals, it is possible to solve for the roots, and hence the α for which there are three solutions, in terms of radicals. We note that $u^6 q(-1/u) = -q(u)$. Thus, if u is a root, so is $-1/u$. We can write $q(u)/u^3$ in the form
$$(u - 1/u)^3 + 8(u - 1/u) - 8.$$
Using Cardano's formula for the roots of a cubic, we can write the one real root as
$$u - 1/u = \sqrt[3]{4 + \frac{4}{9}\sqrt{177}} - \frac{8}{3}\frac{1}{\sqrt[3]{4 + \frac{4}{9}\sqrt{177}}} = \sqrt[3]{4 + \frac{4}{9}\sqrt{177}} + \sqrt[3]{4 - \frac{4}{9}\sqrt{177}}.$$
We then solve the quadratic
$$u^2 - \left(\sqrt[3]{4 + \frac{4}{9}\sqrt{177}} + \sqrt[3]{4 - \frac{4}{9}\sqrt{177}}\right)u - 1 = 0.$$

Problem 140

Given 64 points in the plane which are positioned so that 2001, but no more, distinct lines can be drawn through pairs of points, prove that at least four of the points are collinear.

Solution. First note that there are

$$\binom{64}{2} = \frac{64 \cdot 63}{2} = 2016$$

ways to choose a pair of points from among the 64 given points. Since there are only 2001 distinct lines through pairs of points, it is clear that some three of the points must be collinear.

Now suppose that no four of the points are collinear. Then every line which connects a pair of the points either contains exactly two or exactly three of the points. If we count lines by counting distinct pairs of points, we will count each line containing exactly two of the points once (which is correct) and each line containing exactly three of the points, say A, B, C, three times (as AB, AC, BC). So to get the correct line count we should subtract 2 for every line containing three of the points. That is, if there are n such lines, then $2001 = 2016 - n \cdot 2$. However, this yields $n = 7\frac{1}{2}$, and n must be an integer, a contradiction.

Comment. One way to create an arrangement satisfying the conditions of the problem is to take four points on each of three lines and 52 other points, while making sure that no other collinearity occurs.

Problem 141

Let $f_1, f_2, \ldots, f_n$ be linearly independent, differentiable functions. Prove that some $n-1$ of their derivatives $f_1', f_2', \ldots, f_n'$ are linearly independent.

Solution 1. If $f_1', f_2', \ldots, f_n'$ are linearly dependent, then there is some relation of the form

$$a_1 f_1' + a_2 f_2' + \cdots + a_n f_n' = 0,$$

where we may assume $a_n \neq 0$. We will show that $f_1', f_2', \ldots, f_{n-1}'$ are then linearly independent. First observe that integrating each side of the preceding equation yields

$$a_1 f_1 + a_2 f_2 + \cdots + a_n f_n = C,$$

where the constant C is nonzero by the linear independence of $f_1, f_2, \ldots, f_n$. If

$$b_1 f_1' + b_2 f_2' + \cdots + b_{n-1} f_{n-1}' = 0,$$

then
$$b_1 f_1 + b_2 f_2 + \cdots + b_{n-1} f_{n-1} = D,$$
where D is some constant. We then have
$$(Da_1 - Cb_1) f_1 + (Da_2 - Cb_2) f_2 + \cdots + (Da_{n-1} - Cb_{n-1}) f_{n-1} + Da_n f_n = 0.$$
The linear independence of $f_1, f_2, \ldots, f_n$ forces $Da_n = 0$, hence $D = 0$. Therefore, $b_i = 0$ for all i, and $f'_1, f'_2, \ldots, f'_{n-1}$ are linearly independent.

Solution 2. (William C. Waterhouse, Pennsylvania State University) The map taking f to f' is a linear transformation whose kernel (null space) is the 1-dimensional space consisting of the constant functions. The image of any n-dimensional space of differentiable functions must therefore have dimension at least $n - 1$.

Problem 142

Find all real numbers A and B such that
$$\left| \int_1^x \frac{1}{1+t^2} dt - A - \frac{B}{x} \right| < \frac{1}{3x^3}$$
for all $x > 1$.

Answer. The only pair of real numbers which satisfy the given condition is $(A, B) = (\pi/4, -1)$.

If distinct pairs (A, B) and (A', B') satisfied the condition, we would have
$$\left| A - A' + \frac{B - B'}{x} \right| \le \left| \int_1^x \frac{1}{1+t^2} dt - A - \frac{B}{x} \right| + \left| \int_1^x \frac{1}{1+t^2} dt - A' - \frac{B'}{x} \right|$$
$$< \frac{2}{3x^3}$$
for all $x > 1$, clearly an impossibility. We therefore focus our attention on finding A and B.

Solution 1. We have
$$\int_1^x \frac{1}{1+t^2} dt = \arctan x - \frac{\pi}{4} = \left(\frac{\pi}{2} - \arctan \frac{1}{x} \right) - \frac{\pi}{4} = \frac{\pi}{4} - \arctan \frac{1}{x}.$$
For $|1/x| < 1$, the Taylor series expansion for the inverse tangent implies
$$\arctan \frac{1}{x} = \frac{1}{x} - \frac{1}{3x^3} + \frac{1}{5x^5} - \frac{1}{7x^7} + \cdots .$$

This series meets the conditions of the alternating series test, hence

$$\left|\arctan\frac{1}{x} - \frac{1}{x}\right| < \left|\frac{1}{3x^3}\right|.$$

Combining our results, we obtain

$$\left|\int_1^x \frac{1}{1+t^2}\,dt - \frac{\pi}{4} + \frac{1}{x}\right| < \frac{1}{3x^3}$$

for $x > 1$.

Solution 2. Because $\arctan t$ is an antiderivative of $1/(1+t^2)$, we have

$$\int_1^\infty \frac{dt}{1+t^2} = \frac{\pi}{4},$$

and therefore

$$\int_1^x \frac{1}{1+t^2}\,dt = \frac{\pi}{4} - \int_x^\infty \frac{1}{1+t^2}\,dt$$

$$= \frac{\pi}{4} - \int_x^\infty \frac{1}{t^2}\,dt - \int_x^\infty \left(\frac{1}{1+t^2} - \frac{1}{t^2}\right)dt$$

$$= \frac{\pi}{4} - \frac{1}{x} + \int_x^\infty \frac{1}{(1+t^2)t^2}\,dt.$$

We can estimate this last integral by

$$0 < \int_x^\infty \frac{1}{(1+t^2)t^2}\,dt < \int_x^\infty \frac{1}{t^4}\,dt = \frac{1}{3x^3},$$

and so we have

$$\left|\int_1^x \frac{dt}{1+t^2} - \frac{\pi}{4} + \frac{1}{x}\right| < \frac{1}{3x^3}$$

for all $x > 1$.

Solution 3. As in the second solution, our starting point is

$$\int_1^x \frac{1}{1+t^2}\,dt = \frac{\pi}{4} - \int_x^\infty \frac{1}{1+t^2}\,dt.$$

Converting this latter integral into a series yields

$$\int_x^\infty \frac{1}{1+t^2}\,dt = \int_x^\infty \frac{1}{t^2}\frac{1}{1+1/t^2}\,dt = \int_x^\infty \frac{1}{t^2}\sum_{k=0}^\infty (-1)^k t^{-2k}\,dt.$$

Because $\sum_{k=0}^{\infty}(-1)^k t^{-2k}$ converges absolutely and uniformly on $[x,\infty)$, we can switch integration and summation to get

$$\int_x^\infty \frac{1}{t^2} \sum_{k=0}^{\infty}(-1)^k t^{-2k}\, dt = \sum_{k=0}^{\infty} \frac{(-1)^k}{2k+1} x^{-(2k+1)} = \frac{1}{x} - \frac{1}{3x^3} + \frac{1}{5x^5} - \cdots,$$

and we can proceed as in the first solution.

Problem 143

a. Prove that there is no closed knight's tour of the chessboard which is symmetric under a reflection in one of the main diagonals of the board.
b. Prove that there is no closed knight's tour of the chessboard which is symmetric under a reflection in the horizontal axis through the center of the board.

Solution. a. A closed knight's path, symmetric about a diagonal, can pass through at most two diagonal squares on the axis of reflection. For let A be a diagonal square in the path, and imagine simultaneously traversing the route, in opposite directions, starting at A. The two paths thus traversed are symmetric to each other, so once one of them encounters another diagonal square, so will the other, and the two paths will close to form a closed loop. Since there are eight diagonal squares on the axis of symmetry, there can be no such knight's tour. (In fact, any such tour would require at least four loops.)

b. Consider a closed path of knight moves that is symmetric about the horizontal axis through the center of the board. Let A denote a square in this path, and let α denote the sequence of moves that goes from A to its mirror image A' (in one of the directions of the closed path). A and A' will be squares of opposite colors (since squares on opposite sides of the horizontal center line have opposite colors), so α consists of an odd number of moves.

None of the individual moves in this "half-loop" from A to A' are symmetric to each other (with respect to the horizontal center line). For if there were such a move, then the moves contiguous (immediately preceding and immediately succeeding) to such a move would also be symmetric, and continuing this argument, the mirror image of each move in α would itself be in α. But this would imply that α is made up of an even number of moves, contradicting the conclusion of the preceding paragraph.

Thus, the path α from A to A' will reflect into a path α' from A' to A. The number of moves in the resulting closed path (α followed by α') is twice an odd number. Conclusion: Any closed path, symmetric about the horizontal axis, must contain twice an odd number of moves. Since 64 is not twice an odd number, no such closed path is possible on the ordinary 8×8 chessboard.

Comments. Similarly, there can be no knight's tour symmetric about a vertical axis through the center.

To complete the analysis of symmetric knight's tours, we sketch a proof that there can be no closed knight's tour which is preserved by a 90° rotation about the center. Consider a closed knight's path which is preserved by such a rotation. Pick a square, A, in the path; let A', A'', and A''' be the squares obtained by rotating A through 90°, 180°, and 270°, respectively, about the center. Let α be the sequence of moves from A to A' which does not pass through A''. Since A and A' have opposite color, α consists of an odd number of moves. The result of rotating α through 90°, say α', connects A' to A''; similarly, α'', α''' connect A'' to A''', A''' to A respectively. The four segments α, α', α'', α''' are disjoint and together form the closed knight's path. Thus the total number of moves is four times an odd number, so it cannot be 64 and the path cannot be a knight's tour. (There *are* closed knight's tours on a 6×6 board which are preserved by a 90° rotation; you might enjoy finding one.)

Problem 144

A function f on the rational numbers is defined as follows. Given a rational number $x = \frac{m}{n}$, where m and n are relatively prime integers and $n > 0$, set $f(x) = \frac{3m-1}{2n+1}$. Now, starting with a rational number x_0, apply f repeatedly to get an infinite sequence $x_1 = f(x_0)$, $x_2 = f(x_1)$, ..., $x_{n+1} = f(x_n)$, ... Find all rational x_0 for which that sequence is periodic.

THE SOLUTIONS   251

Answer. The sequence x_0, x_1, x_2, ... is periodic for $x_0 = \frac{2}{7}$, $x_0 = \frac{1}{3}$, $x_0 = \frac{5}{3}$, and $x_0 = 2$, and for no other rational number x_0.

Solution. First, let's compare x and $f(x)$ to see which is larger for various x, and whether x and $f(x)$ might be equal. Because n and $2n+1$ are positive,

$$f(x) > x \iff \frac{3m-1}{2n+1} > \frac{m}{n} \iff n(3m-1) > m(2n+1)$$
$$\iff mn > m+n$$
$$\iff (m-1)(n-1) > 1.$$

Similarly, $f(x) < x$ if and only if $(m-1)(n-1) < 1$, and $f(x) = x$ if and only if $(m-1)(n-1) = 1$. Thus $f(x) = x$ if and only if $m-1 = n-1 = 1$ or $m-1 = n-1 = -1$, that is, $m = n = 2$ or $m = n = 0$. But 0 is unacceptable as a denominator and $2/2$ is not in lowest terms, so $f(x) \neq x$ for all x.

If $x = \frac{m}{n} < 0$, then $m-1 < 0$ and $n-1 \geq 0$, so $(m-1)(n-1) \leq 0$ and $f(x) < x$. So starting with any negative rational x_0, the sequence (x_i) will be decreasing. Also, if $x_0 = 0 = \frac{0}{1}$, then $x_1 = \frac{-1}{3}$ so the sequence decreases. In particular, for $x_0 \leq 0$ the sequence cannot be periodic.

Now suppose that $x = \frac{m}{n} > 0$. Then $m-1 \geq 0$ and $n-1 \geq 0$, so $(m-1)(n-1) < 1$ if and only if at least one of $m-1$, $n-1$ is zero. That is, $f(x) > x$ unless $m = 1$ or $n = 1$, in which case $f(x) < x$.

So if the sequence x_0, x_1, x_2, ... is going to be periodic, then somewhere in the sequence a rational number with $m = 1$ or $n = 1$ has to occur, else the sequence would be increasing.

Suppose $x_i = \frac{1}{n}$. Then $x_{i+1} = f(x_i) = \frac{3 \cdot 1 - 1}{2n+1} = \frac{2}{2n+1}$, and $\frac{2}{2n+1}$ is in lowest terms because $2n+1$ is odd. So the next term in the sequence is $x_{i+2} = \frac{3 \cdot 2 - 1}{2(2n+1)+1} = \frac{5}{4n+3}$. Now $x_i = \frac{1}{n} < \frac{5}{4n+3} = x_{i+2}$ if and only if $4n+3 < 5n$, or $3 < n$. So, if $n > 3$, the sequence x_i, x_{i+2}, x_{i+4}, ... will keep increasing unless an integer m or a number $\frac{1}{n_1}$ with $n_1 \leq 3$ occurs in the sequence. So if x_0, x_1, x_2, ... is to be periodic, for some i we must have that either $x_i = \frac{1}{n}$ with $n \leq 3$ or x_i is a positive integer.

For $n = 3$ we do get a periodic sequence from $x_i = \frac{1}{3}$ on:

$$x_i = \frac{1}{3}, \quad x_{i+1} = \frac{2}{7}, \quad x_{i+2} = \frac{5}{15} = \frac{1}{3}, \quad x_{i+3} = \frac{2}{7}, \quad \ldots .$$

So if $x_0 = \frac{1}{3}$ or $x_0 = \frac{2}{7}$, we get a periodic sequence.

If we have $x_i = \frac{1}{2}$, then $x_{i+1} = \frac{2}{5}$, $x_{i+2} = \frac{5}{11}$, $x_{i+3} = \frac{14}{23}$ and now $x_i < x_{i+3}$, so x_i, x_{i+3}, x_{i+6}, ... is increasing unless we encounter an integer somewhere (because $\frac{14}{23}$ is already greater than $\frac{1}{m}$ for all $m \geq 2$).

Now suppose $x_i = m$. Then $x_{i+1} = \frac{3m-1}{3}$, and this is in lowest terms, so $x_{i+2} = \frac{3(3m-1)-1}{2 \cdot 3 + 1} = \frac{9m-4}{7}$. Comparing x_{i+2} to x_i, we see that $m < \frac{9m-4}{7}$ if and only if $7m < 9m-4$, or $m > 2$. Thus if $m > 2$, the sequence $x_i = m$, x_{i+2}, x_{i+3}, x_{i+4}, ... will be increasing unless and until another integer is encountered, which will be greater than m; therefore, if $m > 2$ the sequence (x_i) cannot be periodic.

If $m = 1$, we have $x_i = 1, x_{i+1} = \frac{2}{3}$, $x_{i+2} = \frac{5}{7}$, $x_{i+3} = \frac{14}{15}$, $x_{i+4} = \frac{41}{31}$, so x_i, x_{i+4}, x_{i+5}, ... is increasing unless and until another integer is encountered. So if the sequence (x_i) is to be periodic for $m = 1$, it must also contain 2. On the other hand, for $m = 2$, $\frac{9m-4}{7} = 2$ also, so we do get a periodic sequence 2, $\frac{5}{3}$, 2, $\frac{5}{3}$, ... This confirms the answer as given.

Comment. It might be interesting to investigate the set of positive rationals x_0 that lead to sequences that eventually become periodic. For example,

$$\frac{17}{107} \longrightarrow \frac{50}{215} = \frac{10}{43} \longrightarrow \frac{29}{87} = \frac{1}{3} \longrightarrow \frac{2}{7} \longrightarrow \frac{1}{3} \longrightarrow \cdots$$

$$\frac{5}{24} \longrightarrow \frac{14}{49} = \frac{2}{7};$$

$$\frac{17}{12} \longrightarrow \frac{50}{25} = 2;$$

$$\frac{23}{36} \longrightarrow \frac{68}{73} \longrightarrow \frac{203}{147} = \frac{29}{21} \longrightarrow \frac{86}{43} = 2;$$

$$\frac{22}{19} \longrightarrow \frac{65}{39} = \frac{5}{3}.$$

Problem 145

a. Find a sequence (a_n), $a_n > 0$, such that

$$\sum_{n=1}^{\infty} \frac{a_n}{n^3} \quad \text{and} \quad \sum_{n=1}^{\infty} \frac{1}{a_n}$$

both converge.

THE SOLUTIONS

b. Prove that there is no sequence (a_n), $a_n > 0$, such that

$$\sum_{n=1}^{\infty} \frac{a_n}{n^2} \quad \text{and} \quad \sum_{n=1}^{\infty} \frac{1}{a_n}$$

both converge.

Solution. a. Let $a_n = n^{3/2}$.

b. Suppose there were such a sequence (a_n). Then, since

$$\sum_{n=1}^{\infty} \frac{a_n}{n^2} \quad \text{and} \quad \sum_{n=1}^{\infty} \frac{1}{a_n}$$

would both converge, so would

$$\sum_{n=1}^{\infty} \left(\frac{a_n}{n^2} + \frac{1}{a_n} \right).$$

Now note that

$$\frac{a_n}{n^2} + \frac{1}{a_n} = \left(\frac{\sqrt{a_n}}{n} - \frac{1}{\sqrt{a_n}} \right)^2 + \frac{2}{n} \geq \frac{2}{n},$$

so by the comparison test for positive series, $\sum \frac{2}{n}$ would converge, a contradiction.

Comment. Alternatively, one could observe that if $a_n \geq n$, then

$$\frac{a_n}{n^2} \geq \frac{1}{n},$$

and if $a_n \leq n$, then

$$\frac{1}{a_n} \geq \frac{1}{n}.$$

Hence, for all n,

$$\frac{a_n}{n^2} + \frac{1}{a_n} \geq \max\left\{ \frac{a_n}{n^2}, \frac{1}{a_n} \right\} \geq \frac{1}{n}.$$

Also, note that the condition $a_n > 0$ is needed, as shown by the example $a_n = (-1)^n n$.

Problem 146

What is the limit of the repeated power $x^{x^{\cdot^{\cdot^{\cdot^{x}}}}}$ with n occurrences of x, as x approaches zero from above?

Answer. The limit is 0 for n odd, 1 for n even.

Solution 1. Clearly, $\lim_{x\to 0^+} x = 0$, and it is a standard calculus exercise to show that $\lim_{x\to 0^+} x^x = 1$. For $n = 3$, we have $\lim_{x\to 0^+} x^{(x^x)} = 0^1 = 0$ and so we meet the first real problem when $n = 4$.

If we use the standard approach and set $z = x^{\left(x^{x^x}\right)}$ and proceed to rewrite

$$\ln z = x^{(x^x)} \ln x = \frac{\ln x}{x^{-(x^x)}},$$

we find that l'Hôpital's rule produces a mess. Instead, we look at the size of $\ln z$. Since $\lim_{x\to 0^+} x^x = 1$, we know that for (positive) x close enough to 0, $\frac{1}{2} < x^x < 1$. From this we get $x^{1/2} > x^{x^x} > x$ for x close enough to 0, and so $x^{1/2} \ln x < x^{x^x} \ln x < x \ln x$ for x close enough to 0. It is not hard to show by l'Hôpital's rule that $\lim_{x\to 0^+} x^{1/2} \ln x = \lim_{x\to 0^+} x \ln x = 0$. Thus, $\ln z = x^{x^x} \ln x$ is "squeezed" between $x^{1/2} \ln x$ and $x \ln x$, and therefore, $\lim_{x\to 0^+} z = e^0 = 1$.

For $n = 5$ we do not have an indeterminate form, and we have

$$\lim_{x\to 0^+} x^{\left(x^{x^{x^x}}\right)} = 0^1 = 0.$$

Now that we have discovered the pattern, we show that it will continue by using induction on k to prove that the limit is 0 for $n = 2k - 1$ and 1 for $n = 2k$. We have seen this for $k = 1$. If it is true for k, then

$$\frac{1}{2} < \underbrace{x^{x^{\cdot^{\cdot^{\cdot^x}}}}}_{2k \ x\text{'s}} < 1$$

for x close enough to 0, hence

$$\lim_{x\to 0^+} \underbrace{x^{x^{\cdot^{\cdot^{\cdot^x}}}}}_{(2k+1) \ x\text{'s}} = 0$$

and even, as in the case $k = 1$ above,

$$\lim_{x\to 0^+} \underbrace{x^{x^{\cdot^{\cdot^{\cdot^x}}}}}_{(2k+1) \ x\text{'s}} \ln x = 0.$$

This implies

$$\lim_{x\to 0^+} \underbrace{x^{x^{\cdot^{\cdot^{\cdot^x}}}}}_{(2k+2) \ x\text{'s}} = 1,$$

completing the proof.

THE SOLUTIONS

Solution 2. We actually prove the following stronger version:

$$\underbrace{x^{x^{\cdot^{\cdot^{\cdot^{x}}}}}}_{n\ x\text{'s}} < x^{1/2} \quad \text{if } n \text{ is odd and } x \text{ is sufficiently small,}$$

while

$$\lim_{x \to 0^+} \underbrace{x^{x^{\cdot^{\cdot^{\cdot^{x}}}}}}_{n\ x\text{'s}} = 1 \quad \text{if } n \text{ is even.}$$

The proof is by induction on n; the case $n = 1$ is clear. Suppose our claim is true for $n - 1$. If n is odd, then

$$\underbrace{x^{x^{\cdot^{\cdot^{\cdot^{x}}}}}}_{(n-1)\ x\text{'s}} > 1/2$$

for x sufficiently small, and the assertion for n follows. If n is even, then, for x sufficiently small,

$$x^{x^{1/2}} < \underbrace{x^{x^{\cdot^{\cdot^{\cdot^{x}}}}}}_{n\ x\text{'s}} < 1.$$

An easy application of l'Hôpital's rule to $\lim_{x \to 0^+} x^{1/2} \ln x$ yields

$$\lim_{x \to 0^+} x^{x^{1/2}} = 1,$$

hence

$$\lim_{x \to 0^+} \underbrace{x^{x^{\cdot^{\cdot^{\cdot^{x}}}}}}_{n\ x\text{'s}} = 1$$

as desired. This completes the induction step, and the proof is complete.

Problem 147

Show that there exist an integer N and a rational number r such that $\sum_{n=2012}^{\infty} \dfrac{(-1)^n}{\binom{n}{2012}} = N \ln 2 + r$, and find the integer N.

Idea. If we were to replace 2012 by 1, by a well-known formula the sum would become $\sum_{n=1}^{\infty} \dfrac{(-1)^n}{n} = -\sum_{n=1}^{\infty} \dfrac{(-1)^{n+1}}{n} = -\ln 2$, which would be a good start if we could find a recurrence relation for $S_k = \sum_{n=k}^{\infty} \dfrac{(-1)^n}{\binom{n}{k}}$.

Answer. $\sum_{n=2012}^{\infty} \frac{(-1)^n}{\binom{n}{2012}} = 1006 \cdot 2^{2012} \ln 2 + r$, where r is a rational number.

Solution. Continuing with the idea above, we note that

$$S_2 = \sum_{n=2}^{\infty} \frac{(-1)^n}{\binom{n}{2}} = \sum_{n=2}^{\infty} \frac{2 \cdot (-1)^n}{n(n-1)} = \sum_{n=2}^{\infty} 2 \cdot (-1)^n \left(\frac{1}{n-1} - \frac{1}{n}\right)$$

$$= 2\left(1 - \frac{1}{2} - \frac{1}{2} + \frac{1}{3} + \frac{1}{3} - \frac{1}{4} - \frac{1}{4} + \frac{1}{5} + \cdots\right)$$

$$= 2\left(-1 + 2\left(1 - \frac{1}{2} + \frac{1}{3} - \frac{1}{4} + \cdots\right)\right)$$

$$= 2(-1 - 2S_1) = 2(-1 + 2\ln 2) = 4\ln 2 - 2.$$

In general, for $k > 1$, we have

$$S_k = \sum_{n=k}^{\infty} \frac{(-1)^n k!}{n(n-1)\cdots(n-k+1)}$$

$$= \sum_{n=k}^{\infty} \frac{(-1)^n k}{k-1}\left[\frac{(k-1)!}{(n-1)\cdots(n-k+1)} - \frac{(k-1)!}{n(n-1)\cdots(n-k+2)}\right]$$

$$= \sum_{n=k}^{\infty} \frac{(-1)^n k}{k-1}\left[\frac{1}{\binom{n-1}{k-1}} - \frac{1}{\binom{n}{k-1}}\right]$$

$$= \frac{k}{k-1}\left[-\sum_{n=k}^{\infty} \frac{(-1)^{n-1}}{\binom{n-1}{k-1}} - \sum_{n=k}^{\infty} \frac{(-1)^n}{\binom{n}{k-1}}\right]$$

$$= \frac{k}{k-1}[-S_{k-1} - (S_{k-1} - (-1)^{k-1})] = \frac{k}{k-1}[(-1)^{k-1} - 2S_{k-1}].$$

It follows by induction that $S_k = a_k \ln 2 + r_k$ with a_k and r_k rational. Specifically, we can start with $a_1 = -1$ and $r_1 = 0$, and for $k > 1$, we can take

$$a_k = -\frac{2k}{k-1} a_{k-1}$$

and

$$r_k = \frac{(-1)^{k-1} k}{k-1} - \frac{2k}{k-1} r_{k-1}.$$

THE SOLUTIONS   257

Clearly, a_k, r_k are rational if a_{k-1}, r_{k-1} are. Furthermore,

$$a_k = \left(-\frac{2k}{k-1}\right)\left(-\frac{2(k-1)}{k-2}\right) a_{k-2} = \cdots$$
$$= \left(-\frac{2k}{k-1}\right)\left(-\frac{2(k-1)}{k-2}\right)\cdots\left(-\frac{2(2)}{1}\right) a_1$$
$$= (-2)^{k-1} k\, a_1 = (-1)^k\, k\, 2^{k-1},$$

an integer. In particular, $a_{2012} = 1006 \cdot 2^{2012}$.

Problem 148

A new subdivision is being laid out on the outskirts of Wohascum Center. There are ten north–south streets and six east–west streets, forming blocks which are exactly square. The Town Council has ordered that fire hydrants be installed at some of the intersections, in such a way that no intersection will be more than two "blocks" (really sides of blocks) away from an intersection with a hydrant. (The two blocks need not be in the same direction.) What is the smallest number of hydrants that could be used?

Answer. The smallest number of hydrants required is seven. In fact, seven hydrants suffice even if an additional north–south road is added, as shown in the following map.

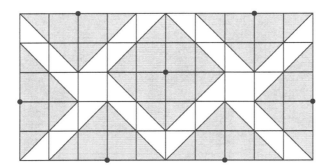

Solution. To show that six hydrants will not do for the 6×10 grid, consider that there are 28 "edge" vertices (vertices on the outside boundary of the grid) and 12 "center" vertices (vertices 2 "blocks" away from the boundary).

The following table indicates how many edge vertices and center vertices are "covered" by hydrants in positions labeled A, B, C.

A	A	A	A	A	A	A	A	A	A
A	A	B	C	C	C	C	B	A	A
A	B							B	A
A	B							B	A
A	A	B	C	C	C	C	B	A	A
A	A	A	A	A	A	A	A	A	A

Position of Hydrant	Number of Edge Vertices Covered	Number of Center Vertices Covered
A	5	0 or 1
B	4	3
C	3	4

A hydrant placed on a center vertex will cover at most two edge vertices. Therefore, if six hydrants are to cover all 28 edge vertices, their positions must consist of either

(1) 6 of type A,
(2) 5 of type A, 1 of type B,
(3) 5 of type A, 1 of type C, or
(4) 4 of type A, 2 of type B.

However, in case (1) they will cover at most 6 center vertices, and in cases (2), (3), and (4) they will cover at most 8, 9, and 10 center vertices, respectively. Thus, when the edges are covered, the center is not covered. This shows that seven hydrants are necessary.

Problem 149

Consider a continuous function $f : \mathbb{R}^+ \longrightarrow \mathbb{R}^+$ with the following properties:
  (i) $f(2) = 3$,
  (ii) For all $x, y > 0$, $f(xy) = f(x)f(y) - f\left(\frac{x}{y}\right)$.

a. Show that if such a function f exists, it is unique.

b. Find an explicit formula for such a function.

Answer. The only such continuous function is given by $f(x) = x^\alpha + \dfrac{1}{x^\alpha}$, where $\alpha = \log_2\left(\dfrac{3+\sqrt{5}}{2}\right)$.

Solution. a. Let f be a function with the given properties. If we substitute $x = y = 1$ into the functional equation we get $f(1) = (f(1))^2 - f(1)$, so $(f(1))^2 = 2f(1)$, or $f(1) = 2$. If we take $y = x$, we get $f(x^2) = (f(x))^2 - f(1)$, so $f(x^2) = (f(x))^2 - 2$. Replacing x by $\sqrt{x}$ in this last equation yields $f(x) = (f(\sqrt{x}))^2 - 2$, so

$$f(\sqrt{x}) = \sqrt{f(x) + 2}. \qquad (*)$$

Using this recurrence we can get $f(x^{1/4})$ from $f(x^{1/2})$, $f(x^{1/8})$ from $f(x^{1/4})$, and so forth. Also, if we have $f(x)$ we can get $f(x^n)$ for any positive integer n, using the recurrence $f(x^n) = f(x^{n-1})f(x) - f(x^{n-2})$. We can also get $f(x^n)$ for negative n, because if we put $x = 1$ in the functional equation we get $f(y) = f(1)f(y) - f\left(\frac{1}{y}\right)$, so $f\left(\frac{1}{y}\right) = 2f(y) - f(y) = f(y)$. So the given value $f(2) = 3$ determines $f(2^{n/k})$ whenever the rational exponent n/k has a denominator which is a power of 2.

Now note that any positive real number x can be written in the form 2^z (specifically, with $z = \log_2 x$) and that the exponent z can be approximated arbitrarily closely by rationals whose denominators are powers of 2. So there will be a sequence $x_1 = 2^{z_1}$, $x_2 = 2^{z_2}$, $x_3 = 2^{z_3}$, ..., where the exponents $z_1, z_2, z_3, \ldots$ are rationals whose denominators are powers of 2, and where $\lim_{i \to \infty} z_i = z$ (and therefore $\lim_{i \to \infty} x_i = x$). Then because f is continuous, $f(x) = \lim_{i \to \infty} f(x_i)$. Thus $f(x)$ is determined by $f(2) = 3$, and if the function f exists it is unique.

b. Start with the function given by $g(x) = x + \frac{1}{x}$. Certainly g is a continuous function from $\mathbb{R}^+$ to itself, and g satisfies the functional equation:

$$g(xy) = xy + \tfrac{1}{xy} = xy + \tfrac{x}{y} + \tfrac{y}{x} + \tfrac{1}{xy} - \left(\tfrac{x}{y} + \tfrac{y}{x}\right)$$
$$= \left(x + \tfrac{1}{x}\right)\left(y + \tfrac{1}{y}\right) - \left(\tfrac{x}{y} + \tfrac{1}{x/y}\right)$$
$$= g(x)g(y) - g\left(\tfrac{x}{y}\right).$$

Unfortunately, $g(2) = 5/2$, not 3. But we can adjust g a little without disturbing the functional equation, by replacing $g(x)$ by $g(x^\alpha)$ for any fixed number α. That is, $f(x) = x^\alpha + \tfrac{1}{x^\alpha}$ will satisfy the functional equation for any choice of the exponent α. We want $f(2) = 2^\alpha + \tfrac{1}{2^\alpha} = 3$; if we put $a = 2^\alpha$, this becomes $a + 1/a = 3$, or $a^2 - 3a + 1 = 0$, or $a = (3 \pm \sqrt{5})/2$. If we take the plus sign, we get $2^\alpha = (3 + \sqrt{5})/2$, or $\alpha = \log_2\left(\tfrac{3+\sqrt{5}}{2}\right)$, as given in the answer. (If we take the minus sign, α is replaced by $-\alpha$, but the function doesn't change.)

Alternatively, setting $y = 2$ in the functional equation, we obtain
$$f(2x) = 3f(x) - f(x/2).$$
If we set $f_n = f(2^n)$, we have
$$f_{n+1} = 3f_n - f_{n-1}.$$
From the characteristic equation $r^2 = 3r - 1$ of this linear recurrence, we get $r = \dfrac{3 \pm \sqrt{5}}{2}$, so
$$f_n = A\left(\frac{3+\sqrt{5}}{2}\right)^n + B\left(\frac{3-\sqrt{5}}{2}\right)^n.$$
From $f_0 = 2$ and $f_1 = 3$, we find $A = B = 1$, so that
$$f(2^n) = \left(\frac{3+\sqrt{5}}{2}\right)^n + \left(\frac{3-\sqrt{5}}{2}\right)^n.$$
It is now easy to check that the natural extension of f to
$$f(x) = \left(\frac{3+\sqrt{5}}{2}\right)^{\log_2 x} + \left(\frac{3-\sqrt{5}}{2}\right)^{\log_2 x} = x^\alpha + \frac{1}{x^\alpha}$$
satisfies the functional equation.

Problem 150

A can is in the shape of a right circular cylinder of radius r and height h. An intelligent ant is at a point on the edge of the top of the can and wants to crawl to the point on the edge of the bottom of the can that is diametrically opposite to its starting point. As a function of r and h, what is the minimum distance the ant must crawl?

Answer. The minimum distance is $2r + h$ if $h/r \leq \pi^2/4 - 1$ and it is $\sqrt{\pi^2 r^2 + h^2}$ if $h/r \geq \pi^2/4 - 1$.

Solution. Let P be the ant's starting point and Q be its end point. On the way from P to Q, the ant may crawl across the top of the can for a while, and it might even crawl down onto the lateral side and then return to the top of the can. Eventually, though, the ant must leave the top of the can for the last time, at some point R along the top edge (it may be that $R = P$). However, the ant could also have crawled from P to R in a straight line and because this is the shortest path from P to R, we may assume that the ant follows this path. Similarly, if S is the point on the bottom edge where the ant first reaches the bottom of the can (possibly $S = Q$), then we can assume that the ant crawls in a straight line from S to Q. So in finding the shortest path for the ant, we need only consider paths of the form $PRSQ$, where PR and SQ are straight line segments and RS is some path along the lateral side, as shown in the figure.

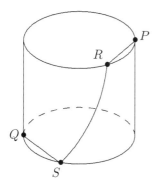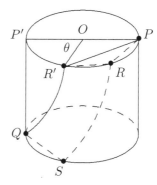

Now note that we could get a path $PRR'Q$ of the same length as $PRSQ$ by rotating RS around the lateral side of the cylinder until S reaches Q and R reaches (say) R', because the straight-line distances RR' on the top and

SQ on the bottom are the same length, while the distances RS and $R'Q$ are the same. On the other hand, if the points R and R' are not the same we can now do even better by going from P to R' in a straight line (rather than by way of R). In other words, we only have to look at the paths of the form $PR'Q$, where PR' is a line segment across the top of the can and $R'Q$ is some path along the lateral side.

To find the shortest distance between R' and Q along the lateral side, cut the side along the vertical line from Q to P' and roll it out into a rectangle. Clearly, the shortest distance joining Q and R' on this rectangle is the straight-line distance. To find an expression for the total length of the path $PR'Q$ it is convenient to let θ denote the central angle $R'OP'$ of the top circle, where O is the center of the top circle and P' is diametrically opposite P (see previous figure). Because $P'R'$ has length $r\theta$, $R'Q = \sqrt{(r\theta)^2 + h^2}$. Also, $PR' = 2r\sin\bigl((\pi - \theta)/2\bigr) = 2r\cos(\theta/2)$. Thus, the total length of the path $PR'Q$ is $2r\cos(\theta/2) + \sqrt{(r\theta)^2 + h^2} = r\Bigl(2\cos(\theta/2) + \sqrt{\theta^2 + (h/r)^2}\Bigr)$. So the problem comes down to finding the minimum value of the function

$$f(\theta) = 2\cos(\theta/2) + \sqrt{\theta^2 + (h/r)^2}, \quad 0 \leq \theta \leq \pi.$$

Using physical intuition or plotting the graph may convince you that the minimum must occur at one of the endpoints, but proving that is another matter. For this, we consider the derivatives of f, and we find

$$f'(\theta) = -\sin(\theta/2) + \frac{\theta}{\sqrt{\theta^2 + (h/r)^2}},$$

$$f''(\theta) = -\frac{1}{2}\cos(\theta/2) + \frac{\sqrt{\theta^2+(h/r)^2} - \theta\frac{\theta}{\sqrt{\theta^2+(h/r)^2}}}{\theta^2 + (h/r)^2}$$

$$= -\frac{1}{2}\cos(\theta/2) + \frac{(h/r)^2}{\bigl(\theta^2 + (h/r)^2\bigr)^{3/2}}.$$

Now suppose there is a critical point θ_0, $0 < \theta_0 < \pi$. From $f'(\theta_0) = 0$ it follows that

$$\sin(\theta_0/2) = \frac{\theta_0}{\sqrt{\theta_0^2 + (h/r)^2}}. \qquad (*)$$

Then

$$\cos(\theta_0/2) = \sqrt{1 - \sin^2(\theta_0/2)} = \frac{h/r}{\sqrt{\theta_0^2 + (h/r)^2}}$$

THE SOLUTIONS

(note that $\cos(\theta_0/2) > 0$ because $0 < \theta_0/2 < \pi/2$). Using this,

$$f''(\theta_0) = -\frac{1}{2}\frac{h/r}{\sqrt{\theta_0^2 + (h/r)^2}} + \frac{(h/r)^2}{\left(\theta_0^2 + (h/r)^2\right)^{3/2}}$$

$$= \frac{h/r}{2\left(\theta_0^2 + (h/r)^2\right)^{3/2}}\left(2(h/r) - \left(\theta_0^2 + (h/r)^2\right)\right).$$

It's enough to show that $2(h/r) - \left(\theta_0^2 + (h/r)^2\right) < 0$, for this will imply that any such critical point θ_0 would give a local maximum, proving that the minimum value must be at an endpoint.

Because $\sin\theta < \theta$ for $0 < \theta < \pi/2$, from $(*)$ we see that

$$\frac{\theta_0}{\sqrt{\theta_0^2 + (h/r)^2}} < \frac{\theta_0}{2},$$

or $\theta_0^2 + (h/r)^2 > 4$. So, if $0 < h/r < 2$, then $2(h/r) - (\theta_0^2 + (h/r)^2) < 0$. On the other hand, if $h/r \geq 2$, then $2(h/r) - (\theta_0^2 + (h/r)^2) = (h/r)(2 - h/r) - \theta_0^2 < 0$. So in either case, $2(h/r) - \left(\theta_0^2 + (h/r)^2\right) < 0$, and the minimum is indeed at an endpoint.

Finally, we can compare $f(0) = 2 + h/r$ and $f(\pi) = \sqrt{\pi^2 + (h/r)^2}$ by squaring both. We see that $f(0)$ is the smaller of the two if and only if

$$4 + 4(h/r) + (h/r)^2 \leq \pi^2 + (h/r)^2,$$

which yields $h/r \leq \pi^2/4 - 1$.

Remembering that the distance was given by $r f(\theta)$, we conclude that the ant must crawl a distance $2r + h$ (straight across the top of the can and down the side) if $h/r \leq \pi^2/4 - 1$, and a distance $\sqrt{\pi^2 r^2 + h^2}$ (entirely along the lateral side) if $h/r \geq \pi^2/4 - 1$.

Problem 151

Consider a triangle ABC whose angles α, β, and γ (at A, B, C respectively) satisfy $\alpha \leq \beta \leq \gamma$. Under what conditions on α, β, and γ can a beam of light placed at C be aimed at the segment AB, reflect to the segment BC, and then reflect to the vertex A?

Answer. The necessary and sufficient conditions are $45° - \frac{1}{2}\alpha < \beta < 60°$.

Solution. To see why, first recall the reflection principle: If points X, Y are on the same side of a line L, then a light beam originating at X, reflecting in L, and ending at Y, arrives at Y from the direction of X', the reflection of the point X in the line L.

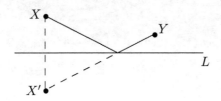

In our problem, a light beam placed at C and reflecting to A as described, will arrive at point D on segment BC (after reflecting in AB at E) from the direction of C' and then arrive at A from the direction of C'', where C' is the reflection of C in the line AB and C'' is the reflection of C' in the line BC (see the following figure).

If AC'' lies within $\angle CAB$, there will be such a light beam, for we can "retrace" the beam from A via D (the intersection point of AC'' and BC) and E (the intersection point of DC' and AB, which exists since α and β are acute) to C. Thus the question becomes: What are the conditions on angles α, β, γ for AC'' to lie within $\angle CAB$?

For AC'' to lie within $\angle CAB$, it is necessary and sufficient that both angles ABC'' and ACC'' (see figure) be less than $180°$. Now, by symmetry, $\angle CBC'' = \angle CBC' = 2\beta$, so $\angle ABC'' = \angle ABC + \angle CBC'' = 3\beta$, which is less than $180°$ when $\beta < 60°$. On the other hand, since $\triangle CBC''$ is isosceles,

$$\angle BCC'' = \frac{1}{2}(180° - \angle CBC'') = 90° - \beta.$$

Therefore,

$$\angle ACC'' = \angle ACB + \angle BCC'' = \gamma + (90° - \beta) = 270° - \alpha - 2\beta,$$

which is less than $180°$ when $\beta > 45° - \frac{1}{2}\alpha$, and we are done.

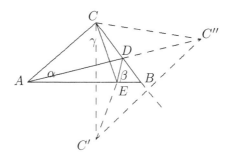

Problem 152

What is the largest possible number of nonzero vectors in n-space so that the angle between any two of the vectors is the same (and not zero)? In that situation, what are the possible values for the angle?

Answer. The largest number of nonzero vectors in n-space that all make equal, nonzero angles with each other is $n+1$. The *only* possible value for the angle is $\arccos(-1/n)$.

Solution 1. Let $f(n)$ be the largest possible number of nonzero vectors in n-space that make equal nonzero angles with each other. Clearly, $f(1) = 2$ (take two vectors in opposite directions), and we've seen that $f(2) = 3$. We'll show that $f(n) = n+1$ by showing that $f(n) = f(n-1) + 1$ for all $n \geq 2$. This equation holds for $n = 2$, so henceforth we'll assume that $n \geq 3$.

Let $k = f(n)$ and suppose that $\mathbf{u}_1, \mathbf{u}_2, \ldots, \mathbf{u}_k$ are nonzero vectors in n-space such that any two of these vectors make the same nonzero angle θ with each other. Note that $k \geq 3$ and therefore, $\theta \neq \pi$. Also, we may assume that the $\mathbf{u}_i$ are unit vectors, so that for $i \neq j$, $\mathbf{u}_i \cdot \mathbf{u}_j = \cos\theta = a$, where $-1 < a < 1$. The (projected) vectors $\mathbf{w}_i = \mathbf{u}_i - a\,\mathbf{u}_1$, $i = 2, 3, \ldots, k$, have equal lengths given by

$$|\mathbf{w}_i|^2 = \mathbf{w}_i \cdot \mathbf{w}_i = (\mathbf{u}_i - a\,\mathbf{u}_1) \cdot (\mathbf{u}_i - a\,\mathbf{u}_1) = 1 - 2a^2 + a^2 = 1 - a^2 > 0,$$

and are each orthogonal to $\mathbf{u}_1$ because

$$\mathbf{w}_i \cdot \mathbf{u}_1 = (\mathbf{u}_i - a\,\mathbf{u}_1) \cdot \mathbf{u}_1 = \mathbf{u}_i \cdot \mathbf{u}_1 - a\,\mathbf{u}_1 \cdot \mathbf{u}_1 = a - a = 0.$$

Therefore, the $\mathbf{w}_i$ lie in an $(n-1)$-dimensional space. Moreover, they all make the same angle with each other, because for $i \neq j$,

$\mathbf{w}_i \cdot \mathbf{w}_j = (\mathbf{u}_i - a\,\mathbf{u}_1) \cdot (\mathbf{u}_j - a\,\mathbf{u}_1) = a - a^2$, so the cosine of the angle is $\dfrac{a-a^2}{1-a^2} = \dfrac{a}{1+a}$. Therefore, $k - 1 \leq f(n-1)$, so $f(n) \leq f(n-1) + 1$.

We will now complete the proof of our claim by induction. Suppose that $f(n-1) = n$ and that the only possible value for the common angle is $\arccos(-1/(n-1))$. We've seen that this is the case for $n = 2$ and $n = 3$ (note that $\arccos((-1)/1) = \pi$ and $\arccos(-1/2) = 2\pi/3$). Let $\mathbf{x}_1, \mathbf{x}_2, \ldots, \mathbf{x}_n$ be nonzero unit vectors in an $(n-1)$-dimensional space which do make angles $\arccos(-1/(n-1))$ with each other, so that for $i \neq j$, $\mathbf{x}_i \cdot \mathbf{x}_j = -1/(n-1) = b$, say.

Consider the n-dimensional space spanned by the given $(n-1)$-dimensional space and an additional unit vector $\mathbf{p}$ orthogonal to it. For a fixed constant λ, define new vectors $\mathbf{y}_i = \mathbf{x}_i + \lambda \mathbf{p}$ for $i = 1, 2, \ldots n$, and let $\mathbf{y}_{n+1} = \mathbf{p}$. We aim to find a value for λ that will make these $n+1$ vectors have equal nonzero angles with each other. This will show that $f(n) \geq f(n-1) + 1 = n + 1$ (so $f(n) = f(n-1) + 1$).

First, for $i = 1, 2, \ldots n$,
$$|\mathbf{y}_i|^2 = \mathbf{y}_i \cdot \mathbf{y}_i = (\mathbf{x}_i + \lambda \mathbf{p}) \cdot (\mathbf{x}_i + \lambda \mathbf{p}) = 1 + \lambda^2$$
so the cosine of the angle between $\mathbf{y}_i, \mathbf{y}_j$, $1 \leq i, j \leq n$, $i \neq j$, is
$$\frac{\mathbf{y}_i \cdot \mathbf{y}_j}{|\mathbf{y}_i||\mathbf{y}_j|} = \frac{1}{1+\lambda^2}\Big((\mathbf{x}_i + \lambda \mathbf{p}) \cdot (\mathbf{x}_j + \lambda \mathbf{p})\Big) = \frac{b + \lambda^2}{1 + \lambda^2},$$
while
$$\frac{\mathbf{y}_i \cdot \mathbf{y}_{n+1}}{|\mathbf{y}_i||\mathbf{y}_{n+1}|} = \frac{1}{\sqrt{1+\lambda^2}}\Big((\mathbf{x}_i + \lambda \mathbf{p}) \cdot \mathbf{p}\Big) = \frac{\lambda}{\sqrt{1+\lambda^2}}.$$

Thus, the $\mathbf{y}_i$ will make equal angles with each other provided
$$\frac{b+\lambda^2}{1+\lambda^2} = \frac{\lambda}{\sqrt{1+\lambda^2}} \iff b + \lambda^2 = \lambda\sqrt{1+\lambda^2}$$
$$\iff (b+\lambda^2)^2 = \lambda^2(1+\lambda^2) \text{ and } b+\lambda^2, \lambda \text{ have the same sign}$$
$$\iff b^2 + 2b\lambda^2 = \lambda^2 \text{ and } b+\lambda^2, \lambda \text{ have the same sign}$$
$$\iff \lambda^2 = \frac{b^2}{1-2b} \text{ and } b+\lambda^2, \lambda \text{ have the same sign.}$$

This will yield a unique λ (recall that $b = -1/(n-1)$, so $1 - 2b > 0$). So we can find $n + 1$ (and no more) nonzero vectors in n-space that make equal angles with each other. We saw earlier that if that angle θ satisfies $\cos\theta = a$, then we can find n nonzero vectors in $(n-1)$-space that make equal angles of cosine $\dfrac{a}{1+a}$. The induction hypothesis tells us that $\dfrac{a}{1+a} = \dfrac{-1}{n-1}$, and we soon see that $a = -1/n$, completing the induction.

Solution 2. As in Solution 1, let $\mathbf{u}_1, \ldots, \mathbf{u}_k$ be nonzero vectors in n-space that make equal, nonzero angles with each other. We may assume that the vectors $\mathbf{u}_i$ have length 1. We proceed by induction, with the case $n = 2$ already given. There is no loss of generality in assuming $\mathbf{u}_k = (0, 0, \ldots, 0, 1)$. Then for $i < k$ the kth coordinate of $\mathbf{u}_i$ is $u_{ik} = \mathbf{u}_i \cdot \mathbf{u}_k = c$, which is independent of i and must be less than 1. Let W be the orthogonal complement of $\mathbf{u}_k$ in $\mathbb{R}^n$. Then the vectors $\mathrm{Proj}_W \mathbf{u}_i$ ($1 \leq i \leq k-1$) all have length $\sqrt{1-c^2}$ and make equal, nonzero angles with each other. Therefore, $k - 1 \leq (n-1) + 1$, or $k \leq n+1$, by induction.

Now assume $k = n+1$. Then, for $i < j < n+1$, we again apply the inductive hypothesis to conclude
$$\mathbf{u}_i \cdot \mathbf{u}_j = \mathrm{Proj}_W \mathbf{u}_i \cdot \mathrm{Proj}_W \mathbf{u}_j + c^2 = (1-c^2)(-1/n) + c^2 = c.$$
This quadratic in c has roots 1 and $-1/(n+1)$, proving the claim.

Problem 153

Let S be a set of numbers which includes the elements 0 and 1. Suppose S has the property that for any nonempty finite subset T of S, the average of all the numbers in T is an element of S. Prove or disprove: S must contain all the rational numbers between 0 and 1.

Solution 1. We'll show that S does indeed contain all the rational numbers between 0 and 1. First observe that $\{0, \tfrac{1}{2}, 1\}$ is a subset of S (because $\tfrac{1}{2}$ is the average of 0 and 1). Then we can expand this subset by taking averages of successive pairs of elements, and we find that $\{0, \tfrac{1}{4}, \tfrac{2}{4}, \tfrac{3}{4}, 1\}$ is a subset of S, and repeating this, so is $\{0, \tfrac{1}{8}, \tfrac{2}{8}, \tfrac{3}{8}, \tfrac{4}{8}, \tfrac{5}{8}, \tfrac{6}{8}, \tfrac{7}{8}, 1\}$. Continuing in this way we conclude that all numbers of the form $\tfrac{m}{2^n}, 0 \leq m \leq 2^n, n \geq 1$, are in S.

Now suppose that $\tfrac{s}{t}$, reduced to lowest terms, is a rational number in $(0, 1)$. We'll be done if we can prove the following:

Claim: There exist an integer n and t distinct positive integers $a_1, a_2, \ldots, a_t$ less than or equal to 2^n such that $a_1 + a_2 + \cdots + a_t = s\, 2^n$.

For then, because the $\dfrac{a_i}{2^n}$ are in S, as we've already seen, so is their average

$$\dfrac{\dfrac{a_1}{2^n}+\dfrac{a_2}{2^n}+\cdots+\dfrac{a_t}{2^n}}{t}=\dfrac{s}{t}.$$

Proof of claim: Begin by setting $N(a)=a+(a{+}1)+(a{+}2)+\cdots+(a+(t{-}1))$, the sum of t consecutive numbers starting at a, and let n be the smallest integer such that $N(0)<2^n$. Note that

$$N(2^n-(t-1))=\left(2^n-(t-1)\right)+\left(2^n-(t-2)\right)+\cdots+2^n$$
$$=t\,2^n-(1+2+\cdots(t{-}1))$$
$$>t\,2^n-2^n=(t-1)2^n.$$

From what we've just shown, $N(0)<s\,2^n<N(2^n-(t{-}1))$. Consequently, because $N(a)$ is an increasing function of a, there is a unique integer a between 1 and $2^n-(t-1)$ such that $N(a-1)<s\,2^n\le N(a)$. Now let $r=N(a)-s\,2^n$, and note that $r<N(a)-N(a-1)=t$. Now, starting with the numbers $a,a+1,\ldots,a+(t{-}1)$, subtract 1 from the first r of these and keep the others the same. The result will be a set of t positive numbers, each less than or equal to 2^n, that add to $s\,2^n$, as desired.

Solution 2. Let S^* be the intersection of *all* the sets S with the given properties; that is, of all the sets S such that $0\in S$, $1\in S$, and the average of any finite subset of S is in S. Note that $0\in S^*$, $1\in S^*$, and the average of any finite subset of S^* is in S^*, so the intersection S^* is the smallest set with the stated properties. It will be enough to show that S^* contains all the rational numbers between 0 and 1.

One advantage of working with S^* is that S^* has the additional property that $x\in S^*$ implies $1-x\in S^*$. To see this, suppose $x\in S^*$, and let S be some set with the given properties. Let S_1 denote the set of all numbers a such that $1-a$ is in S. Then $0\in S_1$, $1\in S_1$, and if T_1 is a finite subset of S_1, say with elements $a_1,a_2,\ldots,a_n$, then $1-a_1,1-a_2,\ldots,1-a_n\in S$, so the average

$$\dfrac{(1-a_2)+(1-a_2)+\cdots+(1-a_n)}{n}=1-\dfrac{a_1+a_2+\cdots+a_n}{n}\in S,$$

which means that $\dfrac{a_1+\cdots+a_n}{n}\in S_1$. This shows that S_1 has all the properties of the sets we're looking at; in particular, $S^*\subseteq S_1$. Therefore, if $x\in S^*$, then $x\in S_1$, so $1-x\in S$. But now we've shown that $1-x$ is in any set S with the given properties, so $1-x\in S^*$.

Now $0 \in S^*$ and $1 \in S^*$, so their average $\frac{1}{2} \in S^*$. From $0 \in S^*$ and $\frac{1}{2} \in S^*$ we get $\frac{1}{4} \in S^*$, and thus $1 - \frac{1}{4} = \frac{3}{4} \in S^*$. Taking the average of the finite set $\{0, \frac{1}{4}, \frac{3}{4}\}$ we get $\frac{1}{3} \in S^*$, and thus $1 - \frac{1}{3} = \frac{2}{3} \in S^*$.

Now suppose that some rational number $\frac{m}{n}$ between 0 and 1 is not in S^*. Of the possible choices of such a number, pick one with the smallest possible value of n, say N, and with the smallest possible m, say M, for that particular $n = N$. Note that $M < \frac{1}{2}N$, because if $M = \frac{1}{2}N$, then $\frac{M}{N} = \frac{1}{2} \in S^*$, and if $M > \frac{1}{2}N$ then $0 < N - M < M$, so $\frac{N-M}{N} \in S^*$ by our assumption. But then $1 - \frac{N-M}{N} = \frac{M}{N} \in S^*$, contradiction. Also, N must be odd, because if N were even, say $N = 2K$, then $M < \frac{1}{2}N = K$, so $\frac{M}{K} \in S^*$ by the minimality of N, and then $\frac{M}{2K} = \frac{M}{N} \in S^*$, as the average of 0 and $\frac{M}{K}$.

Because N is odd, we can write $N = 2K + 1$; because $N \geq 5$ by our previous work, $K \geq 2$. Now $3, 4, \ldots, K+2$ are all less than $2K + 1 = N$, so all rational numbers between 0 and 1 with those denominators are in S^*. In particular, the average of the finite subset

$$T = \left\{ 0, \underbrace{\frac{1}{3}, \frac{2}{3}}_{1}, \underbrace{\frac{1}{4}, \frac{3}{4}}_{1}, \cdots, \underbrace{\frac{1}{K+1}, \frac{K}{K+1}}_{1}, \underbrace{\frac{1}{K+2}, \frac{K+1}{K+2}}_{1} \right\}$$

is in S^*. Because the numbers in this subset consist of K pairs (as shown) that each add to 1, along with the lone 0, the average is $\frac{1+\cdots+1}{2K+1} = \frac{K}{2K+1} = \frac{K}{N}$, so $\frac{K}{N} \in S^*$.

Now consider the intersection S^{**} of all the sets S_1 such that $0 \in S_1$, $\frac{K}{N} \in S_1$, and the average of any finite subset of S_1 is in S_1.

Claim: If $x \in S^*$, then $\frac{K}{N}x \in S^{**}$.

Proof of claim: Suppose $x \in S^*$ and let S_1 be a set as described above. Let S_0 be the set of all numbers b such that $\frac{K}{N}b \in S_1$. Then $0 \in S_0$ because $0 \in S_1$, and $1 \in S_0$ because $\frac{K}{N} \in S_1$. If T is a finite subset of S_0, say with elements $b_1, b_2, \ldots, b_r$, then $\frac{K}{N}b_1, \frac{K}{N}b_2, \ldots, \frac{K}{N}b_r$ are in S_1, so the average

$$\frac{\frac{K}{N}b_1 + \frac{K}{N}b_2 + \cdots + \frac{K}{N}b_r}{r} = \frac{K}{N}\left(\frac{b_1 + b_2 + \cdots + b_r}{r}\right) \in S_1,$$

so $\frac{b_1 + b_2 + \cdots + b_r}{r} \in S_0$. We've now shown that S_0 has the properties of the sets S that we've been considering from the start, so since $x \in S^*$ is in every such set, $x \in S_0$. That is, $\frac{K}{N}x \in S_1$. But now we've shown that $\frac{K}{N}x$ is in *all* the sets S_1 of the type considered in this paragraph, so $\frac{K}{N}x \in S^{**}$, completing the proof of the claim.

Applying the claim to $x = \frac{M}{K}$ (which is in S^* by the minimality of N), we see that $\frac{K}{N} \cdot \frac{M}{K} = \frac{M}{N} \in S^{**}$. But note that S^* is one of the sets S_1 considered in the previous paragraph, because $0 \in S^*$, $\frac{K}{N} \in S^*$ as previously shown, and the average of any finite subset of S^* is in S^*. Therefore, because $\frac{M}{N} \in S^{**}$ and S^* is one of the sets S_1 whose intersection is S^{**}, we finally have $\frac{M}{N} \in S^*$, contrary to our assumption. We conclude that S^*, and hence every S, contains all rational numbers between 0 and 1.

Problem 154

Find $\lim\limits_{n \to \infty} \left(\sum\limits_{k=1}^{n} \dfrac{1}{\binom{n}{k}} \right)^n$, or show that this limit does not exist.

Answer. The limit exists and equals e^2.

Solution. Let $a_n = \sum\limits_{k=1}^{n} \dfrac{1}{\binom{n}{k}}$. For $n \geq 3$, $a_n \geq 1 + \dfrac{2}{\binom{n}{1}} = 1 + \dfrac{2}{n}$, while for $n \geq 6$,

$$a_n = 1 + 2\dfrac{1}{\binom{n}{1}} + 2\dfrac{1}{\binom{n}{2}} + \underbrace{\left(\dfrac{1}{\binom{n}{3}} + \cdots + \dfrac{1}{\binom{n}{n-3}} \right)}_{n-5 \text{ terms}}$$

$$\leq 1 + \dfrac{2}{n} + \dfrac{4}{n(n-1)} + \dfrac{n-5}{\binom{n}{3}}$$

$$= 1 + \dfrac{2}{n} + \dfrac{4}{n(n-1)} + \dfrac{6(n-5)}{n(n-1)(n-2)}$$

$$< 1 + \dfrac{2}{n} + \dfrac{10}{n(n-1)}$$

$$\leq 1 + \dfrac{2}{n} + \dfrac{12}{n^2}.$$

But $\lim\limits_{n \to \infty} \left(1 + \dfrac{2}{n} \right)^n = e^2$ and

$$\lim_{n \to \infty} \left(1 + \dfrac{2}{n} + \dfrac{12}{n^2} \right)^n = \lim_{n \to \infty} \left[\left(1 + \dfrac{2n+12}{n^2} \right)^{n^2/(2n+12)} \right]^{(2n+12)/n} = e^2,$$

so by the squeeze principle, $\lim_{n \to \infty} a_n^n = e^2$.

Problem 155

A person starts at the origin and makes a sequence of moves along the real line, with the kth move being a change of $\pm k$.

a. Prove that the person can reach any integer in this way.

b. If $m(n)$ is the least number of moves required to reach a positive integer n, prove that $\lim_{n\to\infty} m(n)/\sqrt{n}$ exists and evaluate this limit.

Solution. a. A person moving alternately right and left will, after k moves, be at the position given by the kth partial sum of the series

$$1 - 2 + 3 - 4 + 5 - 6 + \cdots.$$

These partial sums are easily seen to be $1, -1, 2, -2, 3, -3, \ldots$; thus every integer can be reached. (If you do not consider 0 reached by its being the starting point, you can write $0 = 1 + 2 - 3$.)

b. We will show that the limit is $\sqrt{2}$ by finding upper and lower bounds for $m(n)/\sqrt{n}$.

First of all, note that after k moves the maximum possible distance from the origin is $1 + 2 + \cdots + k = k(k+1)/2$. Thus for $m = m(n)$ we have $m(m+1)/2 \geq n$, or equivalently, $m(m+1) \geq 2n$. It follows that $(m+1)^2 > 2n$, so

$$m + 1 > \sqrt{2n} \quad \text{and} \quad \frac{m(n)}{\sqrt{n}} > \sqrt{2} - \frac{1}{\sqrt{n}}.$$

We will now show that

$$\frac{m(n)}{\sqrt{n}} < \sqrt{2} + \frac{3}{\sqrt{n}}.$$

Since

$$\lim_{n\to\infty}\left(\sqrt{2} - \frac{1}{\sqrt{n}}\right) = \lim_{n\to\infty}\left(\sqrt{2} + \frac{3}{\sqrt{n}}\right) = \sqrt{2},$$

we will then be done by the squeeze principle.

The quickest way to get at least as far as n is to move to the right until one reaches or passes position n. As above, this will happen once $k(k+1)/2 \geq n$, which is certainly true if $k^2/2 \geq n$, or $k \geq \sqrt{2n}$. Even though $\sqrt{2n}$ may not be an integer, $k(k+1)/2 \geq n$ will first happen for some $k = k_0$ with $k_0 < \sqrt{2n} + 1$. At this point we will have overshot n by somewhere between 0 and $k_0 - 1$, that is, $1 + 2 + \cdots + k_0 = n + r$ with $0 \leq r \leq k_0 - 1$. If $r = 0$, we have reached n. If $r = 2s (s > 0)$ is even, we can go back and change the sth move from s to $-s$. Since this changes the outcome by $-2s = -r$, we now have a way to reach n in k_0 moves. If $r = 2s+1$ is odd, we can reach n in $k_0 + 2$ moves by again changing the sth move but also adding two moves at the end:

$$1 + 2 + \cdots + (s-1) - s + (s+1) + \cdots + k_0 + (k_0 + 1) - (k_0 + 2)$$
$$= 1 + 2 + \cdots + k_0 - 2s - 1$$
$$= n.$$

In each case, we can reach n in at most $k_0 + 2 < \sqrt{2n} + 3$ moves, so

$$\frac{m(n)}{\sqrt{n}} < \sqrt{2} + \frac{3}{\sqrt{n}}$$

and we are done.

Problem 156

Let $S(n)$ be the number of solutions of the equation $e^{\sin x} = \sin(e^x)$ on the interval $[0, 2n\pi]$. Find $\lim_{n \to \infty} \frac{S(n)}{e^{2n\pi}}$, or show that the limit does not exist.

Answer. We'll show that the limit equals $\dfrac{e^\pi}{\pi(e^\pi + 1)}$.

Solution. To begin, note that $x = 0$ is definitely not a solution, because $e^{\sin 0} = e^0 = 1$ and $\sin(e^0) = \sin 1 < 1$. So we really have the interval $(0, 2n\pi]$; divide this interval into $2n$ half-open subintervals of length π.

On the odd-numbered subintervals $(2k\pi, (2k+1)\pi]$ we have $\sin x > 0$ except for $x = (2k+1)\pi$, so $e^{\sin x} > 1$ and so $e^{\sin x} \neq \sin(e^x)$ except possibly for $x = (2k+1)\pi$. So each odd-numbered subinterval contributes at most 1 to $S(n)$, and together they contribute at most n, which will not affect $\lim_{n \to \infty} \frac{S(n)}{e^{2n\pi}}$ (because $\lim_{n \to \infty} \frac{n}{e^{2n\pi}} = 0$). Thus it's enough to consider the number of solutions on the even-numbered subintervals.

Let $f(x) = e^{\sin x}$, $g(x) = \sin(e^x)$. On the even-numbered subintervals $\big((2k+1)\pi, (2k+2)\pi\big]$ we have $f(x) \leq 1$ and thus as $g(x)$ oscillates back and forth between -1 and 1, the graphs of f and g will cross repeatedly. In fact it seems reasonable that they should cross twice for each oscillation, once as $g(x)$ is "on the way up" to the value 1 and once as $g(x)$ is on the way down. To prove this, it will help to look at concavity.

We see that

$$f'(x) = e^{\sin x} \cdot \cos x,$$
$$f''(x) = \left(e^{\sin x} \cos x\right) \cos x + e^{\sin x}(-\sin x)$$
$$= e^{\sin x}(\cos^2 x - \sin x).$$

On the even-numbered subintervals that we're considering, $\sin x \leq 0$; also, $\cos x$ and $\sin x$ are never 0 together, so $\cos^2 x - \sin x > 0$ for all x in our subintervals, and so the graph of f is concave up there. Also, $f(x) \geq e^{-1}$ for all x.

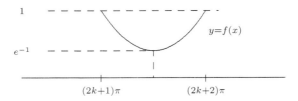

Now consider the concavity of g. We have

$$g'(x) = e^x \cos(e^x)$$
$$g''(x) = e^x \cos(e^x) + e^x(-\sin(e^x))e^x$$
$$= e^x \cos(e^x) - e^{2x} \sin(e^x) = e^x \cos(e^x) - e^{2x} g(x).$$

Claim: $g''(x) < 0$ whenever $g(x) \geq e^{-1}$ on our subintervals.

Proof of Claim: When $g(x) \geq e^{-1}$,

$$g''(x) = e^x \cos(e^x) - e^{2x} \cdot g(x) \leq e^x - e^{2x} \cdot e^{-1} = e^x - e^{2x-1},$$

and on our subintervals, $x < 2x - 1$ (that is, $x > 1$), so $g''(x) < 0$ as claimed.

In particular, we have $g''(x) < 0$ whenever $g(x) \geq f(x)$. This means that when the graph of g crosses the graph of f on its way up to 1, it is concave down, and it remains concave down not only until it crosses the graph of f again, but at least until it falls back below e^{-1} (and is therefore on its way back down to -1, without another opportunity to cross the graph of f). But while the graph of g is concave down and the graph of f is (as on the entire subinterval) concave up, the graphs can only cross twice.

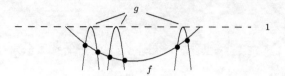

(If the graphs crossed at $x = a$, $x = b$, and $x = c$ with $a < b < c$, then $f(x) - g(x)$ would be 0 at a, b, and c, so by Rolle's theorem, $f'(x) - g'(x)$ would be 0 somewhere between a and b and again somewhere between b and c, so by Rolle's theorem again, $f''(x) - g''(x)$ would be 0 somewhere between those two places and hence between a and c, contradiction.)

It follows that the graphs of f and g cross exactly twice during each oscillation of $g(x)$ from -1 to 1 and back down to -1 in the subinterval, except for a possible "incomplete" oscillation at each end of the subinterval. The number of such oscillations is essentially equal to the number of multiples of 2π in the interval $\left(e^{(2k+1)\pi}, e^{(2k+2)\pi}\right]$. (This is because for x in the interval $\left((2k+1)\pi, (2k+2)\pi\right]$, e^x ranges through the interval $\left(e^{(2k+1)\pi}, e^{(2k+2)\pi}\right]$, and because $g(x) = \sin(e^x)$ equals 1 when $e^x = \pi/2$ or $e^x = 5\pi/2$ or $e^x = 9\pi/2$ or $\ldots$.) So we can write

$$S(n) = \underbrace{2}_{\substack{\text{2 crossings}\\ \text{per oscillation}}} \sum_{k=0}^{n-1} \frac{e^{(2k+2)\pi} - e^{(2k+1)\pi}}{2\pi} + \text{(error term)}$$

where the error term is bounded by a constant multiple of n and will therefore not affect $\lim_{n\to\infty} \dfrac{S(n)}{e^{2n\pi}}$.

Thus the answer is

$$\lim_{n\to\infty} \frac{1}{\pi} \cdot \frac{1}{e^{2n\pi}} \sum_{k=0}^{n-1} \left(e^{(2k+2)\pi} - e^{(2k+1)\pi}\right) = \lim_{n\to\infty} \frac{1}{\pi} \cdot \frac{1}{e^{2n\pi}} \sum_{k=0}^{n-1} (e^\pi - 1) e^{(2k+1)\pi}$$

$$= \lim_{n\to\infty} \frac{e^\pi - 1}{\pi e^{2n\pi}} \sum_{k=0}^{n-1} e^{(2k+1)\pi} = \lim_{n\to\infty} \frac{e^\pi - 1}{\pi e^{2n\pi}} \cdot e^\pi \cdot \frac{1 - e^{2n\pi}}{1 - e^{2\pi}}$$

$$= \frac{(e^\pi - 1)e^\pi}{\pi(1 - e^{2\pi})} \cdot \lim_{n\to\infty} \frac{1 - e^{2n\pi}}{e^{2n\pi}} = \frac{-e^\pi}{\pi(1 + e^\pi)} \cdot \lim_{n\to\infty}\left(\frac{1}{e^{2n\pi}} - 1\right)$$

$$= \frac{-e^\pi}{\pi(1 + e^\pi)} \cdot (-1) = \frac{e^\pi}{\pi(1 + e^\pi)},$$

as claimed.

Problem 157

Let $N > 1$ be a positive integer and consider those functions $\varepsilon\colon \mathbb{Z} \to \{1, -1\}$ having period N. For what N does there exist an infinite series $\sum_{n=1}^{\infty} a_n$ with the following properties: $\sum_{n=1}^{\infty} a_n$ diverges, whereas $\sum_{n=1}^{\infty} \varepsilon(n) a_n$ converges for all nonconstant ε (of period N)?

Solution. Series with the given properties exist only for $N = 2$. One example for $N = 2$ is given by $a_n = 1/n$. To show that there are no such series for $N > 2$, fix N and assume that $\sum_{n=1}^{\infty} a_n$ is such a series. Define the periodic functions $\varepsilon_j\colon \mathbb{Z} \to \{1, -1\}$, $j = 1, 2, \ldots, N$ by $\varepsilon_j = -1$ if and only if $n \equiv j \pmod{N}$. Then since $\sum_{n=1}^{\infty} \varepsilon_j(n) a_n$ converges for all j,

$$\sum_{j=1}^{N} \sum_{n=1}^{\infty} \varepsilon_j(n) a_n = \sum_{n=1}^{\infty} \left(\sum_{j=1}^{N} \varepsilon_j(n) \right) a_n = \sum_{n=1}^{\infty} (N-2) a_n$$

converges as well, so $\sum_{n=1}^{\infty} a_n$ converges.

Problem 158

Suppose you form a sequence of quadrilaterals as follows. The first quadrilateral is the unit square. To get from each quadrilateral to the next, pick a vertex of your quadrilateral and a side that is not adjacent to that vertex, and then connect the midpoint of that side to that vertex. This will divide the quadrilateral into a triangle and a new quadrilateral; discard the triangle, and repeat the process with the new quadrilateral. Show that there are at most $2^{n-1} - n + 1$ possibilities for the area of the nth quadrilateral, and state explicitly what the $2^{n-1} - n + 1$ candidates are.

Solution. It will help to keep track not only of the area of the quadrilaterals, but also of the areas of the triangles into which the diagonals divide the quadrilaterals. If we denote the area of the triangle "opposite" vertex

A by a, etc., then, as shown in the diagram, the area of the quadrilateral is $a + c = b + d$.

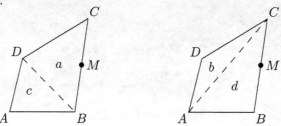

Now suppose we connect A with the midpoint M of BC and discard triangle ABM. Then for the new quadrilateral $AMCD$, we have $a' = \frac{1}{2}a$, $b' = b$, and $d' = \frac{1}{2}d$.

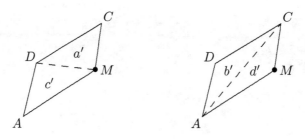

Because $a' + c' = b' + d'$, we have

$$c' = b' + d' - a' = b + \tfrac{1}{2}d - \tfrac{1}{2}a$$
$$= (b+d) - \tfrac{1}{2}d - \tfrac{1}{2}a$$
$$= (a+c) - \tfrac{1}{2}d - \tfrac{1}{2}a = c + \tfrac{1}{2}a - \tfrac{1}{2}d.$$

That is, in this step, the numbers a, b, c, d are replaced by

$$a' = \tfrac{1}{2}a, \quad b' = b, \quad c' = c + \tfrac{1}{2}a - \tfrac{1}{2}d, \quad d' = \tfrac{1}{2}d.$$

Just as the areas a, d of the triangles adjoining BC are halved in this step which involves the midpoint of BC, we can also halve the areas a, b using the midpoint of CD, the areas b, c using the midpoint of AD, or the areas c, d using the midpoint of AB. Also, of the two areas that don't get halved, one of them will stay the same, and we can choose which one by our choice of which vertex to connect to the midpoint.

Because we start with the unit square, we have $a = b = c = d = \frac{1}{2}$ for that first quadrilateral. At the next step, two of these numbers get halved and one stays the same. We still need to have $a+c = b+d$, and so we get

$a = \frac{1}{4}, b = \frac{1}{2}, c = \frac{1}{2}, d = \frac{1}{4}$ up to order. For the third quadrilateral there are different possibilities: We can halve both $\frac{1}{4}$s to get $\frac{1}{8}, \frac{1}{2}, \frac{1}{2}, \frac{1}{8}$, we can halve both $\frac{1}{2}$s to get $\frac{1}{4}, \frac{1}{4}, \frac{1}{4}, \frac{1}{4}$, and we can get either $\frac{1}{8}, \frac{1}{4}, \frac{1}{2}, \frac{3}{8}$ or $\frac{1}{8}, \frac{1}{4}, \frac{3}{8}, \frac{1}{4}$ by halving both a $\frac{1}{4}$ and a $\frac{1}{2}$. As for the areas, the second quadrilateral has area $\frac{1}{4} + \frac{1}{2} = \frac{3}{4}$ while the third quadrilateral has area $\frac{1}{8} + \frac{1}{2} = \frac{5}{8}$ or $\frac{1}{4} + \frac{1}{4} = \frac{1}{2}$.

Continuing in this way we are led to the following conjecture: The only possible areas for the nth quadrilateral are

$$\frac{n+1}{2^n}, \frac{n+2}{2^n}, \ldots, \frac{2^{n-1}+1}{2^n};$$

when the area is $k/2^n$, the four areas formed by two sides and one diagonal of the quadrilateral must be among the values $j/2^n$, $j = 1, 2, \ldots, k-1$.

We will prove this by induction; the first several cases have already been shown. So assume the claim holds for the nth quadrilateral, and assume the vertices of the nth quadrilateral are labeled as in the following diagram so that the $(n+1)$st quadrilateral is $A'BCD$.

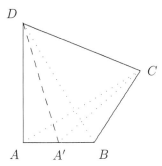

Suppose the area of $ABCD$ is $k/2^n$, the area of ABC is $j_1/2^n$, and the area of ABD is $j_2/2^n$. Then the areas of $A'BCD$, $A'BC$, $A'BD$, $A'CD$, and BCD are

$$\frac{2k - j_2}{2^{n+1}}, \frac{j_1}{2^{n+1}}, \frac{j_2}{2^{n+1}}, \frac{2k - j_1 - j_2}{2^{n+1}}, \text{ and } \frac{2k - 2j_2}{2^{n+1}},$$

respectively. Because it's easy to check that $j_1, j_2, 2k - j_1 - j_2$ and $2k - 2j_2$ are all between 1 and $2k - j_2 - 1$, we only need to establish the bounds on $2k - j_2$. On the one hand,

$$2k - j_2 \leq 2(2^{n-1} + 1) - 1 = 2^n + 1.$$

On the other hand,

$$2k - j_2 = k + (k - j_2) \geq (n+1) + 1 = n + 2,$$

and the induction is complete.

Comment. For $n = 5$, an area of $6/32 = 3/16$ cannot actually occur. Thus, it should be possible to strengthen the conclusion of this problem. We have not investigated this further; as far as we know, finding a complete characterization of the possible areas for the nth quadrilateral is an open problem.

Problem 159

Suppose f is a continuous, increasing, bounded, real-valued function, defined on $[0, \infty)$, such that $f(0) = 0$ and $f'(0)$ exists. Show that there exists $b > 0$ for which the volume obtained by rotating the area under f from 0 to b about the x-axis is half that of the cylinder obtained by rotating $y = f(b)$, $0 \leq x \leq b$, about the x-axis.

Idea. For b near 0, the differentiability condition ensures that the solid is "morally" a cone, having volume $\frac{1}{3}\pi r^2 h$. For b approaching ∞, the boundedness condition ensures that the solid is "morally" a cylinder, with volume $\pi r^2 h$. Somewhere in between, the coefficient will be $\frac{1}{2}$.

Solution. If we rotate $y = f(b)$, $0 \leq x \leq b$, about the x-axis, the cylinder we get has volume $\pi b \bigl(f(b)\bigr)^2$. On the other hand, the volume obtained by rotating the area under f from 0 to b is $\int_0^b \pi \left(f(x)\right)^2 dx$. If we define

$$g(b) = \frac{\int_0^b \pi (f(x))^2\, dx}{\pi b (f(b))^2} = \frac{\int_0^b (f(x))^2\, dx}{b (f(b))^2},$$

then we need to prove the existence of a positive b for which $g(b) = 1/2$.

We first prove the existence of a b with $g(b) > 1/2$. Since f is increasing and bounded, we must have $\lim_{x \to \infty} f(x) = c$ for some nonnegative constant c. Given ε satisfying $0 < \varepsilon < \min\{c, 1\}$, choose x_0 such that $f(x_0) \geq c - \varepsilon$. Then

$$\int_0^{x_0/\varepsilon} (f(x))^2\, dx \geq \int_{x_0}^{x_0/\varepsilon} (f(x))^2\, dx \geq x_0 \left(\frac{1-\varepsilon}{\varepsilon}\right)(c-\varepsilon)^2,$$

and thus for $b = x_0/\varepsilon$, we have

$$g(b) \geq \frac{(1-\varepsilon)(c-\varepsilon)^2}{c^2}.$$

Because ε may be arbitrarily small, $g(b)$ can be made arbitrarily close to 1, and in particular, greater than $1/2$.

To show that g takes on values less than $1/2$, we consider two cases. First, suppose that $f'(0) > 0$. Given $\varepsilon > 0$, there exists a $\delta > 0$ for which $0 \le x < \delta$ implies $|f(x)/x - f'(0)| < \varepsilon$. If, moreover, $\varepsilon < f'(0)$ and $b < \delta$, we then have

$$g(b) \le \frac{\int_0^b (f'(0) + \varepsilon)^2 x^2 \, dx}{b (f'(0) - \varepsilon)^2 b^2} = \frac{1}{3} \frac{(f'(0) + \varepsilon)^2}{(f'(0) - \varepsilon)^2}.$$

We conclude that g can be arbitrarily close to $1/3$, hence takes on values less than $1/2$.

If $f'(0) = 0$, define h on $[0, \infty)$ by $h(x) = f(x)/x$ for $x > 0$ and $h(0) = 0$. Observe that h is continuous, so h attains a maximum on $[0, 1]$, say at $x = b$. We then have $b > 0$, and

$$g(b) \le \frac{\int_0^b \left(\frac{f(b)}{b} x\right)^2 dx}{b (f(b))^2} = \frac{1}{3},$$

as desired.

We apply the Intermediate Value Theorem to conclude that there exists a b for which $g(b) = 1/2$.

Comments. The first part of our argument shows that $\lim_{b \to \infty} g(b) = 1$.

If we had assumed that f is continuously differentiable on $[0, \infty)$ and that $f'(0) > 0$, then we could have used l'Hôpital's rule and the Fundamental Theorem of Calculus to show that $\lim_{b \to 0^+} g(b) = 1/3$.

Problem 160

Suppose that we have three functions f, g, h. Then there are six possible compositions of the three, given by $f(g(h(x))), g(h(f(x))), \ldots$. Give an example of three continuous functions that are defined for all real x and for which exactly five of the six compositions are the same.

Idea. Look for functions f, g, h that are zero most of the time but not all the time, so that five of the six compositions are identically zero but the sixth one is not.

Solution. We construct one example. Let

$$g(x) = \begin{cases} 0 & \text{for } x \leq 2 \text{ and for } x \geq 3, \\ 1 - |2x - 5| & \text{for } 2 \leq x \leq 3, \end{cases}$$

$$h(x) = \begin{cases} 0 & \text{for } x \leq 1 \text{ and for } x \geq 2, \\ |2x - 3| - 1 & \text{for } 1 \leq x \leq 2, \end{cases}$$

with graphs

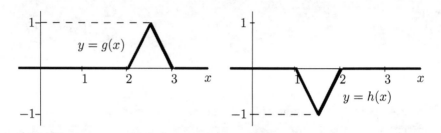

Then it is easy to check that for any x, $g(h(x)) = 0$ and $h(g(x)) = 0$. So regardless of what function f we choose, $f(g(h(x))) = f(h(g(x))) = f(0)$ for all x, while $g(h(f(x))) = h(g(f(x))) = 0$ for all x. So if we want exactly five of the six compositions to be the same, we had better make sure that $f(0) = 0$ and that either $g(f(h(x))) = 0$ for all x or $h(f(g(x))) = 0$ for all x, but not both. We can find another "tent" function to do the trick:

$$f(x) = \begin{cases} 0 & \text{for } x \leq 0 \text{ and for } x \geq 1, \\ \frac{3}{2} - 3\left|x - \frac{1}{2}\right| & \text{for } 0 \leq x \leq 1, \end{cases}$$

with graph

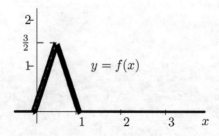

Problem 161

Does the Maclaurin series for e^{x-x^3} have any zero coefficients?

Answer. The Maclaurin series has no zero coefficients.

Solution 1. Let $f(x) = e^{x-x^3}$. The product and chain rules imply that $f^{(n)}(x) = g_n(x)e^{x-x^3}$ where $g_n(x)$ is a polynomial with integer coefficients. A few calculations may lead one to conjecture that $g_n(0)$ is always odd, which would imply the assertion. We prove a stronger statement by induction, namely that $g_n(0)$ is odd and $g'_n(0)$ is even. This is clear for $n = 0$, since $g_0(x) = 1$. Now assume it holds for n. We observe that

$$f^{(n+1)}(x) = g'_n(x)e^{x-x^3} + g_n(x)(1 - 3x^2)e^{x-x^3}.$$

Therefore, $g_{n+1}(x) = g'_n(x) + (1 - 3x^2)g_n(x)$, so $g_{n+1}(0) = g'_n(0) + g_n(0)$ is odd. Differentiating $g_{n+1}(x)$ yields

$$g'_{n+1}(x) = g''_n(x) - 6xg_n(x) + (1 - 3x^2)g'_n(x).$$

Since the second derivative of x^2 is 2, the constant term $g''_n(0)$ of $g''_n(x)$ will be even. Thus, $g'_{n+1}(0) = g''_n(0) + g'_n(0)$ is even, and our proof by induction is complete.

Solution 2. (Bruce Reznick, University of Illinois) Using the power series expansion of the exponential, we have

$$e^{x-x^3} = e^x e^{-x^3} = \sum_{i=0}^{\infty} \frac{x^i}{i!} \sum_{j=0}^{\infty} \frac{(-1)^j x^{3j}}{j!}.$$

Multiplying the series on the right and collecting terms, we see that the coefficient of x^n in the Maclaurin series is given by

$$a_n = \sum_{\substack{i+3j=n \\ i \geq 0, j \geq 0}} \frac{(-1)^j}{i!j!}.$$

Now multiply each side by $(n-1)!$ to get

$$(n-1)! a_n = \sum_{\substack{i+3j=n \\ i \geq 0, j \geq 0}} \frac{(-1)^j (n-1)!}{i!j!}$$

$$= \frac{1}{n} + \sum_{\substack{i+3j=n \\ i \geq 0, j > 0}} (-1)^j (n-1) \cdots (i+j+1) \binom{i+j}{i}.$$

This shows that $(n-1)! a_n$ is an integer plus $1/n$, so $a_n \neq 0$ for $n \geq 2$; it is easily seen that $a_0 = a_1 = 1$, so a_n is nonzero for all n.

Problem 162

Let a and d be relatively prime positive integers, and consider the sequence $a, a+d, a+4d, a+9d, \ldots, a+n^2d, \ldots$. Given a positive integer b, can one always find an integer in the sequence which is relatively prime to b?

Idea. We want $a+n^2d$ to differ from a (mod p) for those primes p which divide both a and b, but not for those which divide b but not a.

Solution. Yes, one can always find a positive integer n such that $a+n^2d$ is relatively prime to b. In fact, let n be the product of those prime factors of b which do not divide a. Then if a prime p divides b and not a, p will divide n^2d and hence not divide $a+n^2d$.

If a prime p divides both b and a, then p will not divide n, and since a, d are relatively prime, p will not divide d. Thus p will not divide n^2d and hence not divide $a+n^2d$. It follows that no prime which divides b can divide $a+n^2d$, and we are done.

Problem 163

A student provided the computation

$$\int_0^\infty \frac{e^{-x} - e^{-2x}}{x}\, dx = \int_0^\infty \frac{e^{-x}}{x}\, dx - \int_0^\infty \frac{e^{-u}}{u}\, du = 0,$$

using the substitution $u = 2x$ in the second part of the "split up" integral. The instructor was not too impressed by this, pointing out that for all positive values of x, $\frac{e^{-x} - e^{-2x}}{x}$ is positive, so how could the integral be zero?

a. Resolve this paradox.

b. *Mathematica* gives the answer $\int_0^\infty \frac{e^{-x} - e^{-2x}}{x}\, dx = \ln 2$. Is *this* answer correct?

Solution. a. The integrals $\int_0^\infty \frac{e^{-x}}{x}\, dx$ and $\int_0^\infty \frac{e^{-2x}}{x}\, dx$ are both divergent (the trouble is at 0, not at ∞; for example, when $0 < x < 1$, $e^{-x}/x > e^{-2x}/x > e^{-2}/x$), so it is not legitimate to split up the original improper integral into a difference of these two separate integrals.

b. *Mathematica* is indeed correct, as we'll show by using a more careful analysis of the same idea. First, split up the interval of integration, say at 1, and use the definition of an improper integral as a limit.

$$\int_0^\infty \frac{e^{-x} - e^{-2x}}{x}\,dx = \int_0^1 \frac{e^{-x} - e^{-2x}}{x}\,dx + \int_1^\infty \frac{e^{-x} - e^{-2x}}{x}\,dx$$

$$= \lim_{t \to 0^+} \int_t^1 \frac{e^{-x} - e^{-2x}}{x}\,dx + \lim_{T \to \infty} \int_1^T \frac{e^{-x} - e^{-2x}}{x}\,dx$$

$$= \lim_{t \to 0^+} \left(\int_t^1 \frac{e^{-x}}{x}\,dx - \int_t^1 \frac{e^{-2x}}{x}\,dx \right) +$$

$$\lim_{T \to \infty} \left(\int_1^T \frac{e^{-x}}{x}\,dx - \int_1^T \frac{e^{-2x}}{x}\,dx \right).$$

Now make the substitution $u = 2x$ in the second and fourth integrals and we get

$$\int_0^\infty \frac{e^{-x} - e^{-2x}}{x}\,dx = \lim_{t \to 0^+} \left(\int_t^1 \frac{e^{-x}}{x}\,dx - \int_{2t}^2 \frac{e^{-u}}{u}\,du \right) +$$

$$\lim_{T \to \infty} \left(\int_1^T \frac{e^{-x}}{x}\,dx - \int_2^{2T} \frac{e^{-u}}{u}\,du \right)$$

$$= \lim_{t \to 0^+} \left(\int_t^{2t} \frac{e^{-x}}{x}\,dx - \int_1^2 \frac{e^{-x}}{x}\,dx \right) + \lim_{T \to \infty} \left(\int_1^2 \frac{e^{-x}}{x}\,dx - \int_T^{2T} \frac{e^{-x}}{x}\,dx \right)$$

$$= \lim_{t \to 0^+} \int_t^{2t} \frac{e^{-x}}{x}\,dx - \lim_{T \to \infty} \int_T^{2T} \frac{e^{-x}}{x}\,dx.$$

The result will follow once we have established the following two claims.

Claim 1: $\displaystyle \lim_{t \to 0^+} \int_t^{2t} \frac{e^{-x}}{x}\,dx = \ln 2.$

The equation of the tangent line to $y = e^{-x}$ at $x = 0$ is $y = 1 - x$. Also, e^{-x} is decreasing and concave up for all x and consequently, $1 - x < e^{-x} < 1$ for $x > 0$. (Alternatively, for $0 < x < 1$, $e^{-x} = 1 - x + x^2/2! - x^3/3! + \cdots$ is an alternating series whose terms are decreasing in absolute value so the partial sums jump across, but ever nearer, the eventual limit.) Therefore, dividing by x and integrating over the interval $[t, 2t]$,

$$\int_t^{2t} \left(\frac{1}{x} - 1 \right) dx < \int_t^{2t} \frac{e^{-x}}{x}\,dx < \int_t^{2t} \frac{1}{x}\,dx$$

$$\ln 2 - t < \int_t^{2t} \frac{e^{-x}}{x}\, dx < \ln 2.$$

Now let $t \to 0^+$; the claim follows by the squeeze principle.

Claim 2: $\lim_{T \to \infty} \int_T^{2T} \frac{e^{-x}}{x}\, dx = 0.$

Because both e^{-x} and $1/x$ are positive and decreasing for $x > 0$, we have

$$0 < \int_T^{2T} \frac{e^{-x}}{x}\, dx < \int_T^{2T} \frac{e^{-T}}{T}\, dx = e^{-T};$$

once again, the claim follows by the squeeze principle.

Problem 164

Find the smallest possible n for which there exist integers $x_1, x_2, \ldots, x_n$ such that each integer between 1000 and 2000 (inclusive) can be written as the sum, without repetition, of one or more of the integers $x_1, x_2, \ldots, x_n$.

Idea. Because a set of 9 elements has 2^9 subsets, including the empty set, 9 integers can form at most $2^9 - 1 = 511$ positive sums, hence there certainly cannot be 1001 distinct, positive sums. On the other hand, it is not hard to construct sets of eleven integers which work, so we must try to rule out sets of 10 integers.

Solution. The smallest possible n is 11. By binary expansion, each integer between 1000 and 2000 can be written as the sum of some of the eleven integers $1, 2, 2^2, \ldots, 2^{10}$.

Now assume that ten integers $x_1, x_2, \ldots, x_{10}$ have the desired property; we will derive a contradiction. First suppose that two of the integers, say x_9 and x_{10}, are greater than 500. Consider a sum s of some of $x_1, x_2, \ldots, x_8$. Since $(s + x_9 + x_{10}) - s \geq 1002$, at most three of the four sums

$$s,\ s + x_9,\ s + x_{10},\ \text{and } s + x_9 + x_{10}$$

can be between 1000 and 2000 (inclusive). There are 2^8 sums s, including 0. Therefore, at most $3 \cdot 2^8 = 768$ sums of $x_1, x_2, \ldots, x_{10}$ can be between 1000 and 2000. Thus no two of $x_1, x_2, \ldots, x_{10}$ can be greater than 500; in other words, at least nine of $x_1, x_2, \ldots, x_{10}$ are less than or equal to 500. There are $\binom{9}{2} = 36$ sums of two of these nine integers; none of these sums is greater than 1000, so at least 35 of them are too small or redundant. Therefore, at most $2^{10} - 35 = 989$ of the integers from 1000 to 2000 can be represented as the sum of some of $x_1, x_2, \ldots, x_{10}$, and we are done.

Comment. The general question of the minimum number of integers needed to represent each integer between M and N (inclusive), where $0 < M < N$, seems to be more difficult. It is clear that at least $\log_2(N - M + 2)$ integers are required. On the other hand, the sets $\{M, 1, 2, 2^2, \ldots, 2^{\lfloor \log_2(N-M) \rfloor}\}$ and $\{1, 2, 2^2, \ldots, 2^{\lfloor \log_2 N \rfloor}\}$ both suffice. Therefore, the smallest possible n satisfies the inequalities

$$\log_2(N - M + 2) \leq n \leq \min\left\{\lfloor \log_2(N - M) \rfloor + 2, \lfloor \log_2 N \rfloor + 1\right\}.$$

Given M and N, at most two integers satisfy the above inequalities. However, the case $M = 1000, N = 2000$ of our problem shows that the upper bound can be attained, while $M = 25$, $N = 67$, $x_1 = 9$, $x_2 = 13$, $x_3 = 16$, $x_4 = 17$, $x_5 = 18$, $x_6 = 19$ shows that the smaller integer can be attained.

Problem 165

Two d-digit integers (with first digit $\neq 0$) are chosen randomly and independently, then multiplied together. Let P_d be the probability that the first digit of the product is 9. Find $\lim_{d \to \infty} P_d$.

Answer. We'll show that $\lim_{d \to \infty} P_d = \dfrac{1}{9} + \ln 9 - \dfrac{80}{81} \ln 10$.

Solution. Write the numbers as $x \cdot 10^{d-1}$ and $y \cdot 10^{d-1}$ ($1 \leq x < 10$, $1 \leq y < 10$). Then their product is $xy \cdot 10^{2d-2}$, so whether the first digit of the product is 9 is determined by xy. More specifically, the first digit of the product is 9 if and only if either $9 \leq xy < 10$ or $90 \leq xy < 100$. Therefore, if we introduce a square grid of "mesh" $1/10^{d-1}$ on the square $1 \leq x < 10$, $1 \leq y < 10$, then the probability P_d is the number of grid points for which $9 \leq xy < 10$ or $90 \leq xy < 100$ divided by the total number of grid points. In the limit as $d \to \infty$, this is just the ratio of the area of the region

$$\{(x, y) \mid 1 \leq x < 10, \ 1 \leq y < 10, \ 9 \leq xy < 10 \text{ or } 90 \leq xy < 100\}$$

(the area of the shaded region in the figure) to the area of the entire square

$$\{(x, y) \mid 1 \leq x < 10, 1 \leq y < 10\}$$

(which is 81).

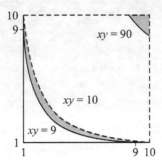

The area of the shaded region is

$$\left[\int_1^9 \left(\frac{10}{x} - \frac{9}{x}\right) dx + \int_9^{10} \left(\frac{10}{x} - 1\right) dx\right] + \int_9^{10} \left(10 - \frac{90}{x}\right) dx$$

$$= \int_1^9 \frac{dx}{x} + \int_9^{10} \left(\left(\frac{10}{x} - 1\right) + \left(10 - \frac{90}{x}\right)\right) dx$$

$$= \ln x \Big|_1^9 + (9x - 80 \ln x)\Big|_9^{10}$$

$$= \ln 9 + (90 - 80 \ln 10) - (81 - 80 \ln 9)$$

$$= 9 + 81 \ln 9 - 80 \ln 10,$$

so dividing this by 81 gives the answer.

Problem 166

Define $(x_n)_{n \geq 1}$ by

$$x_1 = 1, \qquad x_{n+1} = \frac{1}{\sqrt{2}}\sqrt{1 - \sqrt{1 - x_n^2}}.$$

a. Show that $\lim_{n \to \infty} x_n$ exists and find this limit.
b. Show that there is a unique number A for which $L = \lim_{n \to \infty} x_n/A^n$ exists as a finite nonzero number. Evaluate L for this value of A.

Idea. Looking at the first few terms leads us to suspect that the sequence is decreasing. The presence of $\sqrt{1 - x_n^2}$ in the recurrence suggests a "trig substitution."

Solution. a. We show that the limit is 0.

Suppose that $0 < x_n \leq 1$, and let $0 < \theta_n \leq \pi/2$ be such that $\sin \theta_n = x_n$. Then

$$x_{n+1} = \frac{1}{\sqrt{2}} \sqrt{1 - \sqrt{1 - \sin^2 \theta_n}} = \sqrt{\frac{1 - \cos \theta_n}{2}} = \sin(\theta_n/2).$$

This shows that $0 < x_{n+1} < x_n$. Moreover, since $\theta_1 = \pi/2$, it shows that $\theta_n = \pi/2^n$. Thus,

$$\lim_{n \to \infty} x_n = \lim_{n \to \infty} \sin\left(\frac{\pi}{2^n}\right) = 0.$$

b. $L = \pi$, for $A = 1/2$.

Suppose

$$\lim_{n \to \infty} \frac{\sin(\pi/2^n)}{A^n}$$

exists and is nonzero. Since

$$\lim_{x \to 0} \frac{\sin x}{x} = 1,$$

we have

$$\lim_{n \to \infty} \frac{\sin(\pi/2^n)}{A^n} = \lim_{n \to \infty} \frac{\sin(\pi/2^n)}{A^n} \lim_{n \to \infty} \frac{\pi/2^n}{\sin(\pi/2^n)}$$

$$= \lim_{n \to \infty} \frac{\sin(\pi/2^n)}{A^n} \frac{\pi/2^n}{\sin(\pi/2^n)}$$

$$= \pi \lim_{n \to \infty} \frac{1}{(2A)^n}.$$

This latter limit exists and is nonzero if and only if $A = 1/2$, in which case we have

$$\lim_{n \to \infty} \frac{\sin(\pi/2^n)}{A^n} = \pi.$$

Problem 167

Consider the line segments in the xy-plane formed by connecting points on the positive x-axis with x an integer to points on the positive y-axis with y an integer. We call a point in the first quadrant an *I-point* if it is the intersection of two such line segments. We call a point an *L-point* if there is a sequence of distinct I-points whose limit is the given point. Prove or disprove: If (x, y) is an L-point, then either x or y (or both) is an integer.

Solution. The given statement is true: If (x,y) is an L-point, then either x or y is an integer.

Suppose $(\overline{x},\overline{y})$ is a point in the first quadrant with neither $\overline{x}$ nor $\overline{y}$ an integer. We will show that $(\overline{x},\overline{y})$ is not an L-point. Choose positive integers M and N for which the line between $(M,0)$ and $(0,N)$ passes above the point $(\overline{x},\overline{y})$. Let ε be any positive number less than all three of the following: (i) the distance from the above line to $(\overline{x},\overline{y})$, (ii) the distance from $\overline{x}$ to the closest integer, and (iii) the distance from $\overline{y}$ to the closest integer.

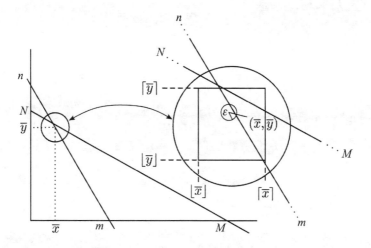

We claim that only finitely many lines whose x- and y-intercepts are both positive integers can pass within ε of $(\overline{x},\overline{y})$. To see this, suppose m and n are positive integers for which the line between $(m,0)$ and $(0,n)$ passes within ε of $(\overline{x},\overline{y})$. Then condition (i) above implies that either $m < M$ or $n < N$. Given m, consider the two tangent lines to the circle with center $(\overline{x},\overline{y})$ and radius ε which pass through $(m,0)$. Since the line connecting $(m,0)$ to $(0,n)$ must lie between these tangent lines, we see from condition (ii) above that there are at most finitely many possibilities for n. Similarly, condition (iii) implies that given n, there are at most finitely many possibilities for m. So, indeed, only finitely many lines whose x- and y-intercepts are both positive integers pass within ε of $(\overline{x},\overline{y})$. Since every I-point within ε of $(\overline{x},\overline{y})$ has to be one of the finitely many intersections of these lines, the point $(\overline{x},\overline{y})$ cannot be an L-point, hence the statement is true.

Comment. One can modify the above proof slightly to show that both coordinates of an L-point must be rational.

Problem 168

a. Find all lines which are tangent to both of the parabolas

$$y = x^2 \quad \text{and} \quad y = -x^2 + 4x - 4.$$

b. Now suppose $f(x)$ and $g(x)$ are any two quadratic polynomials. Find geometric criteria that determine the number of lines tangent to both of the parabolas $y = f(x)$ and $y = g(x)$.

Solution. a. There are two such lines: $y = 0$ and $y = 4x - 4$.

Suppose the line $y = mx + b$ is tangent to the parabola $y = x^2$ at the point $P = (x_1, x_1^2)$ and is also tangent to the parabola $y = -x^2 + 4x - 4$ at the point $Q = (x_2, -x_2^2 + 4x_2 - 4)$. Then, from calculus, the slope of the line is

$$m = 2x_1 = -2x_2 + 4 . \tag{1}$$

First suppose that $x_1 \neq x_2$; then we can also compute the slope from the fact that P and Q are both on the line. This, along with (1), yields

$$m = \frac{-x_2^2 + 4x_2 - 4 - x_1^2}{x_2 - x_1} = \frac{-x_2^2 + 4x_2 - 4 - (2 - x_2)^2}{x_2 - (2 - x_2)} \tag{2}$$
$$= -\frac{(x_2 - 2)^2}{x_2 - 1}.$$

From (1) and (2) we find that $x_2 = 0$ or $x_2 = 2$. For $x_2 = 0$ we get $x_1 = 2$; the tangent line to $y = x^2$ at $(2, 4)$ is $y = 4x - 4$, which is also tangent to $y = -x^2 + 4x - 4$ at $(0, -4)$. For $x_2 = 2$ we get $x_1 = 0$; the tangent line to $y = x^2$ at $(0, 0)$ is $y = 0$, which is also tangent to $y = -x^2 + 4x - 4$ at $(2, 0)$.

On the other hand, if $x_1 = x_2$, equation (1) implies that $x_1 = x_2 = 1$, $P = (1, 1)$, and $Q = (1, -1)$. That is to say, the line joining P and Q is the vertical line $x = 1$, which is not the tangent line at either point, a contradiction. This completes the consideration of all possibilities.

b. If $f(x) = g(x)$, there will obviously be infinitely many common tangent lines to $y = f(x)$ and $y = g(x)$. Otherwise, there will be either zero, one, or two common tangent lines. If the two parabolas open in the same direction (i.e., both upward or both downward), then there will be no common tangent line if the parabolas do not intersect, one common tangent line if they intersect in exactly one point, and two common tangent lines if the two parabolas intersect in two distinct points. On the other hand, if the two parabolas open in opposite directions, then there will be two common tangent lines if the parabolas do not intersect, one if they intersect in exactly one point, and none if they intersect in two distinct points. (Note: The parabolas intersect in at most two points since $f(x) - g(x)$ is a nonzero polynomial of degree ≤ 2, so $f(x) - g(x) = 0$ can have at most two solutions.)

To prove these assertions, note that we can shift the coordinate axes so the vertex of $y = f(x)$ is at the origin; also, replacing $f(x)$ by $-f(x)$ and $g(x)$ by $-g(x)$ if necessary, we can assume that the parabola $y = f(x)$ opens upward. Thus, $f(x) = \alpha x^2$ for some $\alpha > 0$. But now, by replacing x by $x/\sqrt{\alpha}$ (a change of scale along the x-axis), we may assume that $f(x) = x^2$. (Of course, $g(x)$ will change accordingly, but it will still be a quadratic polynomial, and the number of common tangent lines and the number of intersection points will not change.)

So we can assume $f(x) = x^2$, $g(x) = Ax^2 + Bx + C$, $A \neq 0$, $g(x) \neq f(x)$. Let $P = (x_1, x_1^2)$ be a point on $y = f(x)$ and $Q = (x_2, Ax_2^2 + Bx_2 + C)$ be a point on $y = g(x)$. As in (a), if m denotes the slope of a common tangent line, we have

$$m = 2x_1 = 2Ax_2 + B, \tag{1}$$

and if $x_1 \neq x_2$,

$$\begin{aligned} m &= \frac{Ax_2^2 + Bx_2 + C - x_1^2}{x_2 - x_1} \\ &= \frac{(A - A^2)x_2^2 + (1 - A)Bx_2 + (C - B^2/4)}{(1 - A)x_2 - B/2}. \end{aligned} \tag{2}$$

Case 1. Suppose that $x_1 = x_2$. In this case P and Q must be the same point, because otherwise the line connecting them would be vertical and could not be the tangent line. So we have

$$x_1^2 = Ax_2^2 + Bx_2 + C. \tag{3}$$

Substituting $x_1 = x_2$ into equations (1) and (3) yields

$$(A - 1)x_2 + \frac{B}{2} = 0 \tag{4}$$

and

$$(A - 1)x_2^2 + Bx_2 + C = 0.$$

Subtracting x_2 times (4) from this last equation yields

$$\frac{B}{2} x_2 + C = 0. \tag{5}$$

Equations (4) and (5) will have a simultaneous solution if $B = C = 0$ or if $B \neq 0$ and $(A-1)C - B^2/4 = 0$. In the first of these cases we see that $g(x) = Ax^2$, $A \neq 1$, and so we find the unique common tangent line $y = 0$, at $(0, 0)$. In the second case, we have a common tangent line of slope $-4C/B$ at $(-2C/B, 4C^2/B^2)$.

Case 2. If $x_1 \neq x_2$, combining (1) and (2) yields

$$(A^2 - A)x_2^2 + ABx_2 + \left(\frac{B^2}{4} + C\right) = 0. \tag{6}$$

If $A = 1$, then $B \neq 0$ from (1), since $x_1 \neq x_2$. We find that there is a unique solution, with $x_2 = -B/4 - C/B$ and $x_1 = B/4 - C/B$. If $A \neq 1$, then the quadratic equation (6) for x_2 will have two, one, or no solutions depending on whether its discriminant $\Delta = A\left(B^2 - 4C(A-1)\right)$ is positive, zero, or negative. Note that if there is a common tangent line under Case 1, then either $B = C = 0$ or $B^2 = 4C(A-1)$; in both cases $\Delta = 0$. If $A \neq 1$ and $\Delta > 0$, there will be exactly two common tangent lines (corresponding to the two solutions for x_2), while if $\Delta < 0$ there will be none. If $A \neq 1$ and $\Delta = 0$, the double root of (6) will be

$$x_2 = -\frac{AB}{2(A^2 - A)} = \frac{B}{2(1-A)};$$

using (1), this yields

$$x_1 = A\frac{B}{2(1-A)} + \frac{B}{2} = \frac{B}{2(1-A)} = x_2,$$

a contradiction. However, in this case the parabolas have exactly one common tangent line, coming from Case 1.

To summarize, if $A \neq 1$, the parabolas have two, one, or no common tangent lines depending on whether $\Delta = A\left(B^2 - 4C(A-1)\right)$ is positive, zero, or negative. If $A = 1$, the parabolas have one common tangent line when $B \neq 0$ and none when $B = 0$.

On the other hand, the intersections of the parabolas are given by

$$x^2 = Ax^2 + Bx + C, \quad \text{or} \quad (A-1)x^2 + Bx + C = 0.$$

If $A \neq 1$, this has two, one, or no solutions according to whether

$$D = B^2 - 4C(A-1)$$

is positive, zero, or negative. Since $\Delta = AD$, we see that if $A > 0$ (parabolas open the same way) and $A \neq 1$, there are two, one, or zero common tangent lines if there are two, one, or zero intersections, while for $A < 0$ (parabolas open in opposite directions) it is the other way around. Finally, if $A = 1$, there is a single intersection point when $B \neq 0$ and no intersection when $B = 0$; the parabolas open the same way, and there are one or zero common tangent lines if there are one or zero intersections, respectively.

Problem 169

Suppose we are given an m-gon and an n-gon in the plane. Consider their intersection; assume this intersection is itself a polygon.

a. If the m-gon and the n-gon are convex, what is the maximal number of sides their intersection can have?
b. Is the result from (a) still correct if only one of the polygons is assumed to be convex?

Solution. a.   The maximal number of sides that the intersection of a convex m-gon and a convex n-gon can have is $m+n$. To see why more than $m+n$ sides is not possible, first note that any side of the intersection must be part or all of one of the sides of one of the two original polygons, and that there are $m+n$ sides in all of the two original polygons. Thus if the intersection had more than $m+n$ sides, at least two of these would have to be part of the same side of one of the original polygons, which cannot happen because the polygons are convex and hence have convex intersection. Therefore, the intersection can have at most $m+n$ sides.

To show that there really can be $m+n$ sides, we will assume $n \geq m \geq 3$. We will start with a regular $(m+n)$-gon R and show that one can find a convex m-gon X and a convex n-gon Y whose intersection is R. An example will help illustrate the idea. If $m = 3$ and $n = 5$, then the regular octagon R is the intersection of triangle ABC and pentagon $DEFGH$, as shown.

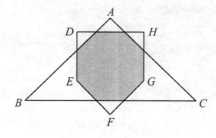

As in the example, the idea in general is to get the m-gon and the n-gon by extending suitable sides of R; we have to make sure that the sides will intersect when we expect them to and that the resulting polygons are convex. So we want to be able to divide the $m + n$ sides of R into two groups: m of the sides will be parts of sides of X while the other n will be parts of sides of Y. If two adjacent sides of R are both parts of sides of X, the vertex where they meet will be a vertex of X, and similarly for Y. On the other hand, if for two adjacent sides of R one is part of a side of X and the other is part of a side of Y, then both these sides get extended beyond the vertex until they meet the extensions of the "next" sides of X, Y respectively.

To make sure that two successive (parts of) sides of X always meet when they are extended, it is enough to make sure that one is less than halfway around the regular $(m + n)$-gon R from the other; the angle at which these sides meet is then certainly less than 180°. So we will have our convex polygons X and Y by choosing m of the sides of R to be sides of X and the others to be sides of Y, provided we can do this in such a way that two successive sides of X or of Y are never halfway or more around the $(m + n)$-gon from each other.

If $m + n$ is even, such an assignment can be carried out by choosing two opposite sides of R, designating one of them as a side of X and the other as a side of Y, and then designating the sides adjacent to each of the two chosen sides as belonging to the polygon which that chosen side *does not* belong to (see figure). This works since $m \geq 3$, $n \geq 3$; the other sides can be designated at random in such a way as to end up with m sides of X and n sides of Y.

If $m + n$ is odd, there are no "opposite sides" of R, but we can designate a side and the two sides adjacent to the opposite vertex as sides of X, then designate the four sides adjacent to these three chosen sides as sides of Y (see figure). This works because $m \geq 3$, $n \geq 4$. Once again, all other sides can be designated arbitrarily so as to end up with m sides of X and n sides of Y.

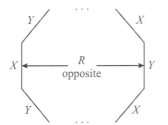 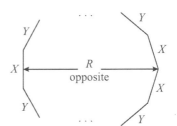

b. No. The following diagram shows that the intersection of a non-convex 4-gon and a convex 3-gon can be an 8-gon.

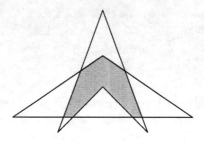

Problem 170

Suppose we start with a Pythagorean triple (a, b, c) of positive integers, that is, positive integers a, b, c such that $a^2 + b^2 = c^2$ and which can therefore be used as the side lengths of a right triangle. Show that it is not possible to have another Pythagorean triple (b, c, d) with the same integers b and c; that is, show that $b^2 + c^2$ can never be the square of an integer.

Solution. Suppose we did have positive integers a, b, c, d such that both $a^2 + b^2 = c^2$ and $b^2 + c^2 = d^2$. If b, c had a common factor k, then k^2 would divide both a^2 and d^2, so k would be a factor of a and d as well, and we could replace a, b, c, d by new integers $a_1 = a/k$, $b_1 = b/k$, $c_1 = c/k$, $d_1 = d/k$ and still have $a_1^2 + b_1^2 = c_1^2$ and $b_1^2 + c_1^2 = d_1^2$. So we may assume that $\gcd(b, c) = 1$. In particular, b and c cannot both be even. If c is even and b is odd, then a has to be odd, but then $a^2 \equiv 1 \pmod 4$ and $b^2 \equiv 1 \pmod 4$, so $c^2 \equiv 1+1 = 2 \pmod 4$. However, all even squares are 0 (mod 4), contradiction. So c is odd. If b is odd also, then $b^2 \equiv 1 \pmod 4$, $c^2 \equiv 1 \pmod 4$, so $d^2 \equiv 1+1 = 2 \pmod 4$, contradiction.

So far we know that $\gcd(b, c) = 1$ and b is even, c is odd (and a is odd). It is known (for a reference, see Problem 118, Solution 2) that under these conditions, the positive integers a, b, c with $a^2 + b^2 = c^2$ must have the form $a = s^2 - t^2$, $b = 2st$, $c = s^2 + t^2$ for some positive integers s, t, where $\gcd(s, t) = 1$ and s, t are of opposite parity. Similarly, because $b^2 + c^2 = d^2$, we have $b = 2vw$, $c = v^2 - w^2$, $d = v^2 + w^2$ for some positive integers v, w, where $\gcd(v, w) = 1$ and v, w are of opposite parity.

In particular, $s^2 + t^2 = v^2 - w^2$, so $v^2 = s^2 + t^2 + w^2$. Because s, t are of opposite parity, $s^2 + t^2 \equiv 1 \pmod 4$ and so w cannot be odd lest $v^2 \equiv 2 \pmod 4$. Consider the equations

$$s^2 + t^2 = c = v^2 - w^2$$

$$2st = b = 2vw.$$

If we remove the condition that $a > 0$, then with respect to these equations, there is no loss of generality in assuming that s is odd and t is even. Let $z = \gcd(s, v)$ and $y = \gcd(t, w)$. Note for later reference that y is even and z is odd. Then from $st = vw$,

$$\frac{s}{z} \cdot \frac{t}{y} = \frac{v}{z} \cdot \frac{w}{y}.$$

Because $\gcd(s/z, v/z) = \gcd(t/y, w/y) = 1$, we can set

$$x = \frac{s}{z} = \frac{w}{y},$$

$$u = \frac{v}{z} = \frac{t}{y}.$$

Note that x and u are odd and relatively prime. Substituting into

$$s^2 + t^2 = c = v^2 - w^2,$$

we get

$$x^2 z^2 + u^2 y^2 = u^2 z^2 - x^2 y^2,$$

or

$$x^2(y^2 + z^2) = u^2(z^2 - y^2).$$

Now $y^2 + z^2$ and $z^2 - y^2$ are relatively prime, because any common factor would be an odd factor of both $2y^2$ and $2z^2$. Therefore,

$$x^2 = z^2 - y^2, \quad \text{that is,} \quad x^2 + y^2 = z^2, \text{ and}$$

$$y^2 + z^2 = u^2.$$

Observe that $u < t < c < d$. Thus we would have a new "double Pythagorean triple" whose largest integer is *less* than the largest integer in the original one. Repeating this process, we arrive at a contradiction, because the sequence of largest (positive) integers cannot decrease indefinitely. We conclude that we can never have had a "double Pythagorean triple" in the first place!

Problem 171

If an insect starts at a random point inside a circular plate of radius R and crawls in a straight line in a random direction until it reaches the edge of the plate, what will be the average distance it travels to the edge?

Solution. The average distance traveled to the edge of the plate is

$$\frac{8}{3\pi} r \approx .8488\, r.$$

We begin by considering the average distance $\overline{D}_X$ to the perimeter if the insect starts at a particular point X inside the circle. Because of symmetry, $\overline{D}_X$ will only depend on the distance from X to the center O of the circle; suppose this distance is a. As shown in the figure, let $f(\alpha)$ be the distance from X to the perimeter along the line at angle α (counterclockwise) from the radius through X.

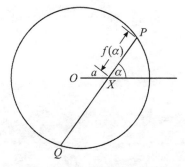

Since the insect crawls off in a random direction, we have

$$\overline{D}_X = \frac{1}{2\pi} \int_0^{2\pi} f(\alpha)\, d\alpha$$

as the average distance it will travel from X to the perimeter. Note that the distance from X to the perimeter at Q (in the opposite direction from P) is $f(\alpha + \pi)$, and that

$$f(\alpha) + f(\alpha + \pi) = PQ = 2\sqrt{r^2 - a^2 \sin^2 \alpha}.$$

THE SOLUTIONS

Thus, we have

$$\overline{D}_X = \frac{1}{2\pi} \int_0^{2\pi} f(\alpha)\, d\alpha = \frac{1}{2\pi} \int_0^{\pi} (f(\alpha) + f(\alpha + \pi))\, d\alpha$$

$$= \frac{1}{\pi} \int_0^{\pi} \sqrt{r^2 - a^2 \sin^2 \alpha}\, d\alpha$$

$$= \frac{2}{\pi} \int_0^{\pi/2} \sqrt{r^2 - a^2 \sin^2 \alpha}\, d\alpha.$$

To get the overall average, we have to average $\overline{D}_X$ over all points X within the circle. This is most easily done using polar coordinates. (Since r is being used for the radius of the circle, we will continue to write a for the radial polar coordinate.) The required average is

$$\frac{1}{\pi r^2} \int_0^{2\pi} \int_0^{r} \left(\frac{2}{\pi} \int_0^{\pi/2} \sqrt{r^2 - a^2 \sin^2 \alpha}\, d\alpha \right) a\, da\, d\theta$$

$$= \frac{4}{\pi r^2} \int_0^{\pi/2} \int_0^{r} a\sqrt{r^2 - a^2 \sin^2 \alpha}\, da\, d\alpha$$

$$= \frac{4}{\pi r^2} \int_0^{\pi/2} \left[-\frac{1}{3} \frac{(r^2 - a^2 \sin^2 \alpha)^{3/2}}{\sin^2 \alpha} \right]_{a=0}^{a=r} d\alpha$$

$$= \frac{4}{\pi r^2} \int_0^{\pi/2} \frac{-1}{3 \sin^2 \alpha} \left[(r^2 - r^2 \sin^2 \alpha)^{3/2} - (r^2)^{3/2} \right] d\alpha$$

$$= \frac{4}{\pi r^2} \int_0^{\pi/2} \frac{-1}{3 \sin^2 \alpha} (r^3 \cos^3 \alpha - r^3)\, d\alpha$$

$$= \frac{4r}{3\pi} \int_0^{\pi/2} \frac{1 - \cos^3 \alpha}{\sin^2 \alpha}\, d\alpha.$$

To continue with this improper integral, we first evaluate the indefinite integral.

$$\int \frac{1 - \cos^3 \alpha}{\sin^2 \alpha}\, d\alpha = \int \frac{1 - (1 - \sin^2 \alpha) \cos \alpha}{\sin^2 \alpha}\, d\alpha$$

$$= \int \left(\csc^2 \alpha - \frac{\cos \alpha}{\sin^2 \alpha} + \cos \alpha \right) d\alpha$$

$$= -\cot \alpha + \frac{1}{\sin \alpha} + \sin \alpha + C$$

$$= \frac{1 - \cos \alpha}{\sin \alpha} + \sin \alpha + C.$$

Therefore,

$$\int_0^{\pi/2} \frac{1-\cos^3\alpha}{\sin^2\alpha}\, d\alpha = \lim_{b\to 0^+}\left[\frac{1-\cos\alpha}{\sin\alpha} + \sin\alpha\right]_b^{\pi/2}$$

$$= \lim_{b\to 0^+}\left(1+1-\frac{1-\cos b}{\sin b} - \sin b\right)$$

$$= 2,$$

and the average distance the insect crawls to the perimeter of the circle is

$$\frac{4r}{3\pi}\int_0^{\pi/2}\frac{1-\cos^3\alpha}{\sin^2\alpha}\,d\alpha = \frac{8}{3\pi}r.$$

Problem 172

Let $ABCD$ be a parallelogram in the plane. Describe and sketch the set of all points P in the plane for which there is an ellipse with the property that the points A, B, C, D, and P all lie on the ellipse.

Answer. The set of points P for which there is such an ellipse can be described by first extending all sides of the parallelogram indefinitely in all directions. Then P is in the set if and only if (0) P is one of the four points A, B, C, D, or (1) P is outside the parallelogram *and* (2) there is a pair of parallel extended sides of the parallelogram such that P is between those sides (see figure).

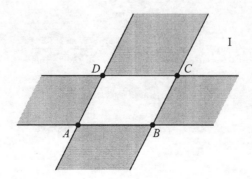

Solution 1. To show why this description is correct, choose coordinates in the plane such that $A = (0,0)$, $B = (1,0)$, and $D = (\lambda, \mu)$. Then $C = (\lambda+1, \mu)$. We may assume that $\mu > 0$. For the four points $(0,0)$, $(1,0)$, (λ, μ), $(\lambda+1, \mu)$ to lie on a conic section $ax^2 + bxy + cy^2 + dx + ey + f = 0$, we must have

$$f = 0,$$
$$a + d + f = 0,$$
$$a\lambda^2 + b\lambda\mu + c\mu^2 + d\lambda + e\mu + f = 0,$$
$$a(\lambda+1)^2 + b(\lambda+1)\mu + c\mu^2 + d(\lambda+1) + e\mu + f = 0.$$

Since $f = 0$, the second, third, and fourth equations simplify to

$$a + d = 0, \tag{1}$$
$$a\lambda^2 + b\lambda\mu + c\mu^2 + d\lambda + e\mu = 0, \tag{2}$$
$$a(\lambda+1)^2 + b(\lambda+1)\mu + c\mu^2 + d(\lambda+1) + e\mu = 0.$$

Subtracting these final two equations, one obtains $2a\lambda + a + b\mu + d = 0$, hence by (1),

$$2a\lambda + b\mu = 0. \tag{3}$$

Combining (1), (2), and (3), we find that $e = (a\lambda^2 - c\mu^2 + a\lambda)/\mu$. Thus, the constraints imply that any conic section through A, B, C, and D has the form

$$ax^2 - 2a\frac{\lambda}{\mu}xy + cy^2 - ax + \frac{a\lambda^2 - c\mu^2 + a\lambda}{\mu}y = 0 \tag{4}$$

for some real numbers a and c. Such a conic section is an ellipse if and only if its discriminant is negative. Therefore, a point $P = (x,y)$ is in the set if and only if there exist real numbers a and c such that (4) holds and $(-2a\lambda/\mu)^2 - 4ac < 0$; this inequality is equivalent to $a^2\lambda^2/\mu^2 < ac$. Clearly, $a \neq 0$, and dividing the equation of the conic section by a we get

$$x^2 - 2\frac{\lambda}{\mu}xy + \frac{c}{a}y^2 - x + \frac{\lambda^2 - (c/a)\mu^2 + \lambda}{\mu}y = 0.$$

If we now put $c_1 = c/a$, we see that P is in the set if and only if there exists a number c_1 such that

$$x^2 - 2\frac{\lambda}{\mu}xy + c_1 y^2 - x + \frac{\lambda^2 - c_1\mu^2 + \lambda}{\mu}y = 0 \quad \text{and} \quad \frac{\lambda^2}{\mu^2} < c_1.$$

The equality can be rewritten as

$$c_1(y^2 - \mu y) = -x^2 + 2\frac{\lambda}{\mu}xy + x - \frac{\lambda^2 + \lambda}{\mu}y. \qquad (5)$$

We now distinguish three cases.

Case 1. $y^2 - \mu y = 0$. Then $y = 0$ or $y = \mu$. If $y = 0$, the equality yields $x = 0$ or $x = 1$, and we find that $P = A$ or $P = B$.

Similarly, for $y = \mu$ we find that $P = C$ or $P = D$.

Case 2. $y^2 - \mu y > 0$. In this case $\lambda^2/\mu^2 < c_1$ is equivalent to

$$(\lambda^2/\mu^2)(y^2 - \mu y) < c_1(y^2 - \mu y),$$

and this together with (5) shows that P is in the set if and only if the following equivalent inequalities hold:

$$\frac{\lambda^2}{\mu^2}(y^2 - \mu y) < -x^2 + 2\frac{\lambda}{\mu}xy + x - \frac{\lambda^2 + \lambda}{\mu}y,$$

$$\lambda^2 y^2 < -\mu^2 x^2 + 2\lambda\mu xy + \mu^2 x - \lambda\mu y,$$

$$(\lambda y - \mu x)(\lambda y - \mu x + \mu) < 0,$$

$$\lambda y - \mu x < 0 \quad \text{and} \quad \lambda y - \mu x + \mu > 0,$$

$$-\mu < \lambda y - \mu x < 0.$$

(Note that since $\mu > 0$, it would be impossible to have $\lambda y - \mu x > 0$ and $\lambda y - \mu x + \mu < 0$.)

Now it is easy to check that $\lambda y - \mu x = 0$ is the equation of line AD, while $\lambda y - \mu x = -\mu$ is the equation of BC. So the condition $-\mu < \lambda y - \mu x < 0$ implies that P is between (the extensions of) AD and BC. On the other hand, $y^2 - \mu y > 0$ is equivalent to ($y < 0$ or $y > \mu$), which says that P is not between AB and CD.

Case 3. $y^2 - \mu y < 0$. The same computation, but with the inequalities reversed, shows that P is not between the lines AD and BC, while P is between (the extensions of) AB and CD. Combining the results of the three cases, we get the answer given above.

Solution 2. Since the set consisting of an ellipse and its interior is convex, it is impossible for a point on an ellipse to lie inside the triangle determined by three other points on that ellipse. We see from this that only the points P

described above can be on the same ellipse as A, B, C, and D. For instance, if P is in the region labeled "I" in the figure, then C is inside triangle BDP, so P cannot be on any ellipse on which B, C, and D lie.

To see the converse, choose a coordinate system for the plane such that $A = (0,0)$, $B = (1,0)$, $C = (\lambda + 1, \mu)$, and $D = (\lambda, \mu)$, where $\mu \neq 0$. The linear transformation T taking (x, y) to $(x - (\lambda/\mu)y, (1/\mu)y)$ takes A, B, C, and D to $(0,0)$, $(1,0)$, $(1,1)$, and $(0,1)$, respectively. Since both T and T^{-1} map straight lines to straight lines and ellipses to ellipses, we see that it suffices to consider the case $A = (0,0)$, $B = (1,0)$, $C = (1,1)$, and $D = (0,1)$. In this case the points $P = (r, s)$, other than A, B, C, D, described above lie in two horizontal and two vertical "half-strips" given by $(r - r^2)/(s^2 - s) > 0$. For such a point P, the curve

$$(x^2 - x) + \frac{r - r^2}{s^2 - s}(y^2 - y) = 0$$

is an ellipse (in fact, the unique ellipse) through A, B, C, D, and P.

Problem 173

Find $\lim\limits_{n \to \infty} \int_0^\infty \dfrac{n \cos\left(\sqrt[4]{x/n^2}\right)}{1 + n^2 x^2} \, dx.$

Idea. First, we note that we cannot switch the limit and the integral, for if we could, we could start by showing that $\lim\limits_{n \to \infty} \dfrac{n \cos\left(\sqrt[4]{x/n^2}\right)}{1 + n^2 x^2} = 0$ for *any* nonzero x, and then conclude that the answer would be 0. This answer is incorrect, and a bit of numerical approximation using technology yields approximate values of the integral as follows:

n	approximate value
3	1.37144
5	1.47507
200	1.5704
2000	1.57078

These values suggest that the limit might be $\pi/2 \approx 1.5708$. To prove this, we note that once n gets large the integrand will be close to zero unless x is small. On the other hand, for values of x reasonably close to zero, $\sqrt[4]{x/n^2}$

should be close to zero, so $\cos\left(\sqrt[4]{x/n^2}\right)$ should be close to 1. Therefore, we can hope that the integral is close to $\displaystyle\int_0^\infty \frac{n\,dx}{1+n^2x^2} = \arctan(nx)\Big|_0^\infty = \pi/2$.

Solution. With the reasoning just presented, we see that it will be enough to show that

$$\lim_{n\to\infty}\left(\int_0^\infty \frac{n\,dx}{1+n^2x^2} - \int_0^\infty \frac{n\cos\left(\sqrt[4]{x/n^2}\right)}{1+n^2x^2}\,dx\right) = 0,$$

so we want to show that

$$\lim_{n\to\infty}\int_0^\infty \frac{n\left(1-\cos\left(\sqrt[4]{x/n^2}\right)\right)}{1+n^2x^2}\,dx = 0.$$

For this, note that

$$0 \leq \int_0^\infty \frac{n\left(1-\cos\left(\sqrt[4]{x/n^2}\right)\right)}{1+n^2x^2}\,dx = \int_0^\infty \frac{2n\sin^2\left(\tfrac{1}{2}\sqrt[4]{x/n^2}\right)}{1+n^2x^2}\,dx$$

$$< \int_0^\infty \frac{\tfrac{1}{2}\sqrt{x}}{1+n^2x^2}\,dx \quad \text{(because } \sin\theta < \theta \text{ for } \theta > 0\text{)}$$

$$= \frac{1}{2n^{3/2}}\int_0^\infty \frac{\sqrt{u}}{1+u^2}\,du \quad \text{(with } u = nx\text{)}.$$

Because the last integral converges to a value independent of n, the result follows by the squeeze principle.

Problem 174

Let x_0 be a rational number, and let $(x_n)_{n\geq 0}$ be the sequence defined recursively by

$$x_{n+1} = \left|\frac{2x_n^3}{3x_n^2 - 4}\right|.$$

Prove that this sequence converges, and find its limit as a function of x_0.

Solution.

$$\lim_{n\to\infty} x_n = \begin{cases} 0 & \text{if } |x_0| < 2/\sqrt{5}, \\ 2 & \text{if } |x_0| > 2/\sqrt{5}. \end{cases}$$

Note that all x_n are actually rational numbers, since when x_n is rational, $3x_n^2 - 4$ will not be zero and so x_{n+1} will again be rational. Also note

THE SOLUTIONS   303

that after the initial term, the sequence starting with x_0 is the same as the sequence starting with $-x_0$, so we can assume $x_0 \geq 0$.

Let $f(x) = 2x^3/(3x^2 - 4)$, so that $x_{n+1} = |f(x_n)|$. We can use standard techniques to sketch the graph of f (shown in the figure for $x \geq 0$ only). From this graph it seems reasonable to suspect that $2 < f(x) < x$ for $x > 2$, and it is straightforward to show this. Thus, if $x_n > 2$ for some n, the sequence will thereafter decrease, but stay above 2. Since any bounded decreasing sequence has a limit, $L = \lim_{n \to \infty} x_n$ will then exist. Taking limits on both sides of $x_{n+1} = f(x_n)$ and using the continuity of f for $x \geq 2$, we then have $L = f(L)$, so $L(3L^2 - 4) = 2L^3$ and $L(L^2 - 4) = 0$. Of the three roots, $L = 0$, $L = \pm 2$, only $L = 2$ is possible in this case, since $x_n > 2$ for large enough n. In particular, we see that $L = 2$ if $x_0 > 2$. We can also see that $L = 2$ if $x_0 = 2$ (in this case, $x_n = 2$ for all n) and even that $L = 2$ if $2/\sqrt{3} < x_0 < 2$, for in that case $x_1 = f(x_0) > 2$.

Now consider what happens if $0 \leq x_0 < 2/\sqrt{3}$. If $x_0 = 0$, then $x_n = 0$ for all n, so $L = 0$. From the graph, we expect that for x close to 0,

$$|f(x)| = \frac{2x^3}{4 - 3x^2}$$

will be even closer to 0, while for x close enough to $2/\sqrt{3}$, $|f(x)|$ becomes large and thus $|f(x)| > x$.

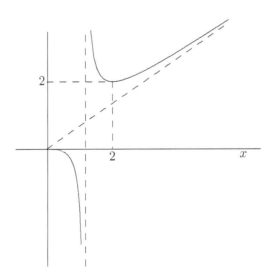

A simple calculation shows that $2x^3/(4 - 3x^2) = x$ for $x = 0$ and for $x = \pm 2/\sqrt{5}$. If $0 < x_n < 2/\sqrt{5}$ at any point in the sequence, then it is easily shown that $0 < x_{n+1} = |f(x_n)| < x_n$, and so forth, so the sequence will decrease (and be bounded) from that point on. Thus, L will exist; as before, we have $L = |f(L)| = 2L^3/(4 - 3L^2)$, which implies $L = 0$ or $L = \pm 2/\sqrt{5}$. However, once $0 < x_n < 2/\sqrt{5}$ and (x_n) is decreasing, we must have $L = 0$.

Since x_0 is rational, $x_0 = 2/\sqrt{5}$ is impossible, so we are left with the case $2/\sqrt{5} < x_0 < 2/\sqrt{3}$. In this case, $x_1 = |f(x_0)| > x_0$, so the sequence increases at first. Suppose that $x_n < 2/\sqrt{3}$ for all n. In this case, the sequence is bounded and increasing, so L exists. But from $L = |f(L)|$ and $0 < L \leq 2/\sqrt{3}$ we again get $L = 0$ or $L = \pm 2/\sqrt{5}$, a contradiction since $x_0 > 2/\sqrt{5}$ and (x_n) is increasing. It follows that there is an n with $x_n > 2/\sqrt{3}$. But then $x_{n+1} \geq 2$, so $L = 2$, and we are done.

Comment. The stipulation that x_0 be rational is stronger than necessary. However, some restriction is needed to ensure that x_{n+1} is defined for all n.

Problem 175

Let f be a continuous function on $[0, 1]$, which is bounded below by 1, but is not identically 1. Let R be the region in the plane given by $0 \leq x \leq 1$, $1 \leq y \leq f(x)$. Let

$$R_1 = \{(x, y) \in R \, | \, y \leq \bar{y}\} \quad \text{and} \quad R_2 = \{(x, y) \in R \, | \, y \geq \bar{y}\},$$

where $\bar{y}$ is the y-coordinate of the centroid of R. Can the volume obtained by rotating R_1 about the x-axis equal that obtained by rotating R_2 about the x-axis?

Answer. Yes, the volumes can be equal. For example, this happens if

$$f(x) = \frac{22}{13} - \frac{9}{13}x.$$

Solution 1. To find an example, we let R be the triangle with vertices $(0, 1)$, $(1, 1)$, and $(0, b)$, where the parameter $b > 1$ will be chosen later. Then

$$f(x) = b - (b - 1)x.$$

There are several expressions for $\bar{y}$ in terms of integrals. For instance,

$$\bar{y} = \frac{\int_1^b y \frac{b-y}{b-1} \, dy}{\int_1^b \frac{b-y}{b-1} \, dy} = \frac{b+2}{3}.$$

THE SOLUTIONS

The horizontal line $y = (b+2)/3$ intersects $y = f(x)$ at $(2/3, (b+2)/3)$.
The volume obtained by rotating R_2 about the x-axis is then

$$\int_0^{2/3} \left(\pi \left(f(x)\right)^2 - \pi \left(\frac{b+2}{3} \right)^2 \right) dx = \frac{20b^2 - 4b - 16}{81} \pi.$$

To find the volume obtained by rotating R_1 about the x-axis, we compute the volume obtained by rotating R, and subtract the volume obtained by rotating R_2. The calculation yields

$$\frac{7b^2 + 31b - 38}{81} \pi.$$

The two volumes are equal when $b = 22/13$ and when $b = 1$, but the solution $b = 1$ is degenerate.

Solution 2. As in the first solution, we let R be the triangle with vertices $(0, 1)$, $(1, 1)$, and $(0, b)$, but this time we avoid computing integrals by using geometry. Recall that by Pappus' theorem, the volume swept out by rotating a region about an exterior axis is the product of the area of the region and the distance traveled by its centroid. Therefore, if $\bar{y}_1$ and $\bar{y}_2$ are the y-coordinates of the centroids of R_1 and R_2 respectively, the two volumes will be equal if and only if

$$(\text{Area of } R_1) \cdot 2\pi \bar{y}_1 = (\text{Area of } R_2) \cdot 2\pi \bar{y}_2.$$

Now R is a triangle, so its centroid, being the intersection of its medians, is two-thirds of the way down from $(0, b)$ to the line $y = 1$; that is, $\bar{y} = (b+2)/3$. Similarly, for the triangle R_2 we have

$$\bar{y}_2 = \frac{b + 2\bar{y}}{3} = \frac{5b + 4}{9}.$$

Also, $\bar{y}$ is the weighted average of $\bar{y}_1$ and $\bar{y}_2$, where the respective weights are the areas of R_1 and R_2. By similar triangles, the area of R_2 is $4/9$ times the area of R, so

$$\bar{y} = \frac{5}{9} \bar{y}_1 + \frac{4}{9} \bar{y}_2;$$

this yields $\bar{y}_1 = (7b+38)/45$. The condition for the volumes to be equal now becomes

$$\frac{5}{9} \cdot \frac{7b+38}{45} = \frac{4}{9} \cdot \frac{5b+4}{9},$$

and we again find $b = 22/13$.

Comment. These solutions illustrate a useful method of constructing examples and counterexamples: take a parametrized family for which computations are feasible, and then choose the parameter(s) so as to get the desired properties.

Problem 176

Let $n \geq 3$ be a positive integer. Begin with a circle with n marks about it. Starting at a given point on the circle, move clockwise, skipping over the next two marks and placing a new mark; the circle now has $n + 1$ marks. Repeat the procedure beginning at the new mark. Must a mark eventually appear between each pair of the original marks?

Answer. Yes, a mark must eventually appear between any pair of original marks.

Solution. To show why, suppose there is a pair m_1, m_2 of consecutive original marks such that no mark will ever appear between them. This implies that as you move around the circle, you always put a mark just before m_1, then skip over both m_1 and m_2 and put the next mark right after m_2, as shown in the figure.

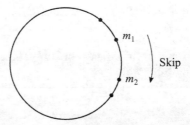

Now suppose that for some particular time around the circle, there are x marks in all as you skip m_1 and m_2. On your next circuit (counting from that moment), you must again arrive just before m_1 in order to skip over m_1 and m_2, so x must be even and you will be adding $x/2$ marks during this next circuit, for a total of $3x/2$. However, not all integers in the sequence $x, (3/2)x, (3/2)^2 x, \ldots$ can be even. Thus on some circuit there will be an odd number of marks as you skip m_1 and m_2, and on the next circuit a mark will appear between m_1 and m_2, a contradiction.

Comments. One can see explicitly when a mark will appear between any two original marks, as follows. Label the original marks consecutively from 1

THE SOLUTIONS

to n, and let I_a be the interval between the mark labeled a and the next mark; suppose that the starting position is in interval I_n. Label the new marks beginning with $n+1$. Then if $n = 2^j k$, where k is odd, mark $(3^{j+1}k+1)/2$ will be the first one to fall in interval I_1, and one must go around the circle $j+1$ times before this happens.

More generally, for $1 \leq a \leq n$, the number of the first mark to fall in interval I_a is $P(n,a)$, where

$$P(n,a) = \begin{cases} n + a/2, & \text{if } a \text{ is even}; \\ (3^{j+1}k+1)/2, & \text{if } a \text{ is odd}, n + (a-1)/2 = 2^j k, k \text{ odd}. \end{cases}$$

Replacing "two marks" by "m marks" (and "$n \geq 3$" by "$n \geq m+1$") in the statement of the problem yields an interesting generalization, which we have not yet been able to solve.

Problem 177

Let

$$c = \sum_{n=1}^{\infty} \frac{1}{n(2^n - 1)} = 1 + \frac{1}{6} + \frac{1}{21} + \frac{1}{60} + \cdots.$$

Show that

$$e^c = \frac{2}{1} \cdot \frac{4}{3} \cdot \frac{8}{7} \cdot \frac{16}{15} \cdots.$$

Solution. Let P_k be the partial product

$$\frac{2}{1} \cdot \frac{4}{3} \cdot \frac{8}{7} \cdots \frac{2^k}{2^k - 1}.$$

We want to show that

$$\lim_{k \to \infty} P_k = e^c,$$

where

$$c = \sum_{n=1}^{\infty} \frac{1}{n(2^n - 1)}.$$

By continuity of the exponential and logarithm functions, this is equivalent to showing that $\lim_{k \to \infty} \ln(P_k) = c$. Now

$$\ln(P_k) = \ln\left(\frac{2}{1} \cdot \frac{4}{3} \cdot \frac{8}{7} \cdots \frac{2^k}{2^k-1}\right)$$

$$= \ln\left(\frac{2}{1}\right) + \ln\left(\frac{4}{3}\right) + \ln\left(\frac{8}{7}\right) + \cdots + \ln\left(\frac{2^k}{2^k-1}\right)$$

$$= -\ln\left(\frac{1}{2}\right) - \ln\left(\frac{3}{4}\right) - \ln\left(\frac{7}{8}\right) - \cdots - \ln\left(\frac{2^k-1}{2^k}\right)$$

$$= -\ln\left(1-\frac{1}{2}\right) - \ln\left(1-\frac{1}{4}\right) - \ln\left(1-\frac{1}{8}\right) - \cdots - \ln\left(1-\frac{1}{2^k}\right)$$

$$= \sum_{i=1}^{k} -\ln\left(1-\frac{1}{2^i}\right).$$

Using the Maclaurin expansion (Taylor expansion about 0)

$$-\ln(1-x) = x + \frac{x^2}{2} + \frac{x^3}{3} + \cdots = \sum_{n=1}^{\infty} \frac{x^n}{n} \quad (|x|<1),$$

we get

$$\ln(P_k) = \sum_{i=1}^{k} \sum_{n=1}^{\infty} \frac{(1/2^i)^n}{n} = \sum_{n=1}^{\infty} \sum_{i=1}^{k} \frac{1}{n2^{in}} = \sum_{n=1}^{\infty} \frac{1}{n} \sum_{i=1}^{k} \frac{1}{(2^n)^i}.$$

Then, because

$$\sum_{i=1}^{k} \frac{1}{(2^n)^i} = \frac{1}{2^n} + \frac{1}{(2^n)^2} + \cdots + \frac{1}{(2^n)^k}$$

is a finite geometric series with sum

$$\frac{1}{2^n} \cdot \frac{1-(1/2^n)^k}{1-1/2^n} = \frac{1-(1/2^n)^k}{2^n-1},$$

we have

$$\ln(P_k) = \sum_{n=1}^{\infty} \frac{1-(1/2^n)^k}{n(2^n-1)} = \sum_{n=1}^{\infty} \frac{1}{n(2^n-1)} - \sum_{n=1}^{\infty} \frac{1}{n(2^n-1)2^{nk}}$$

since the latter two series converge. From the estimate

$$\sum_{n=1}^{\infty} \frac{1}{n(2^n-1)2^{nk}} < \frac{1}{2^k} \sum_{n=1}^{\infty} \frac{1}{n(2^n-1)}$$

we conclude that $\lim_{k\to\infty} \ln(P_k) = \sum_{n=1}^{\infty} \frac{1}{n(2^n-1)} = c$, as claimed.

Problem 178

Let $q(x) = x^2 + ax + b$ be a quadratic polynomial with real roots. Must all roots of $p(x) = x^3 + ax^2 + (b-3)x - a$ be real?

Answer. Yes, the roots of $p(x) = x^3 + ax^2 + (b-3)x - a$ will all be real if the roots of $q(x) = x^2 + ax + b$ are real.

Solution 1. Note that if we replace x by $-x$, we get the same problem with a replaced by $-a$; thus, we may assume that $a \geq 0$. Let $r_1 \leq r_2$ be the roots of $q(x)$. We then have $r_1 + r_2 = -a$, hence $r_1 \leq 0$. If we note that $p(x) = xq(x) - (3x + a)$, we find $p(r_1) = -3r_1 - a = -2r_1 + r_2 \geq 0$. If $a > 0$, then from $p(0) = -a < 0$ and the Intermediate Value Theorem, $p(x)$ has a negative and a positive root, hence a third real root as well. If $a = 0$, then in order for $x^2 + b$ to have real roots, we must have $b \leq 0$. But then all roots of the polynomial $p(x) = x^3 + (b-3)x$ will be real.

Solution 2. It is easy to check that
$$p(x) = \operatorname{Re}\big((x+i)q(x+i)\big) = \big((x+i)q(x+i) + (x-i)q(x-i)\big)/2.$$
If z is a (complex) root of $p(x)$, then, in particular,
$$\big|(z+i)\,q(z+i)\big| = \big|(z-i)\,q(z-i)\big|,$$
or, if we denote the (real) roots of $q(x)$ by r_1 and r_2,
$$\big|(z+i)(z+i-r_1)(z+i-r_2)\big| = \big|(z-i)(z-i-r_1)(z-i-r_2)\big|.$$
If z has positive imaginary part, then each of the three factors on the left-hand side of the equation will be larger in absolute value than the corresponding factor on the right-hand side, a contradiction. Similarly, z cannot have negative imaginary part; hence, z must be real.

Solution 3. Note that
$$p(x) = x(x^2 + ax + b) - (2x + a) - x$$
$$= xq(x) - q'(x) - x;$$
we will use this connection between $p(x)$ and $q(x)$ to show that all roots of $p(x)$ are real. Let r_1 and r_2, with $r_1 \leq r_2$, be the roots of $q(x)$. Since the graph of $q(x)$ opens upward, we have $q'(r_1) \leq 0$ and $q'(r_2) \geq 0$, with equality only when $r_1 = r_2$. We now distinguish three cases.

Case 1. $r_1 \leq 0$, $r_2 \geq 0$. In this case, $p(r_1) = -q'(r_1) - r_1 \geq 0$ and $p(r_2) \leq 0$.

Thus, by the Intermediate Value Theorem, $p(x)$ has a root in the interval $[r_1, r_2]$. Since $\lim_{x\to -\infty} p(x) = -\infty$ and $\lim_{x\to\infty} p(x) = \infty$, $p(x)$ also has roots in each of the intervals $(-\infty, r_1]$ and $[r_2, \infty)$, so we are done with Case 1 unless $p(r_1) = 0$ or $p(r_2) = 0$. If $p(r_1) = 0$, we must have $q'(r_1) = r_1 = 0$, so $r_1 = r_2 = 0$, hence $q(x) = x^2$ and $p(x) = x^3 - 3x$, with all roots real. The case $p(r_2) = 0$ is similar.

Case 2. $r_1 > 0, r_2 > 0$. Because $q(x) = (x-r_1)(x-r_2)$, we have $a = -r_1 - r_2$, so $a < 0$ and $p(0) = -a > 0$. On the other hand, $p(r_2) = -q'(r_2) - r_2 < 0$, so $p(x)$ has roots in each of the intervals $(-\infty, 0)$, $(0, r_2)$, and (r_2, ∞).

Case 3. $r_1 < 0, r_2 < 0$. This time, $a = -r_1 - r_2 > 0$, so $p(0) < 0$, while $p(r_1) = -q'(r_1) - r_1 > 0$. Thus $p(x)$ has roots in each of the intervals $(-\infty, r_1)$, $(r_1, 0)$, and $(0, \infty)$, and we are done.

Problem 179

a. For what numbers α is $\int_0^\infty \left(\frac{\pi}{2} - \arctan(x^\alpha)\right) dx$ convergent?

b. Evaluate $\lim_{\alpha \to \infty} \int_0^\infty \left(\frac{\pi}{2} - \arctan(x^\alpha)\right) dx$.

Answer. a. $\int_0^\infty \left(\frac{\pi}{2} - \arctan(x^\alpha)\right) dx$ converges for $\alpha > 1$ and for no other α.

b. $\lim_{\alpha \to \infty} \int_0^\infty \left(\frac{\pi}{2} - \arctan(x^\alpha)\right) dx = \frac{\pi}{2}$.

Solution. a. First note that $\pi/2 - \arctan(x^\alpha)$ is a positive continuous function of x (for any given α), so the only problem with the integral is the upper bound ∞. If $\alpha \leq 0$, the integrand does not approach 0 as $x \to \infty$, so there is no hope of the integral converging. In fact, for any $x > 1$ we see that as a function of α, x^α is increasing, so $\arctan(x^\alpha)$ is increasing, and $\pi/2 - \arctan(x^\alpha)$ is decreasing (and positive). Therefore, if the integral converges for $\alpha = \alpha_0$, it also converges for all α with $\alpha > \alpha_0$. So it's enough to show that the integral diverges for $\alpha = 1$ and converges for all $\alpha > 1$.

For $\alpha = 1$, we can find an antiderivative, using integration by parts:

$$\int \left(\frac{\pi}{2} - \arctan x\right) dx = \frac{\pi}{2} x - x \arctan x + \int \frac{x}{1+x^2} dx$$

$$= \frac{\pi}{2} x - x \arctan x + \frac{1}{2} \ln(1+x^2) + C.$$

Therefore,

$$\int_0^\infty \left(\frac{\pi}{2} - \arctan x\right) dx = \lim_{b\to\infty} \left[\frac{\pi}{2} x - x\arctan x + \frac{1}{2}\ln(1+x^2)\right]_0^b$$

$$= \lim_{b\to\infty}\left[\frac{\pi}{2}b - b\arctan b + \frac{1}{2}\ln(1+b^2)\right].$$

Now $\pi/2 > \arctan b$ for all b, so this limit is at least $\lim_{b\to\infty} \frac{1}{2}\ln(1+b^2) = \infty$, so the integral diverges.

Now consider $\alpha > 1$. Observe that $\pi/2 - \arctan(x^\alpha) = \arctan(1/x^\alpha)$. Because

$$\frac{d}{d\theta}(\theta - \arctan\theta) = 1 - \frac{1}{1+\theta^2} > 0,$$

for $\theta > 0$ we have the inequality $\theta > \arctan\theta$. Therefore,

$$\int_1^\infty \left(\frac{\pi}{2} - \arctan(x^\alpha)\right) dx = \int_1^\infty \arctan\frac{1}{x^\alpha}\, dx < \int_1^\infty \frac{1}{x^\alpha}\, dx = \frac{1}{\alpha-1},$$

completing the proof.

b. From our work in part (a), we see that

$$\lim_{\alpha\to\infty} \int_1^\infty \left(\frac{\pi}{2} - \arctan(x^\alpha)\right) dx = 0 \quad \text{(by the squeeze principle)},$$

so we can focus on finding $\lim_{\alpha\to\infty} \int_0^1 \left(\frac{\pi}{2} - \arctan(x^\alpha)\right) dx$. The idea is that for large α, $\arctan(x^\alpha)$ will be arbitrarily close to 0 on almost the entire interval $[0,1]$. More specifically, given $\varepsilon > 0$, we will have $\arctan(1-\varepsilon)^\alpha < \varepsilon$ for $\alpha > \ln(\tan\varepsilon)/\ln(1-\varepsilon)$. For such α,

$$\frac{\pi}{2} > \int_0^1 \left(\frac{\pi}{2} - \arctan(x^\alpha)\right) dx > \int_0^{1-\varepsilon} \left(\frac{\pi}{2} - \varepsilon\right) dx = (1-\varepsilon)\left(\frac{\pi}{2} - \varepsilon\right).$$

Then by the squeeze principle,

$$\lim_{\alpha\to\infty} \int_0^1 \left(\frac{\pi}{2} - \arctan(x^\alpha)\right) dx = \frac{\pi}{2},$$

so

$$\lim_{\alpha\to\infty} \int_0^\infty \left(\frac{\pi}{2} - \arctan(x^\alpha)\right) dx = \frac{\pi}{2}.$$

Problem 180

Let $p(x) = x^3 + a_1 x^2 + a_2 x + a_3$ have rational coefficients and have roots r_1, r_2, r_3. If $r_1 - r_2$ is rational, must r_1, r_2, and r_3 be rational?

Answer. Yes, if $r_1 - r_2$ is rational, then r_1, r_2, and r_3 must all be rational.

Solution 1. Let $q = r_1 - r_2$ be rational. Note that if either r_1 or r_2 is rational, then so is the other, and since $r_1 + r_2 + r_3 = -a_1$ is rational, so is r_3. On the other hand, if r_3 is rational, then so is $r_1 + r_2 = -a_1 - r_3$, hence $r_1 = \bigl((r_1+r_2) + (r_1-r_2)\bigr)/2$ and r_2 are rational also. Thus it is enough to show that $p(x)$ has at least *one* rational root. We do this by showing that $p(x)$ is reducible over the rationals; for then, since it is a cubic polynomial, it must have a linear factor, and therefore a rational root.

Using the rational numbers $q = r_1 - r_2$ and $-a_1 = r_1 + r_2 + r_3$, we can express the roots r_2 and r_3 as $r_2 = r_1 - q$, $r_3 = -2r_1 + q - a_1$. We then have

$$a_2 = r_1 r_2 + r_1 r_3 + r_2 r_3$$
$$= r_1(r_1 - q) + r_1(-2r_1 + q - a_1) + (r_1 - q)(-2r_1 + q - a_1)$$
$$= -3r_1^2 + q_1 r_1 + q_2,$$

where q_1 and q_2 are rational. Therefore, r_1 is a root of the rational quadratic polynomial $p_1(x) = 3x^2 - q_1 x + (a_2 - q_2)$. Since $p(x)$ and $p_1(x)$ have the root r_1 in common, their greatest common divisor $d(x)$, which has rational coefficients (and can be found using the Euclidean algorithm), also has r_1 as a root. Hence $d(x)$, which has degree 1 or 2, is a divisor of $p(x)$, so $p(x)$ is reducible over the rationals, and we are done.

Solution 2. (Eugene Luks, University of Oregon) As in Solution 1, it is enough to show that $p(x)$ is reducible over the rationals. Consider the polynomial $p_1(x) = p(x + r_1 - r_2)$. Since $r_1 - r_2$ is rational, $p_1(x)$ has rational coefficients. Also, $p_1(r_2) = p(r_1) = 0$, so $p(x)$ and $p_1(x)$ have r_2 as a common root. Thus the greatest common divisor $d(x)$ of $p(x)$ and $p_1(x)$ is a (monic) rational polynomial of degree at least 1, which shows that $p(x)$ is reducible unless $p(x) = d(x) = p_1(x)$. If $p(x) = p_1(x)$, then $r_1 = r_2$ and $p(x)$ has a multiple root r_1. But then r_1 is also a root of $p'(x)$, so the greatest common divisor of $p(x)$ and $p'(x)$, which is rational, has degree 1 or 2, so once again $p(x)$ is reducible over the rationals.

Solution 3. As in Solution 1, it is enough to show that $p(x)$ is reducible over the rationals. Suppose not. Then by Galois theory, the splitting field $\mathbb{Q}(r_1, r_2, r_3)$ of $p(x)$ has an automorphism σ which takes r_1 to r_2, r_2 to r_3, and r_3 to r_1. Now $r_1 - r_2$ is rational and hence fixed by σ, so we have $r_1 - r_2 = \sigma(r_1) - \sigma(r_2) = r_2 - r_3$, that is, $r_3 = 2r_2 - r_1$. On the other hand, $r_1 + r_2 + r_3 = -a_1$ is rational, so $r_1 + r_2 + (2r_2 - r_1) = 3r_2$ is rational, r_2 is rational, and we are done.

Comment. One natural generalization to polynomials of degree 4 does not hold. That is, there exist fourth-degree rational polynomials with irrational roots r_1, r_2, r_3, r_4 for which two differences of pairs of roots are rational; one example is $p(x) = x^4 + 4$, with roots $1+i$, $-1+i$, $1-i$, $-1-i$ and $r_1 - r_2 = r_3 - r_4 = 2$.

Problem 181

Let $f(x) = x^3 - 3x + 3$. Prove that for any positive integer P, there is a "seed" value x_0 such that the sequence $x_0, x_1, x_2, \ldots$ obtained from Newton's method, given by

$$x_{n+1} = x_n - \frac{f(x_n)}{f'(x_n)},$$

has period P.

Idea. If the "seed" is planted at a point just to the right of the critical point $x = -1$, the following term in the sequence will be a large positive number and the next few terms will decrease (see figure). If the seed is planted exactly right, the sequence might return to the seed after P steps.

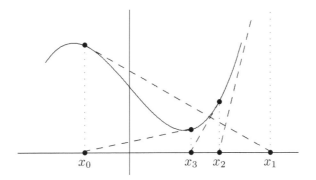

Solution. We first consider the special case $P = 1$. For the sequence to have period 1, we need $f(x_0) = 0$. Since any cubic polynomial with real coefficients has a real root, we can indeed choose x_0 such that $f(x_0) = 0$; we then have $x_1 = x_0$, the sequence has period 1, and we are done.

From now on we assume $P > 1$. Let

$$F(x) = x - \frac{f(x)}{f'(x)} = \frac{2x^3 - 3}{3(x^2 - 1)},$$

so that the sequence is $x_0, x_1 = F(x_0), x_2 = F(F(x_0)), \ldots$. We have

$$F'(x) = \frac{f''(x)f(x)}{(f'(x))^2} = \frac{6x}{(3(x^2-1))^2}(x^3 - 3x + 3).$$

Since $f(x)$ increases on the interval $[1, \infty)$, it is at least 1 there, so we see that $F'(x) > 0$ on $(1, \infty)$. Thus, F is increasing on $(1, \infty)$. Note that

$$\lim_{x \to 1^+} F(x) = -\infty \quad \text{and} \quad \lim_{x \to \infty} F(x) = \infty.$$

Therefore, if F is restricted to $(1, \infty)$, it has a continuous inverse G whose domain is $(-\infty, \infty)$. Thus, for any real number x, $G(x)$ is the unique number in $(1, \infty)$ for which $F(G(x)) = x$. (Note that to go from x to $G(x)$ is "backtracking" in Newton's method.) Let G^k denote the composition of k copies of G; that is,

$$G^k(x) = \underbrace{G \circ G \circ \cdots \circ G}_{k \text{ times}}(x) = \underbrace{G(G(\cdots(G(x))\cdots))}_{k \text{ times}}.$$

Define $H(x)$ on $(-1, 0]$ by $H(x) = F(x) - G^{P-1}(x)$. Since F and G are continuous on $(-1, 0]$, so is H, and since $G^{P-1}(-1)$ is finite,

$$\lim_{x \to -1^+} H(x) = \infty.$$

Meanwhile, $F(0) = 1$ and all values of G are greater than 1, so $H(0) < 0$. Therefore, by the Intermediate Value Theorem there exists an $x_0 \in (-1, 0)$ for which $H(x_0) = 0$. We claim that the sequence starting from this seed value x_0 has period P. First, since $H(x_0) = 0$, we have $F(x_0) = G^{P-1}(x_0)$ and thus

$$x_P = F^P(x_0) = F^{P-1}(G^{P-1}(x_0)) = x_0.$$

To show that the sequence cannot have a period less than P, note that $F(x) < x$ for all x in $(1, \infty)$, so $G(x) > x$ for all x. Therefore,

$$x_1 = G^{P-1}(x_0) > x_2 = G^{P-2}(x_0) > \cdots > x_{P-1} = G(x_0) > x_0,$$

and x_P is the first term of the sequence to return to x_0.

Problem 182

Show that
$$\sum_{k=0}^{n} \frac{(-1)^k}{2n+2k+1} \binom{n}{k} = \frac{\left(2^n (2n)!\right)^2}{(4n+1)!}.$$

Solution. Starting with the left-hand side, we have

$$\sum_{k=0}^{n} \binom{n}{k} \frac{(-1)^k}{2n+2k+1} = \sum_{k=0}^{n} \binom{n}{k}(-1)^k \int_0^1 u^{2n+2k} \, du$$

$$= \int_0^1 u^{2n} \sum_{k=0}^{n} \binom{n}{k}(-u^2)^k \, du$$

$$= \int_0^1 u^{2n}(1-u^2)^n \, du.$$

To evaluate this integral (which is available from tables, see the comments below), we generalize it by defining

$$F(k,m) = \int_0^1 u^k (1-u^2)^m \, du;$$

note that we are looking for $F(2n, n)$. Now $F(k, 0) = 1/(k+1)$, and integration by parts yields

$$F(k,m) = \frac{2m}{k+1} F(k+2, m-1)$$

for $m > 0$. Applying this recursive equation repeatedly, we find

$$F(2n,n) = \frac{2^n \, n!}{(2n+1)(2n+3)\cdots(4n-1)} F(4n, 0)$$

$$= \frac{2^n \, n!}{(2n+1)(2n+3)\cdots(4n-1)(4n+1)}$$

$$= \frac{(2n)!}{(2n)!} \cdot \frac{2^n \, n!}{(2n+1)(2n+3)\cdots(4n-1)(4n+1)}$$

$$\cdot \frac{2^n(n+1)(n+2)\cdots(2n)}{(2n+2)(2n+4)\cdots(4n)}$$

$$= \frac{\left(2^n (2n)!\right)^2}{(4n+1)!},$$

as was to be shown.

Comments. There are several other approaches to finding the integral

$$I = \int_0^1 u^{2n}(1-u^2)^n \, du \, .$$

One is to convert I into a "standard" integral (found in tables) by the substitution $u^2 = x$. This yields

$$I = \tfrac{1}{2} \int_0^1 x^{n-\frac{1}{2}}(1-x)^n \, dx = \tfrac{1}{2} \mathrm{B}(n+\tfrac{1}{2}, n+1),$$

where B is the *beta function* defined by

$$\mathrm{B}(s,t) = \int_0^1 x^{s-1}(1-x)^{t-1} \, dx, \qquad s, t > 0.$$

It is known that

$$\mathrm{B}(s,t) = \frac{\Gamma(s)\Gamma(t)}{\Gamma(s+t)}.$$

Here Γ is the *gamma function*:

$$\Gamma(s) = \int_0^\infty x^{s-1} e^{-x} dx, \qquad s > 0.$$

This last function is known to have the specific values

$$\Gamma(n+\tfrac{1}{2}) = \frac{(2n)!}{4^n \, n!} \sqrt{\pi}, \qquad n = 1, 2, 3, \ldots,$$

$$\Gamma(n) = (n-1)!, \qquad n = 1, 2, 3, \ldots,$$

and so we get

$$I = \tfrac{1}{2} \mathrm{B}(n+\tfrac{1}{2}, n+1)$$

$$= \frac{1}{2} \frac{\Gamma(n+\tfrac{1}{2})\Gamma(n+1)}{\Gamma(2n+\tfrac{3}{2})}$$

$$= \frac{1}{2} \frac{\dfrac{(2n)!}{4^n \, n!} \sqrt{\pi} \, n!}{\dfrac{(4n+2)!}{4^{2n+1}(2n+1)!} \sqrt{\pi}}$$

$$= \frac{\left(2^n \, (2n)!\right)^2}{(4n+1)!}.$$

Another approach, which essentially derives the expression for the beta function in terms of gamma functions, starts with the substitution $u = \sin\theta$,

which yields
$$I = \int_0^{\pi/2} \sin^{2n}\theta \cos^{2n+1}\theta \, d\theta.$$

By considering the double integral
$$\iint_R (r\sin\theta)^{2n} (r\cos\theta)^{2n+1} e^{-r^2} r \, dr \, d\theta,$$
where R is the first quadrant, and converting it to rectangular coordinates, one can then find that
$$I \cdot \int_0^\infty r^{4n+2} e^{-r^2} dr = \iint_R (r\sin\theta)^{2n} (r\cos\theta)^{2n+1} e^{-r^2} r \, dr \, d\theta$$
$$= \int_0^\infty y^{2n} e^{-y^2} dy \cdot \int_0^\infty x^{2n+1} e^{-x^2} dx.$$

Changing all integration variables to t, we have
$$I = \frac{\int_0^\infty t^{2n} e^{-t^2} dt \cdot \int_0^\infty t^{2n+1} e^{-t^2} dt}{\int_0^\infty t^{4n+2} e^{-t^2} dt} = \frac{I_{2n} I_{2n+1}}{I_{4n+2}},$$

where $I_k = \int_0^\infty t^k e^{-t^2} dt$.

Now a straightforward substitution yields $I_1 = \frac{1}{2}$, and integration by parts shows that $I_k = ((k-1)/2) \cdot I_{k-2}$ for $k \geq 2$. Therefore,
$$I = \frac{I_{2n} I_{2n+1}}{I_{4n+2}} = \frac{I_{2n} \cdot n \cdot (n-1) \cdots 1 \cdot I_1}{\frac{4n+1}{2} \cdot \frac{4n-1}{2} \cdots \frac{2n+1}{2} \cdot I_{2n}}$$
$$= \frac{n! \, 2^n}{(4n+1)(4n-1)\cdots(2n+1)},$$

as in the solution above.

Problem 183

Suppose a and b are distinct real numbers such that
$$a-b, \, a^2-b^2, \, \ldots, \, a^k-b^k, \, \ldots$$
are all integers.
a. Must a and b be rational?
b. Must a and b be integers?

Solution. Yes, a and b must be rational, and in fact, a and b must be integers.

a. Since a and b are distinct, we know that $a + b = (a^2 - b^2)/(a - b)$ is a rational number. Therefore, $a = \bigl((a + b) + (a - b)\bigr)/2$ is rational and hence b is also rational.

b. Since $a - b$ is an integer, a and b must have the same denominator, say n, when expressed in lowest terms. If we put $a = c/n$ and $b = d/n$, we see that n^k divides $c^k - d^k$ for $k = 1, 2, \ldots$.

If $n = 1$, we are done; otherwise, let p be a prime factor of n. Since n divides $c - d$, we have $c \equiv d \pmod{p}$. Now note that

$$c^k - d^k = (c - d)(c^{k-1} + c^{k-2}d + \cdots + cd^{k-2} + d^{k-1}),$$

and since $c \equiv d \pmod{p}$, the second factor is

$$c^{k-1} + c^{k-2}d + \cdots + cd^{k-2} + d^{k-1} \equiv kd^{k-1} \pmod{p}.$$

We have $d \not\equiv 0 \pmod{p}$ since $b = d/n$ is in lowest terms, so provided k is not divisible by p, we have $kd^{k-1} \not\equiv 0 \pmod{p}$. Thus p does not divide the second factor above of $c^k - d^k$, and since p^k divides $c^k - d^k$, p^k must divide $c - d$. However, we can choose arbitrarily large k which are not divisible by p, and so we have a contradiction.

Problem 184

The mayor of Wohascum Center has ten pairs of dress socks, ranging through ten shades of color from medium gray (1) to black (10). A pair of socks is unacceptable for wearing if the colors of the two socks differ by more than one shade. What is the probability that if the socks get paired at random, they will be paired in such a way that all ten pairs are acceptable?

Idea. Acceptable pairings can be built up by starting with an extreme color.

Solution. The probability is

$$\frac{683}{19 \cdot 17 \cdot 15 \cdots 3} \approx 10^{-6}.$$

To show this, we first show that there are 683 acceptable ways to pair the socks. Suppose we had n pairs of socks in gradually darkening shades, with the same condition for a pair to be acceptable. Let $f(n)$ be the number of pairings into acceptable pairs. For $n \leq 2$, any pairing is acceptable, and we have $f(1) = 1$, $f(2) = 3$. For $n > 2$, consider one sock from pair n (the darkest pair). It must be paired with its match or with one of the socks in pair $n-1$. If it is paired with its match, there are $f(n-1)$ acceptable ways to pair the remaining $n-1$ pairs. If it is paired with either of the two socks from pair $n-1$, its match must be paired with the other sock from that pair, which leaves $n-2$ matching pairs and $f(n-2)$ acceptable ways to pair them. Since there are two ways to do this, we conclude that, for $n > 2$,

$$f(n) = f(n-1) + 2\,f(n-2).$$

The characteristic equation of this recurrence relation is $r^2 - r - 2 = 0$, which has solutions $r = 2$ and $r = -1$. Therefore, we have

$$f(n) = a\,2^n + b\,(-1)^n$$

for some constants a and b. From $f(1) = 1$ and $f(2) = 3$, we find $a = 2/3, b = 1/3$, so

$$f(n) = \frac{2^{n+1} + (-1)^n}{3}.$$

In particular, the number of acceptable ways to pair 10 pairs of socks is $f(10) = 683$.

To compute the probability that a random pairing is acceptable, we divide $f(10)$ by the total number of pairings. To find the total number of pairings, note that a given sock can be paired in 19 ways. One given sock of the remaining 18 can be paired in 17 ways, and so forth. Thus the total number of pairings is $19 \cdot 17 \cdot 15 \cdots 3 \cdot 1$, and we get the probability given above.

Comment. To solve the recursion $f(n) = f(n-1) + 2f(n-2)$ using generating functions, set

$$G(z) = \sum_{n=1}^{\infty} f(n) z^n.$$

The recursion yields

$$G(z) = z + 3z^2 + \sum_{n=3}^{\infty} f(n-1) z^n + 2 \sum_{n=3}^{\infty} f(n-2) z^n$$
$$= z + 3z^2 + z\bigl(G(z) - z\bigr) + 2z^2 G(z),$$

from which we find
$$G(z) = \frac{z+2z^2}{1-z-2z^2} = z\left(\frac{4/3}{1-2z} - \frac{1/3}{1+z}\right).$$
Expanding $(1-2z)^{-1}$ and $(1+z)^{-1}$ into geometric series, we obtain
$$G(z) = \sum_{n=1}^{\infty} \frac{2^{n+1}+(-1)^n}{3} z^n,$$
and by equating coefficients we find
$$f(n) = \frac{2^{n+1}+(-1)^n}{3}.$$

Problem 185

Let $p(x,y)$ be a real polynomial.
a. If $p(x,y) = 0$ for infinitely many (x,y) on the unit circle $x^2+y^2 = 1$, must $p(x,y) = 0$ on the unit circle?
b. If $p(x,y) = 0$ on the unit circle, is $p(x,y)$ necessarily divisible by x^2+y^2-1?

Answer. "Yes" is the answer to both questions.

Solution 1. a. The unit circle minus the point $(-1,0)$ can be parametrized by
$$x = \frac{1-t^2}{1+t^2}, \quad y = \frac{2t}{1+t^2}, \quad -\infty < t < \infty.$$
(This parametrization is reminiscent of, and related to, Pythagorean triples $m^2-n^2, 2mn, m^2+n^2$.) If the total degree (in x and y) of $p(x,y)$ is d, then
$$(1+t^2)^d p\left(\frac{1-t^2}{1+t^2}, \frac{2t}{1+t^2}\right)$$
is a polynomial in t. By our assumption, this polynomial has infinitely many zeros, hence it is identically zero. But then $p(x,y)$ is identically zero on the unit circle minus the point $(-1,0)$, and continuity forces $p(x,y)$ to be zero at $(-1,0)$ as well.

b. View $p(x,y)$ as a polynomial in x (whose coefficients are polynomials in y). Using long division to divide $p(x,y)$ by x^2+y^2-1, we can write
$$p(x,y) = q(x,y)(x^2+y^2-1) + r(x,y),$$

THE SOLUTIONS

where the degree in x of the remainder $r(x, y)$ is less than 2. We can then write $r(x, y) = f(y)x + g(y)$ for some polynomials $f(y)$ and $g(y)$. Since $p(x, y)$ and $x^2 + y^2 - 1$ are both zero on the unit circle, so is $f(y)x + g(y)$. Furthermore, if (x, y) is on the unit circle, $(-x, y)$ is also, and so we have

$$f(y)(-x) + g(y) = f(y)x + g(y) = 0.$$

Therefore, $f(y) = g(y) = 0$ for all $y \in (-1, 1)$, and thus the polynomials $f(y)$ and $g(y)$ are identically zero. That is, $p(x, y) = q(x, y)(x^2 + y^2 - 1)$, and we are done.

Solution 2. We get the answers to (a) and (b) simultaneously, by showing:

c. If $p(x, y) = 0$ for infinitely many (x, y) on the unit circle, then $p(x, y)$ is divisible by $x^2 + y^2 - 1$.

This clearly answers (b) affirmatively; it also answers (a), because any polynomial which is divisible by $x^2 + y^2 - 1$ is identically zero on the unit circle.

Suppose $p(x, y) = 0$ for infinitely many (x, y) on the unit circle. As in Solution 1, we use long division to write

$$p(x, y) = q(x, y)(x^2 + y^2 - 1) + r(x, y), \qquad r(x, y) = f(y)x + g(y),$$

for some polynomials $f(y)$ and $g(y)$, and it is enough to show that $f(y)$ and $g(y)$ are identically zero.

For the infinitely many points on the unit circle where $p(x, y) = 0$, we have both $x^2 + y^2 - 1 = 0$ and $f(y)x + g(y) = 0$, and we can eliminate x to get

$$\bigl(g(y)\bigr)^2 + y^2\bigl(f(y)\bigr)^2 - \bigl(f(y)\bigr)^2 = 0.$$

Therefore, the polynomial $\bigl(g(y)\bigr)^2 + y^2\bigl(f(y)\bigr)^2 - \bigl(f(y)\bigr)^2$ is zero for infinitely many values of y, so it is identically zero. That is,

$$\bigl(g(y)\bigr)^2 + y^2\bigl(f(y)\bigr)^2 = \bigl(f(y)\bigr)^2$$

as polynomials in y.

Note that because the leading coefficients of $(g(y))^2$ and $y^2(f(y))^2$ are both nonnegative, they cannot cancel when we add these squares together. Therefore, the degree of $(f(y))^2$ is at least the degree of $y^2(f(y))^2$. However, if $f(y)$ is not identically zero, the degree of $y^2(f(y))^2$ is two more than the degree of $(f(y))^2$, a contradiction. So $f(y)$, and hence also $g(y)$, is identically zero, and we are done.

Problem 186

For a real number $x > 1$, we repeatedly replace x by $x - \sqrt[2011]{x}$ until the result is ≤ 1. Let $N(x)$ be the number of replacement steps that is needed. Determine, with proof, whether $\displaystyle\int_1^\infty \frac{N(x)}{x^2}\,dx$ converges.

Solution. We start by investigating the nature of the function $N(x)$. Let $f(x) = x - \sqrt[2011]{x}$, so a replacement step consists of replacing x by $f(x)$. Note that $f(1) = 0$ and $f'(x) = 1 - \frac{1}{2011}x^{-2010/2011} > 0$ for $x \geq 1$. So f is increasing on $[1, \infty)$. Also, $f(x) = x\left(1 - x^{-2010/2011}\right) \to \infty$ and so f takes on every value in the interval $[0, \infty)$ exactly once. In particular, there is a unique number $t_1 > 1$ such that $f(t_1) = 1$, and if $1 \leq x \leq t_1$, we have $f(x) \leq 1$ and thus $N(x) = 1$. There is also a unique number $t_2 > t_1$ such that $f(t_2) = t_1$. If $t_1 < x \leq t_2$, then $1 = f(t_1) < f(x) \leq f(t_2) = t_1$, so one replacement step is not enough but the second replacement step makes the result ≤ 1: $f(f(x)) \leq f(t_1) = 1$. Therefore, if $t_1 < x \leq t_2$, then $N(x) = 2$.

Continuing in this way, we can define a sequence (t_n) of numbers in the interval $[1, \infty)$ by $f(t_n) = t_{n-1}$ $(n \geq 2)$. Because $f(x) \leq x - 1$ for $x \geq 1$, we have $t_{n-1} \leq t_n - 1 \Longrightarrow t_n \geq t_{n-1} + 1$, so $t_n \to \infty$ as $n \to \infty$. If we define $t_0 = 1$, then for each $x > 1$ there is a unique $n \geq 1$ with $t_{n-1} < x \leq t_n$, and n is the number of replacement steps (of x by $f(x)$) needed to make the result ≤ 1, so $N(x) = n$. We've now shown that $N(x)$ is a step function whose value goes up by 1 at each t_n:

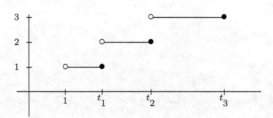

We can now consider the improper integral $\displaystyle\int_1^\infty \frac{N(x)}{x^2}\,dx = \lim_{t\to\infty}\int_1^t \frac{N(x)}{x^2}\,dx$. Because the integrand $\frac{N(x)}{x^2}$ is positive, $\displaystyle\int_1^t \frac{N(x)}{x^2}\,dx$ is an increasing function of t, and so to decide whether the improper integral converges it's enough to look at $t = t_n$, since $t_n \to \infty$ as $n \to \infty$. That is, the improper integral converges if and only if $\lim\limits_{n\to\infty} \int_1^{t_n} \dfrac{N(x)}{x^2}\,dx$ is finite. Using our description of $N(x)$ as a step function, we can calculate $\int_1^{t_n} \dfrac{N(x)}{x^2}\,dx$, as follows:

$$I_n \equiv \int_1^{t_n} \dfrac{N(x)}{x^2}\,dx = \int_1^{t_1} \dfrac{1}{x^2}\,dx + \int_{t_1}^{t_2} \dfrac{2}{x^2}\,dx + \cdots + \int_{t_{n-1}}^{t_n} \dfrac{n}{x^2}\,dx$$

$$= \left(1 - \dfrac{1}{t_1}\right) + \left(\dfrac{2}{t_1} - \dfrac{2}{t_2}\right) + \cdots + \left(\dfrac{n}{t_{n-1}} - \dfrac{n}{t_n}\right)$$

$$= 1 + \dfrac{1}{t_1} + \dfrac{1}{t_2} + \cdots + \dfrac{1}{t_{n-1}} - \dfrac{n}{t_n}.$$

We need to determine whether the sequence (I_n) is bounded. (If it is, then because any bounded, increasing sequence of real numbers has a limit, the improper integral will converge.)

Now $t_n > n$ for all n, because $t_1 > 1$ and $t_n \geq t_{n-1} + 1$, so $n/t_n < 1$ and subtracting n/t_n does not affect boundedness. Thus the improper integral converges if and only if the numbers $1 + \dfrac{1}{t_1} + \dfrac{1}{t_2} + \cdots + \dfrac{1}{t_{n-1}}$ are bounded.

But those are the (increasing) partial sums of the infinite series $\sum\limits_{n=0}^{\infty} \dfrac{1}{t_n}$, so we need to determine whether that series converges; we now show that it does converge.

From $f(t_{n+1}) = t_n$, we have $t_{n+1} - t_n = t_{n+1}^{1/2011}$. Therefore, for any $N \geq 1$,

$$t_N = 1 + \sum_{n=0}^{N-1} (t_{n+1} - t_n)$$

$$= 1 + \sum_{n=0}^{N-1} t_{n+1}^{1/2011}$$

$$> 1 + \sum_{n=0}^{N-1} (n+1)^{1/2011}$$

$$> \int_0^N x^{1/2011}\,dx$$

$$= \dfrac{2011}{2012} N^{2012/2011},$$

and therefore $\dfrac{1}{t_N} < \dfrac{2012}{2011} \cdot \dfrac{1}{N^{2012/2011}}$. It follows, by limit comparison with the convergent "p-series" $\sum_{n=0}^{\infty} \dfrac{1}{n^{2012/2011}}$, that $\sum_{n=0}^{\infty} \dfrac{1}{t_n}$ converges, and by the discussion above, the improper integral converges as well.

Problem 187

Find all real polynomials $p(x)$, whose roots are real, for which
$$p(x^2 - 1) = p(x)p(-x).$$

Idea. If α is a root of $p(x)$, then $\alpha^2 - 1$ is also, and yet $p(x)$ can have only finitely many roots.

Answer. The desired polynomials are 0 and those of the form
$$p_{j,k,l}(x) = (x^2 + x)^j (\varphi - x)^k (\overline{\varphi} - x)^l,$$
where $\varphi = (1+\sqrt{5})/2$, $\overline{\varphi} = (1-\sqrt{5})/2$ are the roots of $x^2 - x - 1 = 0$ and $j, k, l \geq 0$ are integers.

Solution. It is straightforward to check that the $p_{j,k,l}(x)$ do satisfy the functional equation $p(x^2 - 1) = p(x)\,p(-x)$. To see how they were found, observe that if α is a root of $p(x)$, then $p(\alpha^2 - 1) = p(\alpha)p(-\alpha) = 0$, so $\alpha^2 - 1$ is also a root of $p(x)$. But then $(\alpha^2 - 1)^2 - 1$ is, in turn, a root; more generally, each term of the iterative sequence

$$\alpha, \quad \alpha^2 - 1, \quad (\alpha^2 - 1)^2 - 1, \quad \ldots \tag{$*$}$$

is a root of $p(x)$. Since $p(x)$ has only finitely many roots, $(*)$ must eventually become periodic. The easiest way for this to happen is to have $\alpha^2 - 1 = \alpha$ (period 1), which yields $\alpha = \varphi$ or $\alpha = \overline{\varphi}$. In this case $p(x)$ is divisible by $\varphi - x$ or $\overline{\varphi} - x$, respectively.

The next case to consider is $(\alpha^2 - 1)^2 - 1 = \alpha$. Besides φ and $\overline{\varphi}$, this fourth-degree equation for α has roots 0 and -1. Note that if either 0 or -1 is a root of $p(x)$, so is the other, because $0^2 - 1 = -1$ and $(-1)^2 - 1 = 0$. So in this case $p(x)$ is divisible by $x^2 + x$.

We now show that it is not necessary to study any further cases, that is, that every nonzero polynomial $p(x)$ satisfying the given functional equation is one of the $p_{j,k,l}(x)$. Note that if $p(x)$ satisfies the given functional equation and is divisible by $p_{j,k,l}(x)$, then the polynomial $p(x)/p_{j,k,l}(x)$ also satisfies the functional equation. So we can divide out by any factors $\varphi - x$, $\overline{\varphi} - x$, and $x^2 + x$ that $p(x)$ may have, and assume that $p(x)$ does not have any of $-1, 0, \varphi, \overline{\varphi}$ as roots.

If $p(x)$ is nonconstant, let α_0 be the smallest root of $p(x)$. If $\alpha_0 > \varphi$, then $\alpha_0^2 - 1 > \alpha_0$, and the sequence $(*)$ will be strictly increasing, yielding infinitely many roots, a contradiction. If $\overline{\varphi} < \alpha_0 < \varphi$, then $\alpha_0^2 - 1 < \alpha_0$, contradicting our choice of α_0. Thus, we must have $\alpha_0 < \overline{\varphi}$. However, since α_0 is a root of $p(x)$, $p(x)$ has a factor $x - \alpha_0$, and so $p(x^2 - 1)$ has a factor $x^2 - 1 - \alpha_0$. Because $p(x)$ has real roots, $p(x^2 - 1) = p(x)p(-x)$ factors into linear factors, hence $1 + \alpha_0 \geq 0$, and we have $-1 < \alpha_0 < \overline{\varphi}$.

The third term in the sequence $(*)$ is
$(\alpha_0^2 - 1)^2 - 1 = \alpha_0 + ((\alpha_0^2 - 1)^2 - 1 - \alpha_0) = \alpha_0 + \alpha_0(\alpha_0 + 1)(\alpha_0 - \varphi)(\alpha_0 - \overline{\varphi})$.
Because $-1 < \alpha_0 < \overline{\varphi}$, the product on the right is negative, which contradicts our choice of α_0. We conclude that $p(x)$ is constant. The functional equation now shows that $p(x) = 1 = p_{0,0,0}(x)$, and we are done.

Problem 188

Consider sequences of points in the plane that are obtained as follows: The first point of each sequence is the origin. The second point is reached from the first by moving one unit in any of the four "axis" directions (east, north, west, south). The third point is reached from the second by moving $1/2$ unit in any of the four axis directions (but not necessarily in the same direction), and so on. We call a point *approachable* if it is the limit of some sequence of the above type. Describe the set of all approachable points in the plane.

Answer. The set of all approachable points in the plane is the set of all points (x, y) for which $|x| + |y| \leq 2$ (see figure).

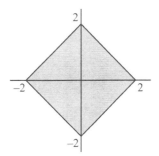

Solution. First we show that $|x| + |y| \leq 2$ for every approachable point (x, y). Let (x, y) be approachable, and let $(x_0, y_0) = (0,0), (x_1, y_1), \ldots,$ $(x_n, y_n), \ldots$ be a sequence, of the type described in the problem, whose limit is (x, y). As we move from one point in the sequence to the next, either the x-coordinate or the y-coordinate changes by $1, \frac{1}{2}, \frac{1}{4}, \ldots$, while the other coordinate does not change. Thus, after n steps we will have
$$|x_n| + |y_n| \leq 1 + \frac{1}{2} + \cdots + \frac{1}{2^n} = 2 - \frac{1}{2^n}.$$
Since (x_n, y_n) approaches (x, y) as $n \to \infty$, it follows that $|x| + |y| \leq 2$.

To show that every point (x, y) with $|x| + |y| \leq 2$ is approachable, begin by dividing the set of such points into four congruent parts, as shown in the following figure.

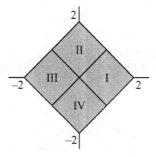

By symmetry, it is enough to show that any point in part I is approachable. To approach a point in part I, take the first "step" (move) to the right. We then have $(x_0, y_0) = (0, 0)$, $(x_1, y_1) = (1, 0)$. Now we have to show that we can approach the point (x, y) from the point $(1, 0)$ by moves of $\frac{1}{2}, \frac{1}{4}, \ldots$ in the axis directions. But this is exactly the original problem scaled down by a factor 2: we have a square whose sides are half the sides of the original square, and we are again at the center of that square. So to decide where to go next, we subdivide square I into four equal squares, as shown in the following figure, and then move right, up, left, or down, depending on whether (x, y) is in A, B, C, or D.

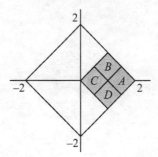

Continuing this process, we obtain a sequence of points (x_n, y_n). To show that the limit of this sequence is (x, y), note that the distance from (x_n, y_n) to (x, y) is at most half the length of a diagonal of the nth new square. Since this length is halved at each stage, $\lim_{n \to \infty}(x_n, y_n) = (x, y)$, and we are done.

Problem 189

A gambling game is played as follows: D dollar bills are distributed in some manner among N indistinguishable envelopes, which are then mixed up in a large bag. The player buys random envelopes, one at a time, for one dollar and examines their contents as they are purchased. If the player can buy as many or as few envelopes as desired, and, furthermore, knows the initial distribution of the money, then for what distribution(s) will the player's expected net return be maximized?

Solution. If $D > (N+1)/2$, the player's expected net return will be maximized if and only if all D dollars are in a single envelope; on the other hand, if $D \leq (N+1)/2$, the distribution does not matter.

If, at some stage in the game, it is in the player's best interest to buy an envelope, then he or she should certainly continue buying envelopes until one of them is found to contain money. (At that point, the player may or may not prefer to stop buying.) In particular, if all D dollar bills are in the same envelope, then the player will either not buy at all or buy envelopes until the one containing the money is purchased. If the latter strategy is followed, the envelope with the money is equally likely to be bought first, second, ..., or Nth, and so the player's expected investment to get the D dollars will be

$$\sum_{j=1}^{N} j \cdot \frac{1}{N} = \frac{N+1}{2}.$$

Therefore, still assuming all the bills are in one envelope, the player will buy envelopes if $D > (N+1)/2$, with an expected net return of $D - (N+1)/2$, and the player will not bother to play at all if $D \leq (N+1)/2$.

Now we will show that from the player's point of view, any other distribution of the bills among the envelopes is worse if $D > (N+1)/2$ and no better (that is, there is still no reason to play) if $D \leq (N+1)/2$.

To see why, let $E(D, N)$ denote the maximum net return, over all possible distributions, when D bills are distributed over N envelopes. (This maximum exists because the number of distributions is finite, as is the number of possible strategies for the player given a distribution.)

Note that $E(D, N - 1) \geq E(D, N)$, because merging the contents of two envelopes and throwing away the now empty envelope can only help the player.

Our claim is that

$$E(D, N) = \begin{cases} D - (N+1)/2 & \text{if } D > (N+1)/2 \\ 0 & \text{if } D \leq (N+1)/2, \end{cases}$$

and that in the case $D > (N+1)/2$, the maximum *only* occurs when all the bills are in the same envelope.

We prove this by induction on N. If $N = 1$, there is only one possible distribution, and the claim follows. Suppose the claim is correct for $N - 1$ and all values of D. If $E(D, N) = 0$, then we must have $D \leq (N+1)/2$ (otherwise, we could do better by putting all bills in the same envelope), and we are done. So assume $E(D, N) > 0$. Suppose the maximum occurs when M of the N envelopes contain money, say $D_1, D_2, \cdots, D_M$ dollars respectively, with $D = \sum_{m=1}^{M} D_m$. Then after the player buys the first envelope, which she or he will do because $E(D, N) > 0$, the player will have lost one dollar with probability $(N - M)/N$ and, for $1 \leq m \leq M$, gained $D_m - 1$ dollars with probability $1/N$. We therefore have the inequality

$$E(D, N) \leq \frac{N - M}{N}(E(D, N - 1) - 1)$$
$$+ \sum_{m=1}^{M} \frac{1}{N}(E(D - D_m, N - 1) + D_m - 1)$$
$$= \frac{N - M}{N} E(D, N - 1) + \frac{1}{N} \sum_{m=1}^{M} E(D - D_m, N - 1) + \frac{D}{N} - 1. \quad (*)$$

Since $E(D, N - 1) \geq E(D, N) > 0$, we know from the induction hypothesis that $E(D, N - 1) = D - N/2$. On the other hand, it is not clear whether $E(D - D_m, N - 1) > 0$, and so we distinguish two cases.

Case 1. $E(D - D_m, N - 1) > 0$ for all m.

In this case inequality $(*)$, together with the induction hypothesis, yields

$$E(D,N) \leq \frac{N-M}{N}\left(D - \frac{N}{2}\right) + \frac{1}{N}\sum_{m=1}^{M}\left(D - D_m - \frac{N}{2}\right) + \frac{D}{N} - 1$$

$$= D - \frac{N}{2} - 1$$

$$< D - \frac{N+1}{2},$$

and we are done.

Case 2. $E(D - D_m, N - 1) = 0$ for at least one m.

Since $E(D - D_m, N - 1) < E(D, N - 1)$ for all m, we then have

$$\sum_{m=1}^{M} E(D - D_m, N - 1) \leq (M-1)E(D, N-1),$$

with equality only for $M = 1$. Inequality $(*)$ then yields

$$E(D,N) \leq \frac{N-M}{N}E(D, N-1) + \frac{1}{N}(M-1)E(D, N-1) + \frac{D}{N} - 1$$

$$= \frac{N-1}{N}E(D, N-1) + \frac{D}{N} - 1$$

$$= \frac{N-1}{N}\left(D - \frac{N}{2}\right) + \frac{D}{N} - 1$$

$$= D - \frac{N+1}{2},$$

with equality only for $M = 1$. That is the situation in which all bills are in one envelope, and we are done.

Problem 190

Let $\alpha = 0.d_1d_2d_3\ldots$ be a decimal representation of a real number between 0 and 1. Let r be a real number with $|r| < 1$.

a. If α and r are rational, must $\sum_{i=1}^{\infty} d_i r^i$ be rational?
b. If α and r are rational, must $\sum_{i=1}^{\infty} i d_i r^i$ be rational?
c. If r and $\sum_{i=1}^{\infty} d_i r^i$ are rational, must α be rational?

Answer. "Yes" to (a) and (b); "no" to (c).

Solution 1. **a.** Suppose that α is rational. Then the decimal representation is eventually periodic; let p be its period, and let k be the number of digits before the periodic behavior starts. We have

$$\alpha = 0.d_1 d_2 \ldots d_k \underbrace{d_{k+1} \ldots d_{k+p}}\, \underbrace{d_{k+1} \ldots d_{k+p}}\, \underbrace{d_{k+1} \ldots d_{k+p}}\, \cdots$$

For any real number r with $|r| < 1$, we can define

$$x = \sum_{i=1}^{k} d_i r^i \quad \text{and} \quad y = \sum_{i=1}^{p} d_{k+i} r^{k+i},$$

and we then have

$$\sum_{i=1}^{\infty} d_i r^i = x + y + y\, r^p + y\, (r^p)^2 + y\, (r^p)^3 + \cdots.$$

Summing the infinite geometric series, we obtain

$$\sum_{i=1}^{\infty} d_i r^i = x + \frac{y}{1 - r^p}. \qquad (*)$$

Now if r is rational, then so are x and y. But then $\sum_{i=1}^{\infty} d_i r^i$ is also rational, and we are done.

b. Again, suppose that α is rational. Note that the sum $\sum_{i=1}^{\infty} i d_i r^i$ can be rewritten as $r \sum_{i=1}^{\infty} i d_i r^{i-1}$; the new sum looks like a derivative.

In fact, we can get the new sum by differentiating both sides of the identity (*), which holds for all r with $|r| < 1$. If we then multiply by r, we find

$$\sum_{i=1}^{\infty} i d_i r^i = \sum_{i=1}^{k} i d_i r^i + \left(\frac{p r^p}{(1 - r^p)^2} \right) \sum_{i=1}^{p} d_{k+i} r^{k+i}$$

$$+ \left(\frac{1}{1 - r^p} \right) \sum_{i=1}^{p} (k+i) d_{k+i} r^{k+i}.$$

If r is a rational number, then the three finite sums on the right are rational; therefore, $\sum_{i=1}^{\infty} i d_i r^i$ is also rational, and we are done.

c. If r has the form $1/b$ where $b \geq 10$ is an integer, then α is rational. For in that case, if $\sum_{i=1}^{\infty} d_i b^{-i}$ is rational, then $0.d_1 d_2 d_3 \ldots$ is the base b expansion of a rational number, hence the sequence $d_1, d_2, d_3, \ldots$ is eventually periodic, and α is also rational.

If r does not have this form, however, α may fail to be rational. To see this, consider the case $r = 1/2$ and the sum

$$\sum_{i=1}^{\infty}\left(\frac{1}{2}\right)^{2i} = \left(\frac{1}{4}\right) + \left(\frac{1}{4}\right)^2 + \left(\frac{1}{4}\right)^3 + \cdots = \frac{1/4}{1-(1/4)} = \frac{1}{3}.$$

This sum can be written as $\sum_{i=1}^{\infty} d_i r^i$ with $d_{2i} = 1$, $d_{2i+1} = 0$ for all i. But we can also, independently for each $i = 1, 2, \ldots$, choose (d_{2i}, d_{2i+1}) to be either $(1, 0)$ or $(0, 2)$; regardless of our choices, if we again set $d_1 = 0$, we will have

$$\sum_{i=1}^{\infty} d_i r^i = \sum_{i=1}^{\infty}\left(d_{2i}\left(\frac{1}{2}\right)^{2i} + d_{2i+1}\left(\frac{1}{2}\right)^{2i+1}\right) = \sum_{i=1}^{\infty}\left(\frac{1}{2}\right)^{2i} = \frac{1}{3}.$$

Furthermore, we can make these choices in such a way that the resulting decimal representation $\alpha = 0.d_1 d_2 d_3 \ldots$ will be nonperiodic; as a result, α will be irrational. (For example, choose $(d_{2i}, d_{2i+1}) = (0, 2)$ precisely when i is a power of 2.)

Solution 2. a. As in Solution 1, let p be the (eventual) period of the decimal representation $0.d_1 d_2 d_3 \ldots$ of α, and suppose that the periodic behavior begins with d_{k+1}, so that $d_{i+p} = d_i$ for all $i > k$. Now recall the technique, which is often used to find the fractional representation of a rational number given its periodic decimal expansion, of "shifting over" an expansion and then subtracting it from the original. Using this technique, we find

$$(1 - r^p)\sum_{i=1}^{\infty} d_i r^i = \sum_{i=1}^{p} d_i r^i + \sum_{i=p+1}^{p+k}(d_i - d_{i-p})r^i + \sum_{i=p+k+1}^{\infty}(d_i - d_{i-p})r^i$$

$$= \sum_{i=1}^{p} d_i r^i + \sum_{i=p+1}^{p+k}(d_i - d_{i-p})r^i$$

for $|r| < 1$. This final expression is a finite sum, so if r is rational, then $\sum_{i=1}^{\infty} d_i r^i$ is also rational.

b. Using the same idea as in (a), we have

$$(1-r^p)\sum_{i=1}^{\infty} id_i r^i = \sum_{i=1}^{p} id_i r^i + \sum_{i=p+1}^{p+k} (id_i - (i-p)d_{i-p})r^i$$

$$+ \sum_{i=p+k+1}^{\infty} (id_i - (i-p)d_{i-p})r^i$$

$$= \sum_{i=1}^{p} id_i r^i + \sum_{i=p+1}^{p+k} (id_i - (i-p)d_{i-p})r^i + \sum_{i=p+k+1}^{\infty} pd_{i-p}r^i.$$

If r is rational, the first two (finite) sums on the right are certainly rational, and the third sum, which may be rewritten as $pr^{p+k}\sum_{i=1}^{\infty} d_{i+k}r^i$, is rational by the result of (a). Therefore, $\sum_{i=1}^{\infty} id_i r^i$ is rational.

c. Let $r = 1/2$, and choose $d_1 = 0$ and (d_{2i}, d_{2i+1}) just as in Solution 1. Now observe that there are uncountably many such choices for the sequence $d_1, d_2, d_3, \ldots$, no two of which yield the same α. Since the rationals are countable, some ("most") of the possible α's are irrational. (In fact, since the set of algebraic numbers is countable, the α's are generally transcendental.)

Problem 191

Let $\mathcal{L}_1$ and $\mathcal{L}_2$ be skew lines in space (that is, straight lines which do not lie in the same plane). How many straight lines $\mathcal{L}$ have the property that every point on $\mathcal{L}$ has the same distance to $\mathcal{L}_1$ as to $\mathcal{L}_2$?

Solution. There are infinitely many lines $\mathcal{L}$ such that every point on $\mathcal{L}$ has the same distance to $\mathcal{L}_1$ as to $\mathcal{L}_2$. To show this, we will first choose convenient coordinates in $\mathbb{R}^3$, then find the set of all points equidistant from $\mathcal{L}_1$ and $\mathcal{L}_2$, and finally find all lines contained in this set of points.

We can set up our coordinate system so that the points on $\mathcal{L}_1$ and $\mathcal{L}_2$ which are closest to each other are $(0,0,1)$ and $(0,0,-1)$. In this case, $\mathcal{L}_1$ lies in the plane $z = 1$ and $\mathcal{L}_2$ lies in the plane $z = -1$. After a suitable rotation about the z-axis, we may assume that for some $m > 0$, $\mathcal{L}_1$ is described by $y = mx$, $z = 1$, while $\mathcal{L}_2$ is given by $y = -mx$, $z = -1$.

The point $(t, mt, 1)$ on $\mathcal{L}_1$ closest to a given point (x, y, z) can be found by minimizing the square of the distance,

$$(t-x)^2 + (mt-y)^2 + (1-z)^2.$$

THE SOLUTIONS

A short computation shows that the minimum occurs for $t = \dfrac{x+my}{1+m^2}$. Therefore, the square of the distance from (x,y,z) to $\mathcal{L}_1$ is

$$\left(\frac{x+my}{1+m^2} - x\right)^2 + \left(\frac{mx+m^2y}{1+m^2} - y\right)^2 + (1-z)^2 = \frac{(y-mx)^2}{1+m^2} + (z-1)^2.$$

In a similar way, or by using the symmetry of the problem, we find that the square of the distance from (x,y,z) to $\mathcal{L}_2$ is

$$\frac{(y+mx)^2}{1+m^2} + (z+1)^2.$$

The point (x,y,z) is equidistant from $\mathcal{L}_1$ and $\mathcal{L}_2$ if and only if the two expressions above are equal. Simplifying, we find that the points (x,y,z) equidistant from $\mathcal{L}_1$ and $\mathcal{L}_2$ form the surface

$$z = -\frac{mxy}{1+m^2}.$$

We can write the equation of a line through a fixed point (x_0, y_0, z_0) on this surface in the form

$$(x_0 + at, y_0 + bt, z_0 + ct),$$

where (a,b,c) represents the direction of the line. For this line to lie entirely on the surface, we must have

$$z_0 + ct = -\frac{m(x_0+at)(y_0+bt)}{1+m^2}$$

for all t. Comparing coefficients of 1, t, and t^2, we see that this comes down to

$$z_0 = -\frac{mx_0y_0}{1+m^2}, \quad c = -\frac{m(bx_0+ay_0)}{1+m^2}, \quad 0 = -\frac{mab}{1+m^2}.$$

The first equation is automatically satisfied, because (x_0, y_0, z_0) is on the surface. The last equation implies $a = 0$ or $b = 0$, so up to constant multiples there are two solutions to the equations: $(a,b,c) = (0, 1, -mx_0/(1+m^2))$ and $(a,b,c) = (1, 0, -my_0/(1+m^2))$. Thus, through the point (x_0, y_0, z_0) on the surface there are exactly two lines $\mathcal{L}$ such that every point on $\mathcal{L}$ has the same distance to $\mathcal{L}_1$ as to $\mathcal{L}_2$. Because x_0 and y_0 can be varied at will, there are, in all, infinitely many such lines.

Problem 192

We call a sequence $(x_n)_{n \geq 1}$ a *superinteger* if (i) each x_n is a nonnegative integer less than 10^n and (ii) the last n digits of x_{n+1} form x_n. One example of such a sequence is $1, 21, 021, 1021, 21021, 021021, \ldots$, which we abbreviate by $\ldots 21021$. We can do arithmetic with superintegers; for instance, if x is the superinteger above, then the product xy of x with the superinteger $y = \ldots 66666$ is found as follows:

$1 \times 6 = 6$: the last digit of xy is 6.
$21 \times 66 = 1386$: the last two digits of xy are 86.
$021 \times 666 = 13986$: the last three digits of xy are 986.
$1021 \times 6666 = 6805986$: the last four digits of xy are 5986, etc.

Is it possible for two nonzero superintegers to have product $0 = \ldots 00000$?

Answer. Yes, it is possible for two nonzero superintegers to have product zero.

Solution. In fact, we will show that there exist nonzero superintegers $x = \ldots a_n \ldots a_2 a_1$ and $y = \ldots b_n \ldots b_2 b_1$ such that for any $k \geq 1$, 2^k divides $a_k \ldots a_2 a_1$ and 5^k divides $b_k \ldots b_2 b_1$. Then 10^k divides the product $a_k \ldots a_2 a_1 \times b_k \ldots b_2 b_1$, so xy ends in k zeros for any k; that is, xy is zero, and we are done.

We start with x, and we show by induction that there exist digits $a_1, a_2, \ldots$ such that for all $k \geq 1$, 2^k divides $a_k \ldots a_2 a_1$. Take $a_1 = 2$ to show this is true for $k = 1$. Now suppose $a_1, a_2, \ldots, a_m$ have been found such that 2^m divides $a_m \ldots a_2 a_1$, so $c_m = (a_m \ldots a_2 a_1)/2^m$ is an integer. We want to find a digit $d = a_{m+1}$ for which 2^{m+1} divides $d \cdot 10^m + a_m \ldots a_2 a_1$, that is, for which 2 divides $d \cdot 5^m + c_m$. Since 5^m is odd, we can take $d = 0$ if c_m is even and $d = 1$ if c_m is odd. (The actual superinteger x constructed by this inductive procedure ends in $\ldots 010112$.)

Similarly, we show that y exists by starting with $b_1 = 5$ and showing that there exist digits $b_2, b_3, \ldots$ such that for all $k \geq 1$, 5^k divides $b_k \ldots b_2 b_1$. This time, assuming 5^m divides $b_m \ldots b_2 b_1$, we want to find a digit d such that 5^{m+1} divides $d \cdot 10^m + b_m \ldots b_2 b_1$, that is, such that 5 divides $d \cdot 2^m + (b_m \ldots b_2 b_1)/5^m$. In other words, we want a solution to $2^m d \equiv -(b_m \ldots b_2 b_1)/5^m \pmod{5}$. Since 2^m has an inverse modulo 5, exactly one of $0, 1, 2, 3, 4$ is a solution for d, and we are done. (The actual superinteger y constructed in this way ends in $\ldots 203125$.)

Comment. (Greg Kuperberg, UC Davis) The "superintegers" described here are commonly known as 10-adic integers. They form a ring $\mathbb{Z}_{10}$ which is isomorphic to the direct sum of the rings $\mathbb{Z}_2, \mathbb{Z}_5$ of 2-adic and 5-adic integers, respectively. (Note that if the "superintegers" are truncated on the left to have n digits, the truncated versions will form the ring $\mathbb{Z}/10^n\mathbb{Z}$, which is isomorphic to $\mathbb{Z}/2^n\mathbb{Z} \oplus \mathbb{Z}/5^n\mathbb{Z}$ by the Chinese Remainder Theorem. These isomorphisms for different n are compatible and give rise to an isomorphism $\mathbb{Z}_{10} \cong \mathbb{Z}_2 \oplus \mathbb{Z}_5$.) The solution above basically constructs explicit zero divisors of the form $(0, x)$ and $(y, 0)$ in $\mathbb{Z}_2 \oplus \mathbb{Z}_5$.

Problem 193

If $\sum a_n$ converges, must there exist a periodic function $\varepsilon : \mathbb{Z} \to \{1, -1\}$ such that $\sum \varepsilon(n)|a_n|$ converges?

Solution. (Greg Kuperberg, UC Davis) No, there exist convergent series $\sum a_n$ such that $\sum \varepsilon(n)|a_n|$ diverges for every periodic function $\varepsilon : \mathbb{Z} \to \{1, -1\}$.

For a specific example, let

$$a_n = \begin{cases} (-1)^k/k & \text{if } n = k! \\ 0 & \text{if } n \text{ is not a factorial.} \end{cases}$$

Note that if we omit the zero terms from $\sum a_n$, we get the series $\sum (-1)^k/k$, which converges by the alternating series test. Omitting the zero terms does not change the values of the partial sums, so the original series $\sum a_n$ is also convergent.

Now suppose that $\varepsilon : \mathbb{Z} \to \{1, -1\}$ is periodic with period P. Then for $k \geq P$, $\varepsilon(k!)$ will not depend on k, because the factorials $P!, (P+1)!, (P+2)!, \ldots$ all differ by multiples of P. Therefore, in the series $\sum \varepsilon(n)|a_n|$, the nonzero terms will eventually all have the same sign. Omitting the zero terms again, we see that this series will have the same convergence/divergence behavior as $\sum_{k \geq P} 1/k$, which diverges because it consists of all but finitely many terms of the (divergent) harmonic series, so we are done.

Comment. For an example with $a_n \neq 0$ for all n, replace a_n by $a_n + 1/2^n$, which will not affect the convergence of either $\sum a_n$ or $\sum \varepsilon(n)|a_n|$.

Problem 194

Let $f(x) = x - 1/x$. For any real number x_0, consider the sequence defined by $x_0, x_1 = f(x_0), \ldots, x_{n+1} = f(x_n), \ldots$, provided $x_n \neq 0$. Define x_0 to be a *T-number* if the sequence terminates, that is, if $x_n = 0$ for some n.
a. Show that the set of all T-numbers is countably infinite (denumerable).
b. Does every open interval contain a T-number?

Solution. a.   Let

$$S_n = \{x_0 \in \mathbb{R} \mid x_n = \underbrace{f(f(\ldots(f(x_0))\ldots))}_{n} = 0\}.$$

We prove by induction that S_n has exactly 2^n elements. This is true for $n = 0$, since 0 is the only element of S_0; assume it is true for n. Note that $x_0 \in S_{n+1}$ if and only if $f(x_0) \in S_n$. Also, the equality $f(x_0) = x_0 - 1/x_0$ can be rewritten as $x_0^2 - f(x_0)x_0 - 1 = 0$, and, given $f(x_0)$, this quadratic equation for x_0 has two real roots. Since there are 2^n possibilities for $f(x_0) \in S_n$, there are $2 \cdot 2^n = 2^{n+1}$ elements $x_0 \in S_{n+1}$, and the induction step is complete.

Now since the set of all T-numbers can be written as $\bigcup_{n=0}^{\infty} S_n$, a countable union of finite sets, it is countable, and since, for any n, it has more than 2^n elements, the set is infinite.

b.   Yes, every open interval does contain a T-number. Suppose that, on the contrary, there exist open intervals which do not contain any T-numbers; we will call such intervals *T-free*. Note that since 0 is a T-number, any T-free interval (a, b) satisfies either $0 \leq a < b$ or $a < b \leq 0$. Also, since f is an odd function, the set of all T-numbers is symmetric about 0, so for any T-free interval (a, b), the interval $(-b, -a)$ is also T-free. Thus there exists some T-free interval (a, b) with $0 < a < b$ (if $a = 0$, shrink the interval slightly). We will show the following:

I. For every T-free interval (a, b) with $0 < a < b$, we can find a T-free interval (c, d) with $0 < c < d \leq 1$ which is *at least* as long as (a, b).

II. For every T-free interval (a, b) with $0 < a < b \leq 1$, we can find a T-free interval (c, d) with $0 < c < d$ which is more than *twice* as long as (a, b).

Repeatedly combining (I) and (II) then yields arbitrarily long T-free intervals that are contained in $(0, 1)$, a contradiction.

Let (a, b) be any T-free interval with $0 < a < b$. Note that f is increasing and continuous on $(0, \infty)$; therefore, the image of (a, b) under f is the interval $(f(a), f(b))$. Because the image of a T-free set is necessarily T-free, this new

THE SOLUTIONS

interval is again T-free, and since its length is

$$f(b) - f(a) = (b-a) + (1/a - 1/b) > b - a,$$

it is longer than the original interval.

To prove (I) above, note that if $b \leq 1$, we can take $(c,d) = (a,b)$. If $b > 1$, consider the successive T-free intervals $(f(a), f(b))$, $(f(f(a)), f(f(b)))$, and so forth. We have $f(b) = b - 1/b$, $f(f(b)) = b - 1/b - 1/f(b) < b - 2/b$, and continuing in this way,

$$\underbrace{f(f(\cdots f(b)\cdots))}_{n} \leq b - \frac{n}{b}.$$

Thus for large enough n, we have

$$\underbrace{f(f(\cdots f(b)\cdots))}_{n} \leq 1.$$

If n_0 is the first value of n for which this happens, then

$$\underbrace{f(f(\cdots f(b)\cdots))}_{n_0} > 0,$$

because f maps any number greater than 1 to a positive number. We can then take

$$(c,d) = \left(\underbrace{f(f(\cdots f(a)\cdots))}_{n_0}, \underbrace{f(f(\cdots f(b)\cdots))}_{n_0}\right),$$

which is a T-free interval with $0 < d \leq 1$ and therefore $0 \leq c < d \leq 1$. By the above, this interval is longer than (a,b), so if $c = 0$, we can shrink the interval slightly; the proof of (I) is now complete.

To prove (II), we now assume $0 < a < b \leq 1$. Thus we have $f(b) \leq 0$. Therefore, using the symmetry mentioned earlier, $(c,d) = (-f(b), -f(a))$ is a T-free interval with $0 \leq c < d$. Its length is

$$f(b) - f(a) = b - a + \frac{b-a}{ab} > 2(b-a),$$

and we are done. (Again, if $c = 0$, we can shrink the new interval slightly.)

Comment. For each n, x_n is a rational function of x_0 with rational coefficients, hence every T-number is an *algebraic number* (a root of a polynomial with rational coefficients). Since the set of all algebraic numbers is countable, we see again that the set of T-numbers is countable. However, this argument does not show that there are infinitely many T-numbers.

Problem 195

For n a positive integer, show that the number of integral solutions (x, y) of $x^2 + xy + y^2 = n$ is finite and a multiple of 6.

Idea. If (x, y) is an integral solution of $x^2 + xy + y^2 = n$, then $(-x, -y)$ is a different solution, so solutions come in pairs. If we can show instead that solutions come in sixes (and that there are only finitely many), we will be done. To see why solutions come in sixes, we can use algebraic manipulation to rewrite $x^2 + xy + y^2$ as $a^2 + ab + b^2$ for suitable $(a, b) \neq (x, y)$.

Solution 1. First note that for any solution (x, y), we have
$$2n = 2x^2 + 2xy + 2y^2 = x^2 + y^2 + (x+y)^2 \geq x^2 + y^2.$$
Therefore, any integral solution is one of the lattice points (points whose coordinates are integers) on or inside a circle of radius $\sqrt{2n}$, and so the number of integral solutions is finite.

Now observe that
$$\begin{aligned} x^2 + xy + y^2 &= (x+y)^2 - xy \\ &= (x+y)^2 - x(x+y) + x^2 \\ &= (x+y)^2 + (x+y)(-x) + (-x)^2. \end{aligned}$$
Thus, if (x, y) is an integral solution of $x^2 + xy + y^2 = n$, then so is $(x+y, -x)$. If we repeat this process with the new solution, we go through a cycle of solutions:
$$(x, y),\ (x+y, -x),\ (y, -x-y),\ (-x, -y),\ (-x-y, x),\ (-y, x+y), \quad (*)$$
after which we get back to (x, y). It can be checked directly that, since x and y cannot both be zero, all six solutions in the cycle $(*)$ are different.

Alternatively, we can use a bit of linear algebra. Because
$$\begin{pmatrix} x+y \\ -x \end{pmatrix} = \begin{pmatrix} 1 & 1 \\ -1 & 0 \end{pmatrix} \begin{pmatrix} x \\ y \end{pmatrix},$$
the solutions in the cycle, rewritten as column vectors, are given by
$$\begin{pmatrix} 1 & 1 \\ -1 & 0 \end{pmatrix}^k \begin{pmatrix} x \\ y \end{pmatrix}$$
for $k = 0, 1, \ldots, 5$. The eigenvalues of the matrix $A = \begin{pmatrix} 1 & 1 \\ -1 & 0 \end{pmatrix}$ are found to be $e^{2\pi i/6}$ and $e^{-2\pi i/6}$, so for $k = 1, 2, \ldots, 5$, A^k cannot have an eigenvalue 1, and hence $A^k \begin{pmatrix} x \\ y \end{pmatrix}$ cannot equal $\begin{pmatrix} x \\ y \end{pmatrix}$.

Because all six solutions in (∗) are different, and since the set of all integral solutions of $x^2 + xy + y^2 = n$ can be partitioned into such cycles, the number of integral solutions is a multiple of 6, and we are done.

Solution 2. (The late Ian Richards, University of Minnesota) Factor $x^2 + xy + y^2$ into (complex) algebraic integers, as follows. Note that

$$x^3 - y^3 = (x-y)(x-\omega y)(x - \omega^2 y),$$

where $\omega \neq 1$ is a cube root of unity, and that for $x \neq y$,

$$x^2 + xy + y^2 = \frac{x^3 - y^3}{x - y}.$$

Therefore,

$$x^2 + xy + y^2 = (x - \omega y)(x - \omega^2 y);$$

this factorization actually holds whether or not $x = y$. Since $\omega^2 + \omega + 1 = 0$, the factorization can be rewritten as follows:

$$(x - \omega y)(x - \omega^2 y) = (x + y + \omega^2 y)(x + y + \omega y)$$
$$= (x + y + \omega y)(x + y + \omega^2 y)$$
$$= \bigl((x+y) - \omega(-y)\bigr)\bigl((x+y) - \omega^2(-y)\bigr).$$

Therefore, if (x, y) is a solution, so is $(x + y, -y)$. At first sight this is not too helpful, since repeating the process gets us right back to (x, y). However, by symmetry, once $(x + y, -y)$ is known to be a solution, so is $(-y, x + y)$. We now find the cycle (∗) in the "backward" direction: (x, y), $(-y, x + y)$, $(-x - y, x)$, and so forth.

Incidentally, the same symmetry $(x, y) \leftrightarrow (y, x)$ can be used to show that the number of integral solutions is often divisible by 12.

However, due to the fact that (y, x) may already be among the solutions in (∗), the number is not always divisible by 12. For instance, there are exactly six integral solutions for $n = 1$.

Solution 3. (Greg Kuperberg, UC Davis) Note that $x^2 + xy + y^2 = x^2 - 2xy\cos(2\pi/3) + y^2$. Therefore, for $x, y > 0$, $x^2 + xy + y^2$ is the square of the length of the long side of a triangle whose shorter sides have lengths x and y and enclose an angle $2\pi/3$.

Now consider the hexagonal lattice shown below, in which each lattice point is a vertex of six different equilateral triangles of side length 1. Every such lattice point can be reached from the origin by a unique displacement

$x(1,0) + y(\frac{1}{2}, \frac{\sqrt{3}}{2})$ that is x units to the right and y units slanting up and to the right (at an angle of $\pi/3$ to the horizontal), where x and y are integers. For $x, y > 0$ this gives rise to a triangle as described above (shown for $x = 2, y = 3$).

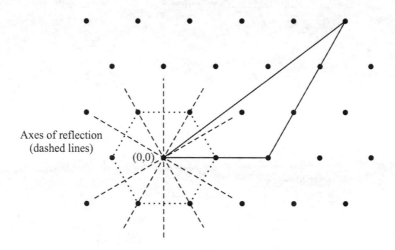

With this motivation, we can check that for *any* integers x, y, the vector $\mathbf{v} = x(1,0) + y\left(\frac{1}{2}, \frac{\sqrt{3}}{2}\right) = \left(x + \frac{1}{2}y, \frac{\sqrt{3}}{2}y\right)$ has length $|\mathbf{v}| = \sqrt{x^2 + xy + y^2}$. Therefore, the number of integral solutions of $x^2 + xy + y^2 = n$ equals the number of lattice points from the hexagonal lattice that lie on the circle with center at the origin and radius $\sqrt{n}$. This number is clearly finite, and by the 6-fold rotational symmetry of the lattice, it must be a multiple of 6.

Comment. Besides rotational symmetry, the lattice also has reflectional symmetry in each of six axes of reflection (see figure). It follows that the number of integral solutions of $x^2 + xy + y^2 = n$, where n is a positive integer, is actually a multiple of 12 *unless* one of the lattice points being counted lies on one of the six axes of reflection. Using the rotational symmetry we can see that this case occurs if and only if we have a point of the form $(x, 0)$ or a point of the form $x(1,0) + x\left(\frac{1}{2}, \frac{\sqrt{3}}{2}\right) = \left(\frac{3}{2}x, \frac{\sqrt{3}}{2}x\right)$. These cases lead to $n = x^2$ and $n = 3x^2$, respectively. Thus, the number of integral solutions of $x^2 + xy + y^2 = n$ is a multiple of 12 unless n can be written as $n = k^2$ or $n = 3k^2$ for some integer k.

Problem 196

For what real numbers x can one say the following?

a. For each positive integer n, there exists an integer m such that
$$\left|x - \frac{m}{n}\right| < \frac{1}{3n}.$$

b. For each positive integer n, there exists an integer m such that
$$\left|x - \frac{m}{n}\right| \leq \frac{1}{3n}.$$

Idea. Consider those integers n which are powers of 2. If the approximations m/n to x for these n are sufficiently close to each other, they cannot be distinct.

Solution. a. The condition holds for x if and only if x is an integer.

If x is an integer, then for any n we can take $m = xn$, so the condition holds for x. Conversely, suppose that the condition holds for x. Let m_k be the numerator corresponding to $n = 2^k$, $k = 0, 1, \ldots$. If, for some k, $m_k/2^k \neq m_{k+1}/2^{k+1}$, then
$$\left|\frac{m_k}{2^k} - \frac{m_{k+1}}{2^{k+1}}\right| = \left|\frac{2m_k - m_{k+1}}{2^{k+1}}\right| \geq \frac{1}{2^{k+1}},$$

and so
$$\frac{1}{2^{k+1}} \leq \left|\frac{m_k}{2^k} - \frac{m_{k+1}}{2^{k+1}}\right| \leq \left|\frac{m_k}{2^k} - x\right| + \left|x - \frac{m_{k+1}}{2^{k+1}}\right| \quad (*)$$
$$< \frac{1}{3 \cdot 2^k} + \frac{1}{3 \cdot 2^{k+1}} = \frac{1}{2^{k+1}},$$

a contradiction. We conclude that $m_k/2^k = m_{k+1}/2^{k+1}$ for all k, so all the approximations $m_k/2^k$ to x are equal to the integer m_0. But then
$$|x - m_0| = \left|x - \frac{m_k}{2^k}\right| \leq \frac{1}{3 \cdot 2^k}$$

for all k; it follows that $x - m_0 = 0$, so x is an integer, and we are done.

b. The condition holds for x if and only if $3x$ is an integer.

Suppose that $3x$ is an integer, and let n be a positive integer. Observe that $|x - \frac{m}{n}| \leq \frac{1}{3n}$ is equivalent to $|3nx - 3m| \leq 1$. Because the distance from the integer $3nx = n(3x)$ to the nearest integer multiple of 3 is at most 1, there exists an integer m for which $|3nx - 3m| \leq 1$; thus, the condition holds for x.

Now suppose that the condition holds for x. As in (a), let m_k be the numerator corresponding to $n = 2^k$. If $m_k/2^k = m_{k+1}/2^{k+1}$ for all k, then we see as in (a) that x is an integer, and we are done. Otherwise, there is an integer $K \geq 0$ such that

$$m_0 = \frac{m_1}{2} = \cdots = \frac{m_K}{2^K} \neq \frac{m_{K+1}}{2^{K+1}}.$$

For $k = K$, the inequalities in $(*)$ of (a) are equalities, and so

$$x = \frac{m_K}{2^K} \pm \frac{1}{3 \cdot 2^K} = m_0 \pm \frac{1}{3 \cdot 2^K}.$$

Now suppose that $K > 0$, and let $n = 3 \cdot 2^{K-1}$. Then for the corresponding numerator m, we have

$$\left| x - \frac{m}{3 \cdot 2^{K-1}} \right| \leq \frac{1}{9 \cdot 2^{K-1}}.$$

On the other hand, since $x = m_0 \pm 1/(3 \cdot 2^K)$, we have

$$x - \frac{m}{3 \cdot 2^{K-1}} = \frac{m_0 \cdot 3 \cdot 2^K \pm 1 - 2m}{3 \cdot 2^K}.$$

Since the numerator of this fraction is odd, $x - m/(3 \cdot 2^{K-1})$ is nonzero and

$$\left| x - \frac{m}{3 \cdot 2^{K-1}} \right| \geq \frac{1}{3 \cdot 2^K} > \frac{1}{9 \cdot 2^{K-1}},$$

a contradiction. We conclude that $K = 0$, $x = m_0 \pm \frac{1}{3}$, and we are done.

Problem 197

Starting with an empty $1 \times n$ board, we successively place 1×2 dominoes to cover two adjacent squares. At each stage, the placement of the new domino is chosen at random; the process continues until no further dominoes can be placed. Find the limit, as $n \to \infty$, of the expected fraction of the board that is covered when the process ends.

Answer. The limit is $1 - e^{-2}$.

Solution. Let $E(n)$ be the expected number of squares that will be covered; we are looking for $\lim_{n \to \infty} \frac{E(n)}{n}$. Note that $E(0) = E(1) = 0$ and $E(2) = E(3) = 2$.

We now find a recursive expression for $E(n+1)$. Starting with an empty $1 \times (n+1)$ board, there are n equally likely placements for the first domino.

THE SOLUTIONS

If the first domino covers squares $k+1$ and $k+2$ (with $0 \le k \le n-1$), then placing that domino leaves us with two separate boards of sizes k and $n-k-1$. Thus, for $n \ge 1$, we have

$$E(n+1) = 2 + \frac{\sum_{k=0}^{n-1}\bigl(E(k) + E(n-k-1)\bigr)}{n} = 2 + \frac{2}{n}\sum_{k=0}^{n-1} E(k).$$

To get rid of the summation, we can multiply by n and subtract the corresponding equation for $n-1$. This yields, for $n \ge 2$,

$$nE(n+1) - (n-1)E(n) = 2 + 2E(n-1). \qquad (*)$$

In fact, from the values above we see that this is still true for $n = 1$. We'll use a generating function to solve this recurrence relation, but first we simplify it slightly by setting $F(n) = n - E(n)$. (Note that $F(n)$ is the expected number of empty squares at the end.) With this notation, $(*)$ becomes

$$n[n+1-F(n+1)] - (n-1)[n-F(n)] = 2 + 2[n-1-F(n-1)],$$
$$nF(n+1) - (n-1)F(n) - 2F(n-1) = 0.$$

Now let $G(x) = \sum_{n=0}^{\infty} F(n+1)x^n$. Because $0 \le F(n) \le n$ for all n, this series is absolutely convergent for $|x| < 1$ by comparison with $\sum_{n=0}^{\infty}(n+1)x^n$. In particular, we can differentiate $G(x)$ on the interval $|x| < 1$, and using $F(0) = 0$ we get

$$G'(x) = \sum_{n=1}^{\infty} nF(n+1)x^{n-1} = \sum_{n=1}^{\infty}\bigl((n-1)F(n) + 2F(n-1)\bigr)x^{n-1}$$
$$= x\sum_{n=1}^{\infty}(n-1)F(n)x^{n-2} + 2x\sum_{n=2}^{\infty} F(n-1)x^{n-2}$$
$$= x\sum_{n=0}^{\infty} nF(n+1)x^{n-1} + 2x\sum_{n=0}^{\infty} F(n+1)x^n = x\,G'(x) + 2x\,G(x).$$

Thus $G(x)$ satisfies the separable differential equation

$$(1-x)\,G'(x) = 2x\,G(x).$$

Solving this equation and using the initial condition $G(0) = F(1) = 1$, we arrive at

$$G(x) = \frac{e^{-2x}}{(1-x)^2}.$$

The Maclaurin series for $G(x)$ is obtained by multiplying the series for e^{-2x} and for $\frac{1}{(1-x)^2}$; the latter series can be found by differentiating the standard geometric series for $\frac{1}{1-x}$. This yields

$$G(x) = \sum_{k=0}^{\infty} \frac{(-2)^k}{k!} x^k \cdot \sum_{k=0}^{\infty} (k+1) x^k.$$

The coefficient of x^{n-1} in this product is $F(n)$, and so

$$E(n) = n - F(n) = n - \sum_{k=0}^{n-1} \frac{(-2)^k}{k!} (n-k)$$

$$= n \left(1 - \sum_{k=0}^{n-1} \frac{(-2)^k}{k!}\right) + \sum_{k=1}^{n-1} \frac{(-2)^k}{(k-1)!}.$$

Finally,

$$\frac{E(n)}{n} = 1 - \sum_{k=0}^{n-1} \frac{(-2)^k}{k!} + \frac{1}{n} \sum_{k=1}^{n-1} \frac{(-2)^k}{(k-1)!}$$

$$= 1 - \sum_{k=0}^{n-1} \frac{(-2)^k}{k!} - \frac{2}{n} \sum_{k=0}^{n-2} \frac{(-2)^k}{k!}.$$

Because both sums on the right are partial sums of the standard series for e^{-2}, in the limit as $n \to \infty$ they both approach e^{-2}, and we get

$$\lim_{n \to \infty} \frac{E(n)}{n} = 1 - e^{-2} - 0 \cdot e^{-2} = 1 - e^{-2},$$

as claimed.

Problem 198

Let $\mathbb{Z}/n\mathbb{Z}$ be the set $\{0, 1, \ldots, n-1\}$ with addition modulo n. Consider subsets S_n of $\mathbb{Z}/n\mathbb{Z}$ such that $(S_n + k) \cap S_n$ is nonempty for every k in $\mathbb{Z}/n\mathbb{Z}$. Let $f(n)$ denote the minimal number of elements in such a subset. Find

$$\lim_{n \to \infty} \frac{\ln f(n)}{\ln n},$$

or show that this limit does not exist.

Answer. The limit is $1/2$.

Solution. First we rephrase the condition that $(S_n + k) \cap S_n$ is nonempty for all k, as follows: For every k in $\mathbb{Z}$, there are elements x and y in S_n such that $x - y \equiv k \pmod{n}$. We call S_n a *difference set* modulo n if this condition is satisfied.

For a difference set S_n with m elements, there are at most m^2 possible differences. This shows that $(f(n))^2 \geq n$, and therefore
$$\frac{\ln f(n)}{\ln n} \geq \frac{1}{2}.$$

On the other hand, let
$$T_n = \{1, 2, 3, \ldots, \lfloor\sqrt{n}\rfloor, 2\lfloor\sqrt{n}\rfloor, 3\lfloor\sqrt{n}\rfloor, \ldots, \lfloor\sqrt{n}\rfloor^2\}.$$

We claim that for $n \geq 16$, T_n is a difference set modulo n. Note that any integer from 0 to $\lfloor\sqrt{n}\rfloor^2 - 1$, inclusive, is a difference of two elements of T_n. When $n \geq 16$, we have $\lfloor\sqrt{n}\rfloor^2 > (\sqrt{n} - 1)^2 \geq n/2 + 1$, so every integer from 0 to $\lfloor n/2 \rfloor$ is a difference of elements of T_n. But then their opposites are also differences, and thus all the integers m satisfying $-n/2 < m \leq n/2$ are differences of elements in T_n. Because every k in $\mathbb{Z}$ is equal (mod n) to such an integer m, T_n is a difference set. The set T_n has $2\lfloor\sqrt{n}\rfloor - 1 < 2\sqrt{n}$ elements, and so we have $f(n) < 2\sqrt{n}$, for $n \geq 16$. This implies
$$\frac{\ln f(n)}{\ln n} < \frac{1}{2} + \frac{\ln 2}{\ln n}.$$

We can now use the squeeze principle to conclude that
$$\lim_{n \to \infty} \frac{\ln f(n)}{\ln n} = \frac{1}{2}.$$

Problem 199

a. If a rational function (a quotient of two real polynomials) takes on rational values for infinitely many rational numbers, prove that it may be expressed as the quotient of two polynomials with rational coefficients.
b. If a rational function takes on integral values for infinitely many integers, prove that it must be a polynomial with rational coefficients.

Solution 1. a. Let $f(x)/g(x)$ be the rational function. We will prove the result by induction on the sum of the degrees, $\deg f + \deg g$. If $f(x) = 0$ or

$\deg f + \deg g = 0$, then the function is constant and the result is immediate. Now suppose it is true for $\deg f + \deg g \le k$, and let $\deg f + \deg g = k+1$. We will soon find it convenient to have $\deg f \ge \deg g$; if necessary, this can be arranged by replacing $f(x)/g(x)$ by its reciprocal $g(x)/f(x)$, which will not affect the rationality of any values except for the finitely many roots of $f(x)$. Assuming $\deg f \ge \deg g$, let r_1 be any rational number for which the value $f(r_1)/g(r_1)$ is also rational, say $f(r_1)/g(r_1) = r_2$. Then the polynomial $f(x) - r_2 g(x)$ has r_1 as a root, so there is a polynomial $h(x)$ for which $f(x) - r_2 g(x) = (x - r_1) h(x)$ and hence

$$\frac{f(x)}{g(x)} - r_2 = (x - r_1) \frac{h(x)}{g(x)}.$$

We see that $h(x)/g(x)$ will be rational whenever x and $f(x)/g(x)$ are both rational, except perhaps for $x = r_1$. Therefore, $h(x)/g(x)$ takes on rational values for infinitely many rational x, so we can apply the induction hypothesis to $h(x)/g(x)$ provided $\deg h + \deg g \le k$. On the other hand, $f(x) - r_2 g(x) = (x - r_1) h(x)$ implies that $\deg h = \deg(f - r_2 g) - 1$, and since $\deg f \ge \deg g$, $\deg(f - r_2 g) \le \deg f$, so that $\deg h \le \deg f - 1$. Therefore, $\deg h + \deg g \le \deg f + \deg g - 1 = k$, the induction hypothesis applies, and $h(x)/g(x)$ is a quotient of polynomials with rational coefficients. But then so is

$$\frac{f(x)}{g(x)} = (x - r_1) \frac{h(x)}{g(x)} + r_2,$$

and we are done.

b. By (a), we can assume that the rational function is in the form $f(x)/g(x)$, where $f(x)$ and $g(x)$ are polynomials with rational coefficients. Using the division algorithm, we can write

$$\frac{f(x)}{g(x)} = q(x) + \frac{r(x)}{g(x)},$$

where $q(x)$, $r(x)$ are polynomials with rational coefficients and either $r(x) = 0$ or $\deg r < \deg g$. Let M be the least common multiple of the denominators of the coefficients of $q(x)$. (If $q(x) = 0$, set $M = 1$.) Then for any integer n, $q(n)$ is either an integer or at least $1/M$ removed from the closest integer. On the other hand,

$$\lim_{|x| \to \infty} \frac{r(x)}{g(x)} = 0.$$

Thus, for integers n with $|n|$ large enough, $|r(n)/g(n)| < 1/M$, so that

$$\frac{f(n)}{g(n)} = q(n) + \frac{r(n)}{g(n)}$$

can only be an integer if $q(n)$ is an integer and $r(n) = 0$. However, this implies that the polynomial $r(x)$ has infinitely many roots, so $r(x)$ is identically zero, $f(x)/g(x) = q(x)$, and we are done.

Solution 2. a. Suppose $f(x)/g(x)$ is a nonzero rational function having rational values for the infinitely many rationals $x_1, x_2, x_3, \ldots$. We can take $f(x)$ and $g(x)$ to be relatively prime. Write $f(x) = a_m x^m + \cdots + a_0$, $g(x) = b_n x^n + \cdots + b_0$, and $y_i = f(x_i)/g(x_i)$, $i = 1, 2, 3, \ldots$. Then the coefficients $a_m, \ldots, a_0, b_n, \ldots, b_0$ satisfy the infinite, homogeneous linear system

$$\sum_{j=0}^{m} x_i^j a_j - \sum_{j=0}^{n} y_i x_i^j b_j = 0, \qquad i = 1, 2, 3, \ldots. \qquad (*)$$

In fact, we claim that the solution $(a_m, \ldots, a_0, b_n, \ldots, b_0)$ of $(*)$ is unique up to scalar multiples. If $(a_m^*, \ldots, a_0^*, b_n^*, \ldots, b_0^*)$ is another solution, then for $f^*(x) = a_m^* x^m + \cdots + a_0^*$, $g^*(x) = b_n^* x^n + \cdots + b_0^*$ we have

$$\frac{f^*(x_i)}{g^*(x_i)} = \frac{f(x_i)}{g(x_i)},$$

so

$$f^*(x_i)g(x_i) - f(x_i)g^*(x_i) = 0$$

for all i. This implies that $f^*(x)g(x) - f(x)g^*(x)$, having infinitely many roots, is identically zero, so $f^*(x)/g^*(x) = f(x)/g(x)$. Since $f(x)$ and $g(x)$ are relatively prime and the degrees of $f^*(x)$ and $g^*(x)$ are at most those of $f(x)$ and $g(x)$, respectively, it follows that for some constant C, $f^*(x) = C f(x)$ and $g^*(x) = C g(x)$. Then the solution $(a_m^*, \ldots, a_0^*, b_n^*, \ldots, b_0^*)$ is C times $(a_m, \ldots, a_0, b_n, \ldots, b_0)$. We now know that the null space (set of solutions) of $(*)$ is a one-dimensional subspace of $\mathbb{R}^{m+n+2}$. Therefore, there exists a positive integer k for which the finite system

$$\sum_{j=0}^{m} x_i^j a_j - \sum_{j=0}^{n} y_i x_i^j b_j = 0, \qquad i = 1, 2, 3, \ldots, k, \qquad (**)$$

has that same one-dimensional null space. Because the x_i and y_i are rational, the Gaussian elimination method applied to $(**)$ then shows that the null space consists of the multiples of a single *rational* vector. Because

$(a_m, \ldots, a_0, b_n, \ldots, b_0)$ is in the null space, it follows that for some nonzero number c, $cf(x)$ and $cg(x)$ have rational coefficients. If we then write

$$\frac{f(x)}{g(x)} = \frac{cf(x)}{cg(x)},$$

we are done.

b. Let $f(x)/g(x)$ take on integer values for infinitely many integers. By (a), we can assume that $f(x)$ and $g(x)$ have rational coefficients; after multiplying both by a suitable integer, we can even assume they have integer coefficients. We can also take $f(x)$ and $g(x)$ to be relatively prime. Using the Euclidean algorithm and clearing denominators then yields polynomials $p_1(x)$ and $p_2(x)$ with integral coefficients and a positive integer D such that

$$p_1(x) f(x) + p_2(x) g(x) = D.$$

We then have

$$p_1(x) \frac{f(x)}{g(x)} + p_2(x) = \frac{D}{g(x)},$$

so by the given, there are infinitely many integers n for which $D/g(n)$ is an integer.

But if $g(x)$ is not constant, we can have $|g(n)| \leq D$ for at most finitely many integers n, a contradiction. Thus $g(x)$ must be a rational number, $f(x)/g(x)$ is a rational polynomial, and we are done.

Comment. It is well known that even if a polynomial takes on integral values for *all* integers, it may not have integral coefficients. In fact, it can be any integral linear combination of the "binomial polynomials"

$$\binom{x}{n} = \frac{x(x-1)\cdots(x-n+1)}{n!}.$$

However, the polynomial $d!\, p(x)$, where d is the degree of $p(x)$, must have integral coefficients.

Problem 200

Can there be a multiplicative $n \times n$ magic square ($n > 1$) with entries $1, 2, \ldots, n^2$? That is, does there exist an integer $n > 1$ for which the numbers $1, 2, \ldots, n^2$ can be placed in a square so that the product of all the numbers in any row or column is always the same?

Solution. No. Suppose that there were such a magic square, with the product of the numbers in each row equal to P. Then the product A of the numbers $1, 2, \ldots, n^2$ would be P^n; in particular, A would be an nth power. Now let m be the number of factors 2 in the prime factorization of A. We will show that m is not divisible by n (provided $n > 1$), so A cannot be an nth power and we will be done.

Note that every second number in the finite sequence $1, 2, \ldots, n^2$ is divisible by 2, which yields $\lfloor n^2/2 \rfloor$ factors 2 in A, every fourth number is divisible by 4 for another $\lfloor n^2/4 \rfloor$ factors 2, and so forth. Therefore, we have

$$m = \lfloor n^2/2 \rfloor + \lfloor n^2/4 \rfloor + \cdots + \lfloor n^2/2^k \rfloor,$$

where k is chosen such that $2^k \leq n^2 < 2^{k+1}$. In particular, for $n = 2, 3, 4, 5, 6$ we find $m = 2 + 1 = 3$, $m = 4 + 2 + 1 = 7$, $m = 8 + 4 + 2 + 1 = 15$, $m = 12 + 6 + 3 + 1 = 22$, $m = 18 + 9 + 4 + 2 + 1 = 34$, respectively, and in each case m is not divisible by n. We now give a general proof that for $n > 6$, we have $n(n-1) < m < n^2$, so that m is between successive multiples of n and hence cannot be divisible by n.

First of all, we see from the expression for m above that

$$m \leq n^2/2 + n^2/4 + \cdots + n^2/2^k = n^2 \left(1/2 + 1/4 + \cdots + 1/2^k\right)$$
$$< n^2 \left(1/2 + 1/4 + \cdots + 1/2^k + \cdots\right)$$
$$= n^2,$$

so we do indeed have $m < n^2$.

On the other hand, $\lfloor x \rfloor > x - 1$ for all x, so we have

$$m > (n^2/2 - 1) + (n^2/4 - 1) + \cdots + (n^2/2^k - 1)$$
$$= n^2(1/2 + 1/4 + \cdots + 1/2^k) - k$$
$$= n^2(1 - 1/2^k) - k.$$

To show that this is greater than $n(n-1)$, it is enough to show $n^2/2^k + k < n$, and, since $n^2/2^k < 2$ (by the definition of k) and k is an integer, it is sufficient to show $k < n - 1$.

From $2^k \leq n^2$ we have $k \leq 2\log_2 n$. It is easy to see, by taking the derivative, that the difference $2\log_2 n - (n-1)$ is decreasing for $n \geq 3$. Because $2\log_2 n < n - 1$ for $n = 7$, we will have $k \leq 2\log_2 n < n - 1$ for any $n > 6$, and we are done.

Comment. One can see much more quickly that A is not an nth power by using Bertrand's postulate, the theorem that for any integer $N > 1$, there is a prime between N and $2N$.

Problem 201

Define a function f by
$$f(x) = x^{1/x^{x^{1/x^{x^{1/x^{\cdots}}}}}} \qquad (x > 0).$$
That is to say, for a fixed x, let
$$a_1 = x, \quad a_2 = x^{1/x}, \quad a_3 = x^{1/x^x} = x^{1/x^{a_1}}, \quad a_4 = x^{1/x^{x^{1/x}}} = x^{1/x^{a_2}}, \quad \ldots$$
and, in general, $a_{n+2} = x^{1/x^{a_n}}$, and take $f(x) = \lim_{n \to \infty} a_n$.

a. Assuming that this limit exists, let M be the maximum value of f as x ranges over all positive real numbers. Evaluate M^M.

b. Prove that $f(x)$ is well defined; that is, that the limit exists.

Solution. a. Because $y = f(x)$ is a limit of positive powers of x, we have $y \geq 1$ for $x > 1$, $y = 1$ for $x = 1$, and $y \leq 1$ for $x < 1$. Thus, to find the maximum value we need only consider $x > 1$. Also, we have
$$y = x^{1/x^y} \implies \ln y = (1/x)^y \ln x \implies x^y \ln y = \ln x.$$
Because, for fixed x, $x^y \ln y$ ($y \geq 1$) is an increasing continuous function of y, which is 0 for $y = 1$ and approaches infinity as $y \to \infty$, there is a unique y for which it takes on the value $\ln x$, and in fact, we have $y > 1$.

Proceeding from $x^y \ln y = \ln x$ and taking logarithms again (which is possible because $\ln x > 0$), we get
$$y \ln x + \ln \ln y - \ln \ln x = 0.$$
From here we can get $\dfrac{dy}{dx}$ by implicit differentiation, or, thinking of this equation as $F(x, y) = 0$, we can differentiate with respect to x (using the multivariable chain rule) to get $\dfrac{\partial F}{\partial x} + \dfrac{\partial F}{\partial y} \dfrac{dy}{dx} = 0$, or
$$\frac{dy}{dx} = -\frac{\partial F/\partial x}{\partial F/\partial y} = -\frac{\dfrac{y}{x} - \dfrac{1}{x \ln x}}{\ln x + \dfrac{1}{y \ln y}} = \frac{\dfrac{1}{\ln x} - y}{x \ln x + \dfrac{x}{y \ln y}}.$$

THE SOLUTIONS

Because $y = x^{1/x^y}$ is less than x (the exponent $(1/x)^y$ is less than 1), we have $\frac{dy}{dx} > 0$ for x close enough to 1 (note that $\frac{1}{\ln x} \to \infty$ as $x \to 1^+$). On the other hand, for $x > e$ we have $y > 1 > \frac{1}{\ln x}$, so $\frac{dy}{dx} < 0$. Thus, if we can show that $\frac{dy}{dx} = 0$ only once, that will be at the maximum value for y.

Now $\frac{dy}{dx} = 0 \Longrightarrow y \ln x = 1 \Longrightarrow x^y = e$. Combining this with our earlier equation $x^y \ln y = \ln x$, we see that when $\frac{dy}{dx} = 0$, we must have $e \ln y = \ln x$, or $x = y^e$. Substituting this into $x^y = e$ gives us

$$(y^e)^y = e \Longrightarrow y^{ey} = e \Longrightarrow y^y = e^{1/e}.$$

This is equivalent to $y \ln y = 1/e$, which has a unique solution for y (because for $y \geq 1$, $y \ln y$ increases from 0 at $y = 1$ and approaches infinity as $y \to \infty$). So there is only one possible value of $x = y^e$ for which $\frac{dy}{dx}$ can be zero, and as we saw above, the corresponding y is the maximum value for $f(x)$. As we've seen, this maximum value M is such that $M^M = e^{1/e}$.

b. First, fix $x > 1$, and consider the sequence (a_n) given in the problem. Note that we have both $a_1 > a_2$ (because $1 > 1/x$) and $a_2 > a_3$ (because $1/x > (1/x)^x$). Also, $a_{n+2} = x^{1/x^{a_n}}$, and the function $t \mapsto x^{1/x^t}$ is decreasing (because $t \mapsto (1/x)^t$ is decreasing; remember that $x > 1$ is fixed). So, for example, $a_1 > a_3$ (because $a_1 > a_2 > a_3$) implies $a_3 = x^{1/x^{a_1}} < a_5 = x^{1/x^{a_3}}$. But we also know how a_1 and a_5 compare, because a_5 is x raised to an exponent less than 1. So

$$a_1 > a_5 \Longrightarrow a_3 < a_7 \Longrightarrow a_5 > a_9 \Longrightarrow a_7 < a_{11} \Longrightarrow \cdots.$$

Thus, $a_1, a_5, a_9, \ldots$ is a decreasing subsequence of our sequence (a_n). This decreasing subsequence is bounded below by 1, so it has a limit L_1. Meanwhile, $a_3, a_7, a_{11}, \ldots$ is an increasing subsequence, bounded above by x, so it has a limit L_2. Once we show that $L_1 = L_2$, the even-numbered terms of the sequence will have that same limit, because they are "trapped": $a_1 > a_2 > a_3$, $a_3 < a_4 < a_5$, etc., so the entire sequence $a_1, a_2, a_3, \ldots$ will have that limit $L = L_1 = L_2$ and $f(x)$ will be well defined. But we also note that the sequence $\log_x a_2, \log_x a_3, \log_x a_4, \ldots = \frac{1}{x}, (1/x)^x, (1/x)^{x^{1/x}}, \ldots$ will have limit $\log_x L$, which shows that $f(1/x)$ is also well defined (and equals $\log_x f(x)$). As x ranges through all numbers greater than 1, $1/x$ ranges through all numbers between 0 and 1, so $f(x)$ is well defined for $0 < x < 1$. Because, obviously, $f(1) = 1$, it's enough to show that $L_1 = L_2$ for $x > 1$.

Because $a_1 > a_3$ implies $a_5 > a_7$ (in two steps) which implies $a_9 > a_{11}$, etc., in the limit we have $L_1 \geq L_2$. Also, from $a_3 = x^{1/x^{a_1}}$, $a_7 = x^{1/x^{a_5}}$, ... we have $L_2 = x^{1/x^{L_1}}$, while from $a_5 = x^{1/x^{a_3}}$, $a_9 = x^{1/x^{a_7}}$, ... we have $L_1 = x^{1/x^{L_2}}$. Now suppose that $L_1 \neq L_2$, so that $L_1 > L_2$. We'll work toward a contradiction by applying the Mean Value Theorem to the function $g(t) = x^{1/x^t} = x^{x^{-t}}$ (for fixed $x > 1$). Note that $L_1 = g(L_2)$ and $L_2 = g(L_1)$, so on the assumption that $L_1 > L_2$ we have

$$L_1 - L_2 = g(L_2) - g(L_1) = g'(c)(L_2 - L_1)$$

for some c with $L_2 < c < L_1$. Dividing by $L_2 - L_1$, we get $g'(c) = -1$.
On the other hand, the derivative of $g(t) = x^{x^{-t}} = e^{x^{-t}\ln x}$ is

$$g'(t) = e^{x^{-t}\ln x} \cdot \frac{d}{dt}(x^{-t}\ln x) = x^{x^{-t}} \cdot \frac{d}{dt}(e^{-t\ln x}\ln x)$$
$$= x^{x^{-t}} \cdot e^{-t\ln x} \cdot (-\ln x)\ln x = -(\ln x)^2 \, x^{x^{-t}} \cdot x^{-t},$$

so $g'(c) = -1$ yields $(\ln x)^2 \, x^{x^{-c}} \cdot x^{-c} = 1$, or $(\ln x)^2 = x^{c-x^{-c}}$. We'll show that this is impossible for $L_2 < c < L_1$.

Note that $c - x^{-c} = c - 1/x^c$ is an increasing function of c, so for $L_2 < c$, because $1 < L_2$, we have

$$x^{c-x^{-c}} > x^{L_2 - x^{-L_2}} > x^{1-1/x}.$$

Then $g'(c) = -1$ would imply $(\ln x)^2 > x^{1-1/x}$. In particular, because $x > 1$ and thus $1 - 1/x > 0$ and $x^{1-1/x} > 1$, this can only happen if $(\ln x)^2 > 1$, that is (as $\ln x > 0$ for $x > 1$) if $\ln x > 1$, or equivalently, if $x > e$.
On the other hand, from $(\ln x)^2 > x^{1-1/x}$ (and $\ln x > 0$) we get

$$\ln x > x^{\frac{1}{2}(1-1/x)} \implies \ln(\ln x) > \frac{1}{2}(1-1/x)\ln x \implies \frac{\ln(\ln x)}{\ln x} > \frac{1}{2}\left(1-1/x\right).$$

Now consider the function $h(y) = (\ln y)/y$. Because $h'(y) = (1 - \ln y)/y^2$, which is positive for $0 < y < e$ and negative for $y > e$, so $h(y)$ has its maximum value for $y = e$, namely $h(e) = (\ln e)/e = 1/e$. In particular, we have $\frac{\ln(\ln x)}{\ln x} \leq \frac{1}{e}$ for all x, so from the inequality above for $(\ln(\ln x))/\ln x$

we get

$$\frac{1}{e} > \frac{1}{2}\left(1 - \frac{1}{x}\right) \Longrightarrow \frac{2}{e} > 1 - \frac{1}{x} \Longrightarrow \frac{1}{x} > 1 - \frac{2}{e}$$

$$\Longrightarrow x < \frac{1}{1 - 2/e} = \frac{e}{e-2} \Longrightarrow \ln x < \ln\left(\frac{e}{e-2}\right) < 1.34.$$

Recall that $h(y) = (\ln y)/y$ is increasing for $y < e$, so $\ln x < 1.34$ implies

$$\frac{\ln(\ln x)}{\ln x} < \frac{\ln(1.34)}{1.34} < 0.22.$$

Thus, for our original inequality for $\frac{\ln(\ln x)}{\ln x}$ we now have

$$0.22 > \frac{1}{2}\left(1 - \frac{1}{x}\right) \Longrightarrow 0.44 > 1 - \frac{1}{x} \Longrightarrow \frac{1}{x} > 0.56 \Longrightarrow x < \frac{1}{0.56} < e,$$

which is a contradiction since we found earlier that $x > e$. Therefore, it is impossible to have $g'(c) = -1$, so we must have $L_1 = L_2$, and this establishes that f is well defined.

Problem 202

Note that if the edges of a regular octahedron have length 1, then the distance between any two of its vertices is either 1 or $\sqrt{2}$. Are there other possible configurations of six points in $\mathbb{R}^3$ for which the distance between any two of the points is either 1 or $\sqrt{2}$? If so, find them.

Answer. There is one other such configuration, consisting of the vertices of a rectangular prism whose triangular faces are equilateral triangles of side 1 and whose rectangular faces are squares. One way of positioning this prism in $\mathbb{R}^3$ is shown below.

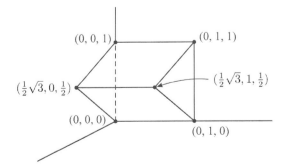

Solution. Let X be any configuration of six points in $\mathbb{R}^3$ such that the distance between any two points of X is either 1 or $\sqrt{2}$. Our proof that the points of X must form the vertices either of a prism as described above or of a regular octahedron will consist of three parts, as follows. In Part 1 we show by contradiction that there are two points of X whose distance is 1. Part 2 is the heart of the proof; here we show that any two points of X whose distance is 1 are vertices of a square whose other two vertices are also in X. Finally, in Part 3 we combine the results of the earlier parts and complete the proof.

Suppose that no two points of X have distance 1, so all distances are $\sqrt{2}$. Then any three points of X form an equilateral triangle. Given such a triangle, there are only two points in $\mathbb{R}^3$ whose distance to each of the vertices is $\sqrt{2}$. But there are three points of X besides the vertices of the triangle, a contradiction. So there are two points of X whose distance is 1, and Part 1 is done.

Now suppose we have any two points of X whose distance is 1. We can position these points at $(0,0,0)$ and $(0,0,1)$. Then in order to have distances from each of these that are either 1 or $\sqrt{2}$, each of the other four points of X must lie on one of the following circles.

C_1: $z = 0$, $x^2 + y^2 = 1$ (distance 1 from $(0,0,0)$, $\sqrt{2}$ from $(0,0,1)$),

C_2: $z = \frac{1}{2}$, $x^2 + y^2 = \frac{3}{4}$ (distance 1 from $(0,0,0)$, 1 from $(0,0,1)$),

C_3: $z = \frac{1}{2}$, $x^2 + y^2 = \frac{7}{4}$ (distance $\sqrt{2}$ from $(0,0,0)$, $\sqrt{2}$ from $(0,0,1)$),

C_4: $z = 1$, $x^2 + y^2 = 1$ (distance $\sqrt{2}$ from $(0,0,0)$, 1 from $(0,0,1)$).

In order to show that $(0,0,0)$ and $(0,0,1)$ are vertices of a square with two other points of X, it will be helpful to consider the projections of the four other points of X onto the xy–plane. In fact, we will show that two of these projections must coincide. This can only happen if one of the points is on C_1 and the other is directly above it on C_4, and we then have our square.

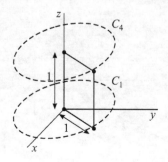

THE SOLUTIONS

Let O be the origin, and let P and Q, with $P \neq Q$, be any two of the projections described above. We can use the law of cosines in $\triangle OPQ$ to get bounds on the angle between the rays OP, OQ from the origin to P, Q respectively. Specifically, if $\theta = \angle POQ$, then we have

$$PQ^2 = OP^2 + OQ^2 - 2 \cdot OP \cdot OQ \cos \theta, \quad \text{so}$$

$$\cos \theta = \frac{OP^2 + OQ^2 - PQ^2}{2 \cdot OP \cdot OQ}.$$

Now PQ is at most the distance between the two points of X whose projections are P and Q, so $PQ \leq \sqrt{2}$. On the other hand, those points of X each lie on one of the circles C_1, C_2, C_3, C_4, and so OP and OQ can only have the values 1, $\frac{1}{2}\sqrt{3}$, $\frac{1}{2}\sqrt{7}$. Therefore,

$$OP^2 + OQ^2 - PQ^2 \geq \frac{3}{4} + \frac{3}{4} - 2 = -\frac{1}{2}$$

and thus

$$\cos \theta \geq \frac{-1}{4 \cdot OP \cdot OQ} \geq \frac{-1}{4 \cdot \frac{1}{2}\sqrt{3} \cdot \frac{1}{2}\sqrt{3}} = -\frac{1}{3}.$$

In particular, $\theta < 120°$.

To get a bound in the other direction, we distinguish a number of cases, depending on the location of the points of X whose projections are P and Q. If both these points of X have the same z-coordinate, then PQ equals their distance, so $PQ \geq 1$ and

$$\cos \theta \leq \frac{OP^2 + OQ^2 - 1}{2 \cdot OP \cdot OQ}.$$

We can then have $OP = OQ = 1$, $OP = OQ = \frac{1}{2}\sqrt{3}$, $OP = OQ = \frac{1}{2}\sqrt{7}$, or OP, OQ equal to $\frac{1}{2}\sqrt{3}$, $\frac{1}{2}\sqrt{7}$ in some order. These cases lead to the respective estimates $\cos \theta \leq \frac{1}{2}$, $\cos \theta \leq \frac{1}{3}$, $\cos \theta \leq \frac{5}{7}$, $\cos \theta \leq \frac{3}{\sqrt{21}}$. Therefore, if both the points of X have the same z-coordinate, then $\cos \theta \leq \frac{5}{7}$.

If the points of X whose projections are P and Q have different z-coordinates, then either one of them is on C_1 and the other is on C_4, or their z-coordinates differ by $\frac{1}{2}$. In the former case, since $P \neq Q$, the distance between the points of X must be $\sqrt{2}$ rather than 1, and by the Pythagorean theorem we must then have $PQ = 1$. Then it follows, as above, that $\cos \theta \leq \frac{5}{7}$. In the latter case, since the points of X are at least 1 apart, by the Pythagorean theorem we have $PQ \geq \frac{1}{2}\sqrt{3}$ and thus

$$\cos \theta \leq \frac{OP^2 + OQ^2 - \frac{3}{4}}{2 \cdot OP \cdot OQ}.$$

Meanwhile, OP and OQ are equal to $1, \frac{1}{2}\sqrt{3}$ or to $1, \frac{1}{2}\sqrt{7}$ in some order, which yields the estimates $\cos\theta \leq \frac{1}{\sqrt{3}}$, $\cos\theta \leq \frac{2}{\sqrt{7}}$ respectively. Because $\frac{2}{\sqrt{7}} > \frac{5}{7} > \frac{1}{\sqrt{3}}$, we can conclude that $\cos\theta \leq \frac{2}{\sqrt{7}}$ in all cases.

Now consider the four rays from the origin to the projections of the points of X. We have just seen that if the projections of two points of X are distinct, the corresponding rays make an angle θ for which $-\frac{1}{3} \leq \cos\theta \leq \frac{2}{\sqrt{7}}$. Suppose that as we go around the origin, the angles between successive rays are α, β, γ, δ, as shown in the following figure.

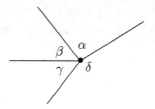

We can assume that $\alpha + \beta \leq 180°$ and $\beta + \gamma \leq 180°$ without loss of generality. But then, by the above, we actually have $\alpha + \beta < 120°$ and $\beta + \gamma < 120°$, so that

$$\delta = 360° - (\alpha + \beta + \gamma) \geq 360° - (\alpha + \beta) - (\beta + \gamma) > 120°.$$

Therefore, the smallest angle between the rays that form δ must be $\alpha+\beta+\gamma$ rather than δ, and we have $\alpha + \beta + \gamma < 120°$. On the other hand, if all four rays are distinct, then $\alpha, \beta, \gamma \geq \cos^{-1}\left(\frac{2}{\sqrt{7}}\right)$, and using the formula

$$\cos 3\theta = 4\cos^3\theta - 3\cos\theta,$$

we find

$$\cos(\alpha+\beta+\gamma) \leq \cos\left(3\cos^{-1}\left(\frac{2}{\sqrt{7}}\right)\right) = \frac{-10}{7\sqrt{7}} < -\frac{1}{2},$$

a contradiction. We have now shown that two of the projections must coincide, so $(0,0,0)$ and $(0,0,1)$ are vertices of a square with two other points of X, and Part 2 is done.

By Parts 1 and 2, we know that there are four points of X which form the vertices of a square of side 1. Let P_1 and P_2 be the remaining two points of X. P_1 and P_2 cannot both have distance $\sqrt{2}$ to each vertex of the square, since the only two points in $\mathbb{R}^3$ with this property lie at distance $\sqrt{6}$ to each other, one on either side of the plane that the square is in. So we can assume that P_1 has distance 1 to one of the vertices of the square, say to P_3. But then by Part 2 of our proof, there is another square whose vertices are points of X and include P_1 and P_3. Because two distinct squares cannot have three vertices in common, the new square can have at most two of the vertices of the old square, and since X has only six points in all, it follows that *both* P_1 and P_2 must be vertices of the new square.

First suppose that P_1 and P_2 are adjacent vertices of the new square. Then the two squares have a side in common, so they are "hinged" as shown below on the left. By considering the distances from P_1 to P_4 and P_5, we see that $P_1 P_3 P_4$ must be an equilateral triangle, and so we have the prism.

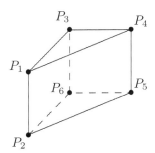 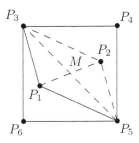

On the other hand, if P_1 and P_2 are opposite vertices of the new square, then the two squares have a diagonal in common (shown above on the right). Because the distances from P_2 and from P_4 to the midpoint M of this diagonal are both $\frac{1}{2}\sqrt{2}$, the distance from P_2 to P_4 cannot be $\sqrt{2}$, so it must be 1. Then $P_2 M$ and $P_4 M$ are perpendicular, and we have the regular octahedron.

Finally, the prism and the regular octahedron yield distinct configurations, because for the prism each vertex has distance $\sqrt{2}$ from two of the other vertices, while for the octahedron each vertex has distance $\sqrt{2}$ from only one other vertex.

Problem 203

Let a and b be positive real numbers, and define a sequence (x_n) by

$$x_0 = a, \quad x_1 = b, \quad x_{n+1} = \frac{1}{2}\left(\frac{1}{x_n} + x_{n-1}\right).$$

a. For what values of a and b will this sequence be periodic?

b. Show that given a, there exists a unique b for which the sequence converges.

Solution. **a.** The sequence is periodic if and only if $ab = 1$. To show this, we can rewrite the given recurrence relation as

$$x_n x_{n+1} = \frac{1}{2}(1 + x_{n-1} x_n).$$

If we then put $y_n = x_n x_{n+1}$, we see that the sequence $(y_n)_{n \geq 0}$ satisfies the recurrence relation

$$y_{n+1} = \frac{1}{2}(1 + y_n).$$

Also, if (x_n) is periodic, then so is (y_n). On the other hand, if $y_n < 1$, then $y_n < y_{n+1} < 1$, and if $y_n > 1$, then $y_n > y_{n+1} > 1$, so (y_n) is not periodic if $y_n \neq 1$. If $y_n = 1$, then $y_{n+1} = 1$. Therefore, (y_n) is periodic if and only if $y_0 = 1$, that is, $ab = 1$. In this case, the original sequence (x_n) has the form $a, 1/a, a, 1/a, \ldots$, so it is periodic as well.

b. First note that if (x_n) converges to L, then the relation

$$x_n x_{n+1} = \frac{1}{2}(1 + x_{n-1} x_n)$$

implies $L^2 = \frac{1}{2}(1 + L^2)$, and so $L = 1$ (since $L \geq 0$).

To find out when this happens, we use the sequence (y_n) from (a) to get explicit expressions for the x_n. Rewriting the recurrence relation for (y_n) as

$$(y_{n+1} - 1) = \frac{1}{2}(y_n - 1),$$

we find by induction that

$$y_n - 1 = \frac{1}{2^n}(y_0 - 1),$$

so

$$y_n = 1 + \frac{y_0 - 1}{2^n} = 1 + \frac{ab - 1}{2^n}.$$

Now

$$\frac{x_{n+2}}{x_n} = \frac{y_{n+1}}{y_n} = \frac{1 + \dfrac{ab-1}{2^{n+1}}}{1 + \dfrac{ab-1}{2^n}}.$$

Therefore, by induction,

$$x_{2n} = \left(\prod_{j=0}^{n-1} \frac{1 + \dfrac{ab-1}{2^{2j+1}}}{1 + \dfrac{ab-1}{2^{2j}}} \right) a$$

and
$$x_{2n+1} = \left(\prod_{j=1}^{n} \frac{1 + \frac{ab-1}{2^{2j}}}{1 + \frac{ab-1}{2^{2j-1}}} \right) b .$$

If (x_n) converges, then
$$\lim_{n \to \infty} x_{2n} = \left(\prod_{j=0}^{\infty} \frac{1 + \frac{ab-1}{2^{2j+1}}}{1 + \frac{ab-1}{2^{2j}}} \right) a = L = 1,$$

so
$$\prod_{j=0}^{\infty} \frac{1 + \frac{ab-1}{2^{2j+1}}}{1 + \frac{ab-1}{2^{2j}}} = \frac{1}{a} .$$

Conversely, if
$$\prod_{j=0}^{\infty} \frac{1 + \frac{ab-1}{2^{2j+1}}}{1 + \frac{ab-1}{2^{2j}}} = \frac{1}{a} ,$$

then we not only have
$$\lim_{n \to \infty} x_{2n} = 1 ,$$

but also
$$\lim_{n \to \infty} x_{2n+1} = \left(\frac{1 + \frac{ab-1}{2^2}}{1 + \frac{ab-1}{2}} \cdot \frac{1 + \frac{ab-1}{2^4}}{1 + \frac{ab-1}{2^3}} \cdots \right) b$$

$$= \frac{1}{1 + (ab-1)} \left(\frac{1 + \frac{ab-1}{2}}{1 + ab-1} \cdot \frac{1 + \frac{ab-1}{2^3}}{1 + \frac{ab-1}{2^2}} \cdots \right)^{-1} b$$

$$= \frac{1}{ab} \left(\prod_{j=0}^{\infty} \frac{1 + \frac{ab-1}{2^{2j+1}}}{1 + \frac{ab-1}{2^{2j}}} \right)^{-1} b = 1,$$

and hence (x_n) converges.

For fixed a, the value of the convergent infinite product

$$\prod_{j=0}^{\infty} \frac{1 + \frac{ab-1}{2^{2j+1}}}{1 + \frac{ab-1}{2^{2j}}}$$

is a continuous function of b for $b > 0$. As b increases from 0 to ∞, the individual factors

$$\frac{1 + \frac{ab-1}{2^{2j+1}}}{1 + \frac{ab-1}{2^{2j}}} = \frac{2^{2j+1} - 1 + ab}{2^{2j+1} - 2 + 2ab}$$

of the product decrease from

$$\begin{cases} (2^{2j+1} - 1)/(2^{2j+1} - 2), & j > 0 \\ \infty, & j = 0 \end{cases}$$

to $1/2$. Thus the product is a decreasing function of b, whose limit as $b \to \infty$ is 0, and whose limit as $b \to 0^+$ is ∞. Therefore, given a, there is a unique b for which the product equals $1/a$, and we are done.

Comment. One systematic way to find the general formula for y_n is to introduce a generating function, as follows. Let

$$g(z) = \sum_{n=0}^{\infty} y_n z^n.$$

Then from the recurrence relation $y_{n+1} = \frac{1}{2}(1 + y_n)$, we find that

$$g(z) = y_0 + z \sum_{n=0}^{\infty} \frac{1}{2}(1 + y_n) z^n$$

$$= y_0 + \frac{z/2}{1-z} + \frac{z}{2} g(z) .$$

Solving for $g(z)$ yields

$$g(z) = \frac{y_0}{1 - z/2} + \frac{z/2}{(1 - z/2)(1 - z)}$$

$$= \frac{y_0 - 1}{1 - z/2} + \frac{1}{1 - z}$$

$$= \sum_{n=0}^{\infty} \left[\frac{1}{2^n}(y_0 - 1) + 1 \right] z^n ,$$

and we can now read off that
$$y_n = \frac{1}{2^n}(y_0 - 1) + 1.$$

Problem 204

Consider the equation $x^2 + \cos^2 x = \alpha \cos x$, where α is some positive real number.
a. For what value or values of α does the equation have a unique solution?
b. For how many values of α does the equation have precisely four solutions?

Idea. Solutions of the equation correspond to intersection points of the circle $x^2 + y^2 = \alpha y$ and the graph of $y = \cos x$.

Solution. a. The equation has a unique solution if and only if $\alpha = 1$.

To see this, note that if x is a solution, so is $-x$; therefore, if the solution is unique, it must be $x = 0$. If $x = 0$ is a solution to $x^2 + \cos^2 x = \alpha \cos x$, then $\alpha = 1$.

It remains to show that when $\alpha = 1$, $x = 0$ is the *only* solution. If $x^2 + \cos^2 x = \cos x$, then $x^2 = \cos x - \cos^2 x \leq 1 - \cos^2 x = \sin^2 x$, so $|x| \leq |\sin x|$, and since this is true only for $x = 0$, we are done.

b. There is exactly one $\alpha > 0$ for which the equation has precisely four solutions.

If we put $y = \cos x$, the given equation becomes $x^2 + y^2 = \alpha y$ or, completing the square, $x^2 + (y - \frac{1}{2}\alpha)^2 = \frac{1}{4}\alpha^2$. Since $\alpha > 0$, this represents a circle with center $(0, \frac{1}{2}\alpha)$ and radius $\frac{1}{2}\alpha$. Thus the solutions of our equation $x^2 + \cos^2 x = \alpha \cos x$ correspond to intersection points of the graph of $y = \cos x$ with the circle $x^2 + (y - \frac{1}{2}\alpha)^2 = \frac{1}{4}\alpha^2$. The figure shows such circles for various values of α, along with the graph of $y = \cos x$. Our problem can be reformulated as: "How many of these circles intersect the graph in precisely four points?"

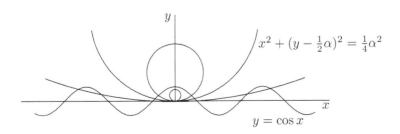

In part (a) we saw that the circle for $\alpha = 1$ lies underneath the graph of $y = \cos x$ except for $x = 0$, where it is tangent to the graph. Therefore, for any $\alpha < 1$ the circle $x^2 + (y - \frac{1}{2}\alpha)^2 = \frac{1}{4}\alpha^2$, which lies inside the circle for $\alpha = 1$, does not intersect $y = \cos x$ at all. Hence it is enough to consider $\alpha > 1$.

The preceding figure suggests that for any $\alpha > 1$, there are two intersection points with $-\pi/2 < x < \pi/2$. In fact, since the point $(0, 1)$ on the graph is inside the circle while the points $(-\pi/2, 0)$ and $(\pi/2, 0)$ are outside, there must be at least two intersection points. On the other hand, if there were more than two, the function $f(x) = x^2 + \cos^2 x - \alpha \cos x$ would have at least three zeros with $-\pi/2 < x < \pi/2$. Applying Rolle's theorem to $f(x)$ and then to $f'(x)$, we see that $f''(x) = 0$ would have a solution in this same interval. But

$$f''(x) = 2 - 2\cos 2x + \alpha \cos x \geq \alpha \cos x > 0$$

for $-\pi/2 < x < \pi/2$, a contradiction. Thus we have shown that if $\alpha > 1$, there are precisely two intersection points with $-\pi/2 < x < \pi/2$.

If, for some α, there are also two intersection points for which $3\pi/2 < x < 5\pi/2$, the next interval where $\cos x > 0$, then by symmetry there will be two more for $-5\pi/2 < x < -3\pi/2$, making at least six intersection points altogether. On the other hand, suppose there are no intersection points for $3\pi/2 < x < 5\pi/2$. Then the circle must pass above the point $(2\pi, 1)$ (if, indeed, the circle extends that far to the right) and thus there will be no intersection points for $|x| > 2\pi$ either. This leaves us with only two intersection points in all.

Thus the only possible way to get precisely four intersection points is to have a *unique* intersection point with $3\pi/2 < x < 5\pi/2$. We now show that this happens for exactly one value of α.

Note that for each point on the graph of $y = \cos x$ with $3\pi/2 < x < 5\pi/2$, there is exactly one of the circles $x^2 + (y - \frac{1}{2}\alpha)^2 = \frac{1}{4}\alpha^2$ which passes through that point; specifically, the one with

$$\alpha = \frac{x^2 + y^2}{y} = \frac{x^2 + \cos^2 x}{\cos x}.$$

If we put

$$F(x) = \frac{x^2 + \cos^2 x}{\cos x},$$

then we want to show that there is exactly one value of α for which the equation $F(x) = \alpha$ has a unique solution with $3\pi/2 < x < 5\pi/2$.

Now we have $F(x) > 0$ for $3\pi/2 < x < 5\pi/2$, and $F(x) \to \infty$ as $x \to (3\pi/2)^+$ and as $x \to (5\pi/2)^-$. Therefore, $F(x)$ has a minimum value α_0 on the interval $3\pi/2 < x < 5\pi/2$; by the Intermediate Value Theorem, $F(x)$ takes on all values greater than α_0 at least *twice* on that interval.

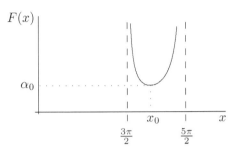

Hence the only value of α for which $F(x) = \alpha$ can actually have a unique solution in the interval is $\alpha = \alpha_0$. To show that this is indeed the case, it is enough to show that $F'(x)$ is zero only once on the interval, because then the minimum value can occur only once.

Now

$$F'(x) = \frac{(2x - 2 \sin x \cos x) \cos x - (x^2 + \cos^2 x)(-\sin x)}{\cos^2 x}$$

can only be zero when its numerator is, that is, when

$$x^2 \sin x + 2x \cos x - \sin x \cos^2 x = 0.$$

On the other hand, the derivative of $g(x) = x^2 \sin x + 2x \cos x - \sin x \cos^2 x$ is

$$g'(x) = (x^2 + 2 + 2 \sin^2 x - \cos^2 x) \cos x = (x^2 + 4 - 3 \cos^2 x) \cos x,$$

which is positive on our interval. So $g(x)$ is increasing there and can only be zero once, and it follows that there is a unique x_0 in the interval with $F(x_0) = \alpha_0$.

Finally, since the circle $x^2 + (y - \frac{1}{2}\alpha_0)^2 = \frac{1}{4}\alpha_0^2$ lies above the graph of $y = \cos x$ at both end points $3\pi/2$, $5\pi/2$ of the interval (provided the circle extends that far to the right), it must be above the graph everywhere in the interval except at the point of tangency $(x_0, \cos x_0)$. But then by the periodicity of cosine, the circle will certainly lie above the graph on all "future" intervals $7\pi/2 < x < 9\pi/2, \ldots$ where $\cos x > 0$. Thus for $\alpha = \alpha_0$ there are precisely four intersection points, for any other value of α there is a different number of intersection points, and we are done.

Problem 205

Fast Eddie needs to double his money; he can only do so by playing a certain win-lose game, in which the probability of winning is p. However, he can play this game as many or as few times as he wishes, and in a particular game he can bet any desired fraction of his bankroll. The game pays even money (the odds are one-to-one). Assuming he follows an optimal strategy if one is available, what is the probability, as a function of p, that Fast Eddie will succeed in doubling his money?

Solution. The probability is 1 for $p > 1/2$ and p for $p \leq 1/2$.

Suppose that $p > 1/2$; we will show a way for Fast Eddie to double his money with probability 1. First, suppose that at some point his bankroll has reached x, and that he proceeds to bet $x/2^n, 2x/2^n, 4x/2^n, \ldots$ until he wins once or until he has lost n of these bets. Then if he wins once, his bankroll will become $x(1 + \frac{1}{2^n})$, and this will occur with probability $1 - (1-p)^n$. If he does win once and he continues with a similar round of up to n bets, but with x replaced by the new bankroll $x(1 + \frac{1}{2^n})$, and if Eddie proceeds in this way for a total of 2^n rounds (winning once in each round), his bankroll will reach

$$x\left(1 + \frac{1}{2^n}\right)^{2^n} > 2x.$$

In particular, if x was Eddie's original bankroll, he will have doubled it after the 2^n rounds. Meanwhile, the probability that he will indeed win once in each round is

$$\left(1 - (1-p)^n\right)^{2^n}.$$

Note that since $p > 1/2$, this approaches 1 as $n \to \infty$. (This can be seen, for instance, by taking the logarithm, rewriting the resulting indeterminate form of type $\infty \cdot 0$ as a quotient, and using l'Hôpital's rule.) So by choosing a large n, we can find a strategy for which Eddie's probability of doubling his bankroll (after at most $N = n2^n$ bets) is as close to 1 as desired. However, since n has to be chosen in advance, we cannot actually achieve probability 1 this way, so we do not yet have an optimal strategy.

To get an optimal strategy, we use a refinement of the same idea. If we replace "2^n rounds" in the discussion above by "$2^n M$ rounds," where M is a positive integer, Eddie's bankroll will grow from x to an amount $> 2^M x$ with probability $(1 - (1-p)^n)^{2^n M}$, which still approaches 1 as $n \to \infty$. So even if x is *not* Eddie's original bankroll, he can make his bankroll reach twice its original size (using at most $N = M n 2^n$ bets) with probability as close to 1 as desired.

Now we can see that the following is an optimal strategy for Fast Eddie (still for $p > 1/2$). Let him choose a number N_1 such that the bankroll will double after at most N_1 bets, as described in the first paragraph, with some probability $p_1 > 1/2$, and let him make those bets. If he loses in some round, he will still have at least $1/2^{N_1}$ of his original bankroll. In this case, let him choose a number N_2 such that this amount will grow to twice the original amount after at most N_2 bets, as described in the second paragraph, with some probability $p_2 > 1/2$. Let him then make those bets, and so forth. The total probability of success (in doubling his original bankroll) will then be

$$p_1 + (1 - p_1)p_2 + (1 - p_1)(1 - p_2)p_3 + \cdots.$$

The partial sums of this series are

$$1 - (1 - p_1), \ 1 - (1 - p_1)(1 - p_2), \ 1 - (1 - p_1)(1 - p_2)(1 - p_3), \ldots,$$

so they have limit 1 (since $p_i > 1/2$ for all i), and we are done with the case $p > 1/2$.

For the case $p \leq 1/2$, an optimal strategy for Fast Eddie is to bet all his money at once; thus, his probability of success will be p. To show this, suppose that he follows another strategy, for which the probability of success is q. We will show that $q \leq p$. In carrying out this strategy, the cumulative amount Eddie must wager is at least his bankroll (otherwise he cannot possibly double his money). Because all his bets have probability $p \leq 1/2$, his expected net return is at most $p - (1 - p) = 2p - 1$ times his bankroll. On the other hand, the worst that can happen to Eddie is that he loses his entire bankroll, and so his expected net return is at least $q - (1 - q) = 2q - 1$ times his bankroll. Thus $2q - 1 \leq 2p - 1$, so $q \leq p$, and we are done.

Comments. An alternative betting strategy in the case $p > 1/2$ was suggested by Greg Kuperberg (UC Davis). Eddie chooses $\alpha > 0$ such that

$$p \ln (1 + \alpha) + (1 - p) \ln (1 - \alpha) > 0$$

and bets fraction α of his bankroll each time. In particular, Eddie will not go broke, so we need only show that his probability of doubling his bankroll is 1. Choose ε between 0 and p such that

$$(p - \varepsilon) \ln (1 + \alpha) + (1 - p + \varepsilon) \ln (1 - \alpha) > 0.$$

Let X be the number of times out of n bets that Eddie wins. By the law of large numbers, for n sufficiently large, the probability that X/n is greater than $p - \varepsilon$ is arbitrarily close to 1. Furthermore, take n sufficiently large so that

$$n(p-\varepsilon)\ln(1+\alpha) + n(1-p+\varepsilon)\ln(1-\alpha) \geq \ln 2.$$

For such n, the probability is arbitrarily close to 1 that Eddie's bankroll has increased by a factor of at least

$$(1+\alpha)^{n(p-\varepsilon)}(1-\alpha)^{n(1-p+\varepsilon)} = e^{n(1-\varepsilon)\ln(1+\alpha)+n(1-p+\varepsilon)\ln(1-\alpha)} \geq e^{\ln 2} = 2.$$

In practice, there will be a minimum bet (a penny, say), and then the probability will *not* be 1, even for $p > 1/2$.

This problem may be viewed as a one-dimensional random walk problem, which has unequal probabilities and variable step sizes.

Problem 206

Define a *die* to be a convex polyhedron. For what n is there a fair die with n faces? By fair, we mean that, given any two faces, there exists a symmetry of the polyhedron which takes the first face to the second.

Solution. There is a fair die with n faces if and only if n is an even integer with $n > 2$.

Any convex polyhedron has at least four faces, so we can assume $n \geq 4$. A regular tetrahedron is a fair die with 4 faces. If n is even and $n > 4$, we can get a fair die with n faces by constructing a "generalized regular octahedron," as follows. Begin with a regular $n/2$-gon in a horizontal plane. Place two points, one on either side of the plane, on the vertical line through the center of the $n/2$-gon, at equal distances from the center. Connect these two points to the $n/2$ vertices of the $n/2$-gon to obtain a fair die with n triangular faces.

Now suppose n is odd. We use Euler's formula, $V - E + F = 2$, where V is the number of vertices, E is the number of edges, and F is the number of faces of a simple (no "holes") polyhedron. In our case, $F = n$. Since all faces are congruent, they all have the same number, say s, of sides. Because each edge bounds two faces, $E = sn/2$, and so s is even. Let $v_1 \leq v_2 \leq \cdots \leq v_s$ denote the numbers of edges emanating from the s vertices of a single face. Since a vertex with v_i edges emanating from it is a vertex for v_i different

THE SOLUTIONS

faces, we have
$$V = n\left(\frac{1}{v_1} + \frac{1}{v_2} + \cdots + \frac{1}{v_s}\right).$$
Substituting all this into Euler's formula yields
$$n\left(\frac{1}{v_1} + \frac{1}{v_2} + \cdots + \frac{1}{v_s} - \frac{s}{2} + 1\right) = 2.$$
Because $v_i \geq 3$ for all i, this implies
$$2 \leq n\left(\frac{s}{3} - \frac{s}{2} + 1\right) = n\left(1 - \frac{s}{6}\right),$$
and since s is even, it follows that $s = 4$. Euler's formula then becomes
$$n\left(\frac{1}{v_1} + \frac{1}{v_2} + \frac{1}{v_3} + \frac{1}{v_4} - 1\right) = 2.$$
Because $v_1 \geq 4$ would imply
$$\frac{1}{v_1} + \frac{1}{v_2} + \frac{1}{v_3} + \frac{1}{v_4} \leq 1,$$
we conclude that $v_1 = 3$.

At this point, our goal is to show that, for odd n, the equation
$$\frac{1}{v_2} + \frac{1}{v_3} + \frac{1}{v_4} = \frac{2}{3} + \frac{2}{n}$$
has no solution in positive integers $v_2 \leq v_3 \leq v_4$ with $v_2 \geq 3$.

Observe that
$$\frac{1}{3} + \frac{1}{6} + \frac{1}{6} = \frac{2}{3} < \frac{2}{3} + \frac{2}{n} \quad \text{and} \quad \frac{1}{4} + \frac{1}{5} + \frac{1}{5} = \frac{13}{20} < \frac{2}{3},$$
which leaves us with the following four cases:

i. $v_2 = 3$, $v_3 = 3$;
ii. $v_2 = 3$, $v_3 = 4$;
iii. $v_2 = 3$, $v_3 = 5$; and
iv. $v_2 = 4$, $v_3 = 4$.

If v_2 and v_3 are both odd, then
$$\frac{1}{v_2} + \frac{1}{v_3} + \frac{1}{v_4} = \frac{v_2 v_3 + v_2 v_4 + v_3 v_4}{v_2 v_3 v_4}$$
has odd numerator, so it cannot equal
$$\frac{2}{3} + \frac{2}{n} = \frac{2(n+3)}{3n};$$

this rules out cases (i) and (iii). Case (ii) reduces to
$$\frac{1}{v_4} = \frac{n+24}{12n}.$$
In this case, then, $n+24$ is an odd factor of $12n$, hence of $3n$. Since $n+24$ also divides $3n+72$, we find that $n+24$ is an odd factor of 72, which is impossible. Finally, case (iv) reduces to
$$\frac{1}{v_4} = \frac{n+12}{6n},$$
and by a similar argument, this is impossible. Thus there is no fair die with an odd number of faces, and we are done.

Problem 207

Prove that
$$\det\begin{pmatrix} 1 & 4 & 9 & \cdots & n^2 \\ n^2 & 1 & 4 & \cdots & (n-1)^2 \\ \vdots & \vdots & \vdots & \vdots & \vdots \\ 4 & 9 & 16 & \cdots & 1 \end{pmatrix}$$
$$= (-1)^{n-1} \frac{n^{n-2}(n+1)(2n+1)\bigl((n+2)^n - n^n\bigr)}{12}.$$

Solution. We begin by sketching a proof of the known formula for the determinant of a circulant matrix:
$$\det\begin{pmatrix} a_1 & a_2 & a_3 & \cdots & a_n \\ a_n & a_1 & a_2 & \cdots & a_{n-1} \\ \vdots & \vdots & \vdots & \vdots & \vdots \\ a_3 & a_4 & a_5 & \cdots & a_2 \\ a_2 & a_3 & a_4 & \cdots & a_1 \end{pmatrix} = (-1)^{n-1} \prod_{j=0}^{n-1}\left(\sum_{k=1}^{n} \zeta^{jk} a_k\right),$$
where $\zeta = e^{2\pi i/n}$.

View the formula above as a polynomial identity in $a_1, \ldots, a_n$. Note that the determinant is zero if $a_1 + a_2 + \cdots + a_n = 0$, because then the sum of all the columns of the matrix is zero. Therefore, the determinant has a factor $a_1 + a_2 + \cdots + a_n$, and this is the factor on the right for $j = 0$. Similarly, if for any j we multiply the first column of the matrix by ζ^j, the second column by ζ^{2j}, and so forth, and add all the columns, we get a multiple of

THE SOLUTIONS

$\sum_{k=1}^{n} \zeta^{jk} a_k$ in each position. Thus if $\sum_{k=1}^{n} \zeta^{jk} a_k = 0$, the original columns are linearly dependent and the determinant is zero. It follows that the determinant has a factor $\sum_{k=1}^{n} \zeta^{jk} a_k$ for every j. Furthermore, no two of these n polynomials are constant multiples of each other. Both sides of the formula above are homogeneous polynomials of total degree n in $a_1, \ldots, a_n$, so there are no nonconstant factors other than the ones we have accounted for already. Finally, we can get the factor $(-1)^{n-1}$ by looking at the case of the identity matrix.

Applying the formula to our special case $a_k = k^2$, we find

$$\det \begin{pmatrix} 1 & 4 & 9 & \cdots & n^2 \\ n^2 & 1 & 4 & \cdots & (n-1)^2 \\ \vdots & \vdots & \vdots & \vdots & \vdots \\ 4 & 9 & 16 & \cdots & 1 \end{pmatrix} = (-1)^{n-1} \prod_{j=0}^{n-1} \left(\sum_{k=1}^{n} \zeta^{jk} k^2 \right).$$

The next step is to derive a "closed" expression for $\sum_{k=1}^{n} k^2 x^k$, after which we will substitute $x = \zeta^j$ for $j = 0, 1, \ldots, n-1$. Specifically, we claim that

$$\sum_{k=1}^{n} k^2 x^k = \begin{cases} \dfrac{n^2 x^{n+3} - (2n^2 + 2n - 1)x^{n+2} + (n^2 + 2n + 1)x^{n+1} - x^2 - x}{(x-1)^3} & \text{if } x \neq 1 \\ \dfrac{n(n+1)(2n+1)}{6} & \text{if } x = 1. \end{cases}$$

To obtain this formula for $x \neq 1$, start with the finite geometric series

$$\sum_{k=0}^{n} x^k = \frac{x^{n+1} - 1}{x - 1}.$$

Then differentiate each side and multiply by x, to find

$$\sum_{k=1}^{n} k x^k = \frac{n x^{n+2} - (n+1) x^{n+1} + x}{(x-1)^2}.$$

Differentiating and multiplying by x a second time yields the formula we want. For the well-known special case $x = 1$, we can apply l'Hôpital's rule to the general case. For either case, one could also use induction on n.

We can now make some progress toward the desired formula:

$$\det \begin{pmatrix} 1 & 4 & 9 & \cdots & n^2 \\ n^2 & 1 & 4 & \cdots & (n-1)^2 \\ \vdots & \vdots & \vdots & \vdots & \vdots \\ 4 & 9 & 16 & \cdots & 1 \end{pmatrix} = (-1)^{n-1} \prod_{j=0}^{n-1} \left(\sum_{k=1}^{n} k^2 \zeta^{jk} \right)$$

$$= (-1)^{n-1} \frac{n(n+1)(2n+1)}{6} \prod_{j=1}^{n-1} \frac{n^2 \zeta^{3j} - (2n^2+2n)\zeta^{2j} + (n^2+2n)\zeta^j}{(\zeta^j - 1)^3}$$

$$= (-1)^{n-1} \frac{n(n+1)(2n+1)}{6} \prod_{j=1}^{n-1} \frac{n^2 \zeta^j (\zeta^j - \frac{n+2}{n})}{(\zeta^j - 1)^2}.$$

To conclude, we use a technique reminiscent of the standard method used to prove the irreducibility of cyclotomic polynomials. To find

$$\prod_{j=1}^{n-1} (\zeta^j - a)$$

(we will later take $a = 0$, $a = 1$, and $a = (n+2)/n$, note that the $\zeta^j - a$ are the roots of the polynomial

$$\frac{(y+a)^n - 1}{(y+a) - 1} = \sum_{k=0}^{n-1} (y+a)^k,$$

so their product is $(-1)^{n-1}$ times the constant term of the polynomial. Thus we have

$$\prod_{j=1}^{n-1} (\zeta^j - a) = (-1)^{n-1} \sum_{k=0}^{n-1} a^k = \begin{cases} (-1)^{n-1} \dfrac{a^n - 1}{a - 1} & \text{if } a \neq 1, \\ (-1)^{n-1} n & \text{if } a = 1. \end{cases}$$

In particular,

$$\prod_{j=1}^{n-1} \zeta^j = (-1)^{n-1}, \qquad \prod_{j=1}^{n-1} (\zeta^j - 1) = (-1)^{n-1} n,$$

and

$$\prod_{j=1}^{n-1} \left(\zeta^j - \frac{n+2}{n} \right) = (-1)^{n-1} \frac{(n+2)^n - n^n}{2n^{n-1}}.$$

Substituting these results into our most recent expression for the determinant and simplifying, we get

$$\det \begin{pmatrix} 1 & 4 & 9 & \cdots & n^2 \\ n^2 & 1 & 4 & \cdots & (n-1)^2 \\ \vdots & \vdots & \vdots & \vdots & \vdots \\ 4 & 9 & 16 & \cdots & 1 \end{pmatrix}$$

$$= (-1)^{n-1} \frac{n^{n-2}(n+1)(2n+1)\big((n+2)^n - n^n\big)}{12},$$

as claimed.

Comment. If we rewrite the determinant as

$$\frac{(-1)^{n-1} n^{2n} \left(1 + \dfrac{1}{n}\right)\left(2 + \dfrac{1}{n}\right)\left[\left(1 + \dfrac{2}{n}\right)^n - 1\right]}{12},$$

we see that its absolute value is asymptotic to

$$\frac{n^{2n}(e^2 - 1)}{6}.$$

Problem 208

Let $a_0 = 0$ and let a_n be equal to the smallest positive integer that cannot be written as a sum of n numbers (with repetitions allowed) from among $a_0, a_1, \ldots, a_{n-1}$. Find a formula for a_n.

Answer. The general term is $a_n = F_{2n}$, where F_k is the kth term of the Fibonacci sequence given by $F_0 = 0$, $F_1 = 1$, $F_{k+2} = F_k + F_{k+1}$.

Solution 1. (Paul Fjelstad, St. Olaf College) Let $(c_k c_{k-1} \ldots c_2 c_1)$ denote the *even-subscripted Fibonacci sum* $c_k F_{2k} + c_{k-1} F_{2(k-1)} + \cdots + c_2 F_4 + c_1 F_2$. For example,

$$(1010) = F_8 + F_4 = 21 + 3 = 24 = 2(8 + 3 + 1) = 2(F_6 + F_4 + F_2) = (0222).$$

We will make use of the following lemmas.

Lemma 1. If an integer can be represented as an even-subscripted Fibonacci sum, then it can be represented as an even-subscripted sum using only the digits $0, 1, 2$.

Proof. If the string 030 occurs in an even-subscripted Fibonacci sum, it can be replaced with the string 101. This is because

$$030 = 3F_{2k} = (F_{2k+2} - F_{2k+1}) + (F_{2k-1} + F_{2k-2}) + (F_{2k+1} - F_{2k-1})$$
$$= F_{2k+2} + F_{2k-2} = 101.$$

Because of this, for $y \geq 3$, the string $x\,y\,z$ can be replaced by the string $(x+1)\,(y-3)\,(z+1)$.

Similarly, because $3F_2 = 3 = F_4$, if $y \geq 3$, a terminal string $x\,y$ can be replaced by $(x+1)\,(y-3)$. For example, the terminal string 03 can be replaced by 10.

The proof now follows by repeated use of these rules, because the sum of the digits in the even-subscripted Fibonacci sum decreases at each step of this substitution procedure. For example, $(4132) = (11232) = (11303) = (12013) = (12020)$.

Lemma 2. For $n \geq 3$, we have $F_{2n} = (\underbrace{100\ldots0}_{n}) = (2\underbrace{11\ldots1}_{n-3}2)$ and $(\underbrace{022\ldots2}_{n}) = (10\underbrace{11\ldots1}_{n-3}0)$.

Proof. For the first equality,

$$\begin{aligned}
(\underbrace{100\ldots0}_{n}) &= F_{2n} = F_{2n-1} + F_{2n-2} = 2F_{2n-2} + F_{2n-3} \\
&= 2F_{2(n-1)} + F_{2n-4} + F_{2n-5} \\
&= 2F_{2(n-1)} + F_{2(n-2)} + F_{2n-6} + F_{2n-7} \\
&\quad \vdots \\
&= 2F_{2(n-1)} + F_{2(n-2)} + F_{2(n-3)} + \cdots + F_4 + F_3 \\
&= 2F_{2(n-1)} + F_{2(n-2)} + F_{2(n-3)} + \cdots + F_4 + 2F_2 \\
&= (2\underbrace{11\ldots1}_{n-3}2).
\end{aligned}$$

THE SOLUTIONS

For the second equality we use the fact, easily established by induction, that for $k \geq 0$, $F_0 + F_1 + F_2 + \cdots + F_k = F_{k+2} - 1$. Thus,

$$(0\underbrace{22\ldots2}_{n}) = 2F_{2(n-1)} + 2F_{2(n-2)} + \cdots + 2F_2$$

$$= \left(F_{2(n-1)} + (F_{2n-3} + F_{2n-4})\right)$$
$$\quad + \left(F_{2(n-2)} + (F_{2n-5} + F_{2n-6})\right) + \cdots + \left(F_2 + (F_1 + F_0)\right)$$

$$= \left(F_{2(n-1)} + F_{2(n-2)} + \cdots + F_4 + F_2\right)$$
$$\quad + \left(F_{2n-3} + F_{2n-4} + \cdots + F_1 + F_0\right)$$

$$= \left(F_{2(n-1)} + F_{2(n-2)} + \cdots + F_4 + F_2\right) + \left(F_{2n-1} - 1\right)$$

$$= \left(F_{2(n-1)} + F_{2(n-2)} + \cdots + F_4 + F_2\right) + \left((F_{2n} - F_{2(n-1)}) - 1\right)$$

$$= F_{2n} + F_{2(n-2)} + \cdots + F_4$$

$$= (10\underbrace{11\ldots1}_{n-3}0).$$

Lemma 3. For $n \geq 2$, $(0\underbrace{22\ldots2}_{n}) < (\underbrace{111\ldots12}_{n}) < (\underbrace{200\ldots0}_{n})$.

Proof. For $n = 2$, $(02) = 2$, $(12) = 5$, $(20) = 6$. For $n \geq 3$, by Lemma 2,

$$(0\underbrace{22\ldots2}_{n}) = (10\underbrace{11\ldots1}_{n-3}0) < (\underbrace{111\ldots12}_{n})$$

and

$$(\underbrace{200\ldots0}_{n}) = (\underbrace{100\ldots0}_{n}) + (\underbrace{100\ldots0}_{n}) = (\underbrace{100\ldots0}_{n}) + (02\underbrace{11\ldots1}_{n-3}2)$$

$$= (\underbrace{1211\ldots12}_{n}) > (\underbrace{111\ldots12}_{n}).$$

Lemma 4. For $n \geq 3$, $(2\underbrace{11\ldots1}_{n-3}2)$ is the only $(n-1)$-digit representation of F_{2n} as a sum of even-subscripted Fibonacci numbers with digits among $0, 1, 2$.

Proof. If there is another representation there will be a first digit (from left to right) that is changed. If it is raised, then changing all the rest to 0's won't subtract enough to reach F_{2n}. This is a consequence of Lemma 3, for

$$(21\underbrace{1\ldots 1}_{n-1}2\underbrace{00\ldots 0}_{k}) > (21\underbrace{1\ldots 11}_{n-1}\underbrace{\ldots 12}_{k}) = F_{2n}.$$

If it is lowered, then changing all the rest to 2's won't add enough to reach F_{2n}. This also follows from Lemma 3, for

$$(1\underbrace{22\ldots 2}_{n-2}) = (1\underbrace{00\ldots 0}_{n-1}) + (0\underbrace{22\ldots 2}_{n-1})$$
$$< (1\underbrace{00\ldots 0}_{n-1}) + (\underbrace{11\ldots 12}_{n-1}) = (2\underbrace{11\ldots 12}_{n-1}) = F_{2n}$$

and

$$(2\underbrace{1\ldots 1}_{n-1}0\underbrace{22\ldots 2}_{k}) < (2\underbrace{1\ldots 11}_{n-1}\underbrace{1\ldots 12}_{k}) = F_{2n}.$$

We're now in a position to complete the solution, so suppose that $a_i = F_{2i}$ for $i = 0, 1, 2, \ldots, n-1$, where $n \geq 3$ ($n = 0, 1, 2$ are easy to check).

First, F_{2n} cannot be equal to a sum of n numbers from $a_0, a_1, \ldots, a_{n-1}$, for if it could, it could be so expressed with no a_i occurring more than 2 times, by Lemma 1. But we know, by Lemma 4, that there is only one way to write F_{2n} as an even-subscripted sum of lower Fibonacci numbers using only the digits $0, 1, 2$, and that is as

$$F_{2n} = 2F_{2(n-1)} + F_{2(n-2)} + F_{2(n-3)} + \cdots + F_4 + 2F_2,$$

which requires $n+1$ terms from $a_0, a_1, \ldots, a_{n-1}$.

Now suppose that s is a nonnegative number smaller than F_{2n}. In order to show that s is a sum of n numbers from $a_0, a_1, \ldots, a_{n-1}$, we'll distinguish different cases, by comparing s to the following integers. Using $(n-1)$-digit even-subscripted notation, let $s_0 = (100\ldots 0)$, $s_1 = (200\ldots 0)$, $s_2 = (210\ldots 0)$, $s_3 = (2110\ldots 0)$, ..., $s_{n-2} = (211\ldots 10)$, and finally $s_{n-1} = (211\ldots 11) = F_{2n} - 1$.

If $0 \leq s < s_0 = F_{2(n-1)} = a_{n-1}$, then, by induction, it is equal to a sum of $n-1$ numbers from among $a_0, a_1, \ldots, a_{n-2}$, so adding $a_0 = 0$ to this sum gives n numbers from this set that add to s. If $s_0 \leq s < s_1$, then

$$0 \leq s - s_0 < s_1 - s_0 = (100\ldots 0) = a_{n-1}.$$

By induction, $s - s_0$ is equal to a sum of $n-1$ numbers from $a_0, a_1, \ldots, a_{n-2}$, which means that s is a sum of n numbers from $a_0, a_1, \ldots, a_{n-1}$.

THE SOLUTIONS

Now assume that $s_k \leq s < s_{k+1}$ with $1 \leq k \leq n-2$. Then we have $0 \leq s - s_k < s_{k+1} - s_k = F_{2(n-(k+1))} = a_{n-(k+1)}$, so by induction, $s - s_k$ is equal to a sum of $n-(k+1)$ numbers from among $a_0, a_1, \ldots, a_{n-(k+1)-1}$. We can combine this with

$$s_k = (\underbrace{21\ldots1}_{k}00\ldots0)$$
$$= 2F_{2(n-1)} + F_{2(n-2)} + \cdots + F_{2(n-k)}$$
$$= 2a_{n-1} + a_{n-2} + \cdots + a_{n-k},$$

a sum of $k+1$ numbers, to get s as a sum of n numbers from among $a_0, a_1, \ldots, a_{n-1}$. Thus every number smaller than s_{n-1} is equal to a sum of n numbers from among $a_0, a_1, \ldots, a_{n-1}$.

Finally, s_{n-1} itself is a sum of n numbers from among $a_0, a_1, \ldots, a_{n-1}$. Putting all of this together, we conclude that $a_n = F_{2n}$, and our proof by induction is complete.

Solution 2. In the course of the proof we'll make repeated use of the fact that

$$F_2 + F_4 + \cdots + F_{2n} = F_1 + F_2 + F_4 + \cdots + F_{2n} - 1$$
$$= F_3 + F_4 + F_6 + \cdots + F_{2n} - 1$$
$$= F_5 + F_6 + \cdots + F_{2n} - 1 = \cdots$$
$$= F_{2n+1} - 1. \qquad (*)$$

We proceed to show that $a_n = F_{2n}$ by induction. The equality holds for $n = 0, 1, 2, 3, 4$, so assume that it holds through n, where $n \geq 4$.

We first show that $a_{n+1} \geq F_{2n+2}$ by showing that one can express every smaller integer as a sum of $n+1$ numbers from among $F_0, F_2, F_4, \ldots, F_{2n}$. By the inductive hypothesis, every integer between F_{2n} and $2F_{2n} - 1$, inclusive, is the sum of F_{2n} and n numbers from among $F_0, F_2, F_4, \ldots, F_{2(n-1)}$. Now consider j between $2F_{2n}$ and $F_{2n+2} - 1$, inclusive. Choose the smallest positive integer k such that

$$F_{2n} + (F_{2n} + F_{2n-2} + \cdots + F_{2k}) \leq j.$$

If $k = 1$, then $(*)$ implies that $j \geq F_{2n} + (F_{2n+1} - 1) = F_{2n+2} - 1$, so $j = F_{2n+2} - 1$, and j is indeed the sum of $n+1$ numbers from among

$F_0, F_2, F_4, \ldots, F_{2n}$. If $k > 1$, then

$$j - (2F_{2n} + F_{2n-2} + \cdots + F_{2k}) < F_{2k-2}$$

is the sum of $k{-}1$ numbers from among $F_0, F_2, F_4, \ldots, F_{2k-4}$. Consequently, j is the sum of $(n{-}k{+}2)+(k{-}1) = n{+}1$ numbers from among $F_0, F_2, F_4, \ldots, F_{2n}$.

We now show that F_{2n+2} cannot be expressed as a sum of $n{+}1$ numbers from among $F_0, F_2, F_4, \ldots F_{2n}$. For suppose this could be done, and choose such a sum having the *smallest* possible number of nonzero terms. From

$$3F_{2k} = 2F_{2k} + (F_{2k+2} - F_{2k+1}) = F_{2k+2} + F_{2k} - F_{2k-1} = F_{2k+2} + F_{2k-2}$$

we see that no nonzero term F_{2k}, $1 \leq k \leq n$ occurs *more* than twice in such a minimal expression. Next, from $(*)$, we see that

$$2(F_2 + F_4 + \cdots + F_{2n-2}) + F_{2n} = (2F_{2n-1} - 2) + F_{2n} = F_{2n-1} + F_{2n+1} - 2$$
$$< F_{2n} + F_{2n+1} - 2 = F_{2n+2} - 2 < F_{2n+2},$$

so F_{2n} must appear twice. However, by $(*)$,

$$F_2 + \cdots + F_{2n-2} + 2F_{2n} = (F_{2n-1} - 1) + 2F_{2n} = F_{2n} + F_{2n+1} - 1 = F_{2n+2} - 1,$$

so some F_{2k} with $k < n$ also appears twice in the expression for F_{2n+2}. Let k be the largest integer less than n for which F_{2k} either appears twice or does not appear in the sum. If F_{2k} appears twice, then

$$F_{2n+2} \geq 2F_{2k} + F_{2k+2} + \cdots + F_{2n-2} + 2F_{2n}$$
$$\geq (F_{2k-1} + F_{2k}) + F_{2k+2} + \cdots + F_{2n-2} + 2F_{2n} \text{ (with equality } \Leftrightarrow k = 1)$$
$$= F_{2k+1} + F_{2k+2} + \cdots + F_{2n-2} + 2F_{2n} = \cdots$$
$$= F_{2n} + F_{2n+1} = F_{2n+2},$$

a contradiction unless equality holds, forcing $k = 1$. But if $k = 1$, the expression for F_{2n+2} requires at least $n{+}2$ terms. If F_{2k} does not appear, using $(*)$ we have

$$F_{2n+2} \le 2(F_2 + \cdots + F_{2k-2}) + F_{2k+2} + \cdots + F_{2n-2} + 2F_{2n}$$
$$= (2F_{2k-1} - 2) + F_{2k+2} + \cdots + F_{2n-2} + 2F_{2n}$$
$$\le (F_{2k-1} + F_{2k}) + F_{2k+2} + \cdots + F_{2n-2} + 2F_{2n} - 2$$
$$= F_{2k+1} + F_{2k+2} + \cdots + F_{2n-2} + 2F_{2n} - 2 = \cdots$$
$$= F_{2n} + F_{2n+1} - 2 = F_{2n+2} - 2,$$

a contradiction, and the proof by induction is complete.

APPENDIX 1
PREREQUISITES BY PROBLEM NUMBER

1. Basic properties of the integers
2. Elementary algebra
3. Differential calculus
4. Precalculus
5. Geometric series
6. Elementary algebra
7. Precalculus
8. Elementary geometry
9. Elementary algebra
10. Infinite series
11. Determinants
12. Elementary algebra
13. Basic counting methods
14. Integral calculus

15. Elementary probability
16. Basic properties of the integers
17. Limits of sequences
18. Basic properties of the integers; elementary algebra
19. Differential calculus
20. Basic counting methods
21. Basic properties of the integers
22. Inclusion-exclusion principle
23. Analytic geometry
24. Basic properties of the integers
25. Vector geometry
26. Basic properties of the integers; precalculus
27. Precalculus
28. Elementary number theory; geometric series
29. Elementary geometry
30. None
31. Elementary geometry
32. Elementary algebra
33. None
34. Calculus
35. Basic properties of the integers
36. Permutations
37. Elementary algebra
38. Elementary geometry
39. Calculus
40. Elementary algebra; mathematical induction

PREREQUISITES BY PROBLEM NUMBER

41. Trigonometry

42. Infinite series

43. Elementary algebra

44. Differential calculus

45. Inverse trig. functions; vector geometry

46. Elementary algebra

47. Analytic geometry

48. Mathematical induction

49. Trigonometry

50. Factorization of polynomials

51. Matrix algebra

52. Elementary geometry

53. Elementary number theory; mathematical induction

54. Differential calculus

55. Elementary algebra

56. Elementary geometry; law of cosines

57. Elementary geometry

58. Calculus

59. Factorization of polynomials; complex numbers

60. Trigonometry

61. Binomial coefficients; mathematical induction

62. Basic properties of the integers; binomial theorem

63. None

64. Differential calculus

65. Mathematical induction

66. Differential calculus

67. Basic counting methods
68. Mathematical induction; elementary probability; geometric series
69. Algebra of complex numbers
70. Differential calculus
71. Basic properties of the integers
72. Elementary geometry; elementary algebra
73. Permutations; elementary algebra
74. Taylor series
75. Elementary algebra
76. Linear algebra
77. Basic properties of the integers
78. Trigonometry
79. Basic properties of the integers
80. Differential calculus
81. Precalculus
82. Elementary geometry
83. Linear algebra
84. Integral calculus; infinite series
85. Elementary number theory; one-to-one and onto functions
86. Elementary geometry
87. Limits of sequences; mathematical induction
88. Elementary number theory
89. Elementary algebra
90. Mathematical induction
91. Basic counting methods
92. Infinite series

PREREQUISITES BY PROBLEM NUMBER

93. Vector arithmetic
94. Trigonometry; geometric series; limits
95. Elementary geometry
96. Matrix algebra
97. Factorization of polynomials
98. Basic counting methods
99. Basic properties of the integers; mathematical induction
100. Basic properties of the integers
101. Calculus; mathematical induction
102. Analytic geometry
103. Limits of sequences
104. Mathematical induction
105. Vector geometry
106. None
107. Precalculus
108. Elementary probability
109. Limits of sequences; Fibonacci numbers
110. Polar coordinates
111. Linear differential equations
112. Probability; basic counting methods
113. Elementary number theory
114. Infinite series
115. Probability; basic counting methods; infinite series
116. Differential calculus
117. Vector geometry
118. Elementary number theory

119. Calculus

120. Precalculus

121. Basic counting methods

122. Separable differential equations; arc length; polar coordinates

123. Elementary number theory

124. Analytic geometry

125. Limits of sequences

126. Linear algebra; probability; modular arithmetic

127. Improper integrals

128. Limits; mathematical induction

129. Multivariable calculus

130. Calculus

131. Basic properties of the integers; complex numbers

132. Analytic geometry; elementary number theory

133. Infinite series

134. Elementary algebra; elementary geometry

135. Abstract algebra

136. Analytic geometry

137. Elementary number theory

138. Limits of sequences; mathematical induction

139. Differential calculus

140. Basic counting methods

141. Linear algebra; calculus

142. Taylor series

143. None

144. Precalculus

PREREQUISITES BY PROBLEM NUMBER

145. Infinite series
146. Indeterminate forms; mathematical induction
147. Infinite series; mathematical induction
148. None
149. Precalculus; limits
150. Differential calculus
151. Elementary geometry
152. Vector geometry in n-space; mathematical induction
153. Basic properties of the integers
154. Indeterminate forms
155. Limits of sequences
156. Calculus
157. Infinite series
158. Mathematical induction
159. Calculus; definition of limit
160. Precalculus
161. Taylor series
162. Elementary number theory
163. Improper integrals
164. Basic counting methods
165. Calculus
166. Limits of sequences
167. Limits of sequences in the plane
168. Differential calculus
169. Elementary geometry
170. Pythagorean triples

171. Multivariable calculus; probability

172. Conic sections

173. Improper integrals

174. Limits of sequences

175. Integral calculus; centroid

176. Basic properties of the integers

177. Taylor series

178. Factorization of polynomials

179. Improper integrals

180. Factorization of polynomials

181. Differential calculus

182. Calculus; binomial theorem

183. Elementary number theory

184. Recurrence relations; elementary probability

185. Polynomials in two variables

186. Improper integrals; infinite series

187. Factorization of polynomials; sequences

188. Limits of sequences in the plane

189. Expectation; mathematical induction

190. Infinite series

191. Analytic geometry in 3-space

192. Elementary number theory; mathematical induction

193. Infinite series

194. Continuous functions; countable sets

195. Basic properties of the integers

196. Basic properties of the integers

PREREQUISITES BY PROBLEM NUMBER

197. Generating functions; separable differential equations; Taylor series
198. Modular arithmetic; limits of sequences
199. Factorization of polynomials; elementary number theory
200. Elementary number theory
201. Limits of sequences; calculus
202. Analytic geometry; trigonometry
203. Limits of sequences; infinite products
204. Differential calculus
205. Probability theory; differential calculus
206. Euler's formula for polyhedra
207. Roots of unity; factorization of polynomials; theory of determinants; closed forms for finite sums
208. Fibonacci numbers; mathematical induction

APPENDIX 2
PROBLEM NUMBERS BY SUBJECT

ALGEBRA AND TRIGONOMETRY

    Complex numbers   47, 69, 82; see also LINEAR AND ABSTRACT ALGEBRA

    Equations   2, 7, 9, 22, 26, 27, 69, 89, 97, 110, 180; see also CALCULUS

    Functions   4, 85, 107, 144, 160

    Inequalities   5, 12, 35, 37, 43, 46, 196

    Trigonometry   41, 45, 49, 56, 60, 78, 94, 110, 202, 204

CALCULUS

    Chain rule   66, 129

    Differential equations   111, 122

    Equations   19, 80, 139, 178, 204; see also ALGEBRA AND TRIGONOMETRY

    Estimation   39, 119, 142, 155, 198; see also Limits

    Integration   14, 34, 39, 58, 84, 119, 122, 127, 130, 142, 159, 163, 171, 175, 179, 182, 186

    Intermediate Value Theorem   58, 86, 159, 181

    Infinite series (convergence)   10, 42, 115, 133, 145, 157, 186, 193

    Infinite series (evaluation)   74, 92, 94, 114, 147, 190

    Infinite series (Taylor series)   74, 84, 142, 161, 177, 197

    Limits   87, 94, 103, 111, 128, 130, 146, 154, 155, 156, 159, 167, 173, 179, 188, 198, 201

    Limits (iteration, recursion)   17, 101, 109, 125, 136, 138, 149, 166, 174, 181, 194, 203

    Maximum-minimum problems   3, 78, 150, 201

    Multivariable calculus   129, 171

    Polynomials   80, 116, 178; see also LINEAR AND ABSTRACT ALGEBRA

    Probability   165, 171, 205; see also DISCRETE MATHEMATICS

    Tangent and normal lines   44, 54, 64, 70, 168

DISCRETE MATHEMATICS

Basic counting techniques   13, 15, 20, 22, 67, 81, 91, 98, 112, 115, 120, 121, 164
Combinatorial geometry   56, 81, 95, 104, 140, 158, 206
Games   68, 121, 189, 205
Generating functions   184, 197, 203
Mathematical induction, pattern finding   16, 40, 48, 61, 65, 67, 68, 81, 90, 99, 104, 158, 208
Probability   15, 68, 108, 112, 115, 184, 189, 205; see also CALCULUS
Recreational mathematics   2, 9, 20, 22, 33, 35, 36, 46, 48, 55, 61, 63, 89, 90, 98, 99, 104, 106, 112, 143, 148, 176, 200 (see also Games)
Recurrence relations   67, 109, 128, 147, 182, 184, 189, 197, 203

GEOMETRY

Analytic geometry   23, 47, 57, 102, 110, 124, 132, 167, 172
Geometry in 3-space   93, 105, 175, 191, 202
Plane geometry   8, 29, 31, 38, 52, 57, 60, 72, 78, 82, 86, 134, 151, 169
Vector geometry   25, 45, 93, 105, 117, 152

LINEAR AND ABSTRACT ALGEBRA

Abstract algebra   135, 180, 207
Complex numbers   178, 195, 207; see also ALGEBRA AND TRIGONOMETRY
Linear independence   76, 83, 126, 141
Matrix algebra   11, 51, 83, 96, 195, 207
Permutations   30, 36, 55, 73, 76, 106
Polynomials   7, 40, 50, 59, 69, 97, 180, 185, 187, 199; see also CALCULUS

NUMBER THEORY

Base b representation   6, 21, 28, 35, 53, 62, 71, 79, 87, 99, 190, 192
Diophantine problems   1, 18, 24, 26, 32, 65, 75, 113, 118, 123, 131, 170, 195
Modular arithmetic   21, 62, 137, 183, 192, 198
Primes and divisibility   9, 11, 21, 53, 77, 85, 88, 100, 113, 137, 162, 176, 183, 192, 200
Rational/irrational numbers   132, 144, 153, 180, 183, 190, 196, 199

INDEX

n : occurs in the statement or the solution(s) of problem n
n^C : occurs only in the comment of problem n
n^I : occurs only in the idea of problem n
n^m : occurs in solution m of problem n

Abel's Limit Theorem 74^2
affine transformation 29^1
algebraic integer 195^2
algebraic number $190^2, 194^C$
alternating series 74^1
alternating series test 142^1
Andrews, George 14^4
angle, between planes   105
    between vectors $45^1, 152$
    for icosahedron   105
    maximal   38
    obtuse   117
angle bisector   134
ant   150
approachable points   188
approximation(s)   79, 96
    by rational numbers   149, 196
arc length   39, 122, 130
area $47, 94, 158, 165, 175^2$
    equal   52, 86
    minimal   3, 78
array   12, 61, 99
astroid   110
automorphism 180^3
average(s)   89, 153
    batting   37
    of squares   118
    of subsets   77
    weighted $116, 175^2$
average distance   57, 171

Babe Ruth   37
ballot   2
bankroll   205
barrels   48
base b representation $6^2, 87$
    $b = 2$ 71, 99, 164
    $b = 8$ 75^2

$b = 9$ 28
$b = 10$ 79, 99, 190
baseball   37
batting average   37
beam of light   25, 151
Becker, Amy 82^1
Bertrand's postulate 200^C
beta function 182^C
bijective proof $67^{3,4}$
binary expansion   71, 99, 164
Binet's formula   109
binomial coefficients   61, 112, 115
binomial expansion $62^1, 69, 133^C$
"binomial polynomials" 199^C
Boucher, Robert 73^2
bridge (game)   63
Buhler, Joe   138

can   150
Cardano's formula 139^C
center, of configuration   95
Centigrade   108
centroid $29^{2,3}, 47, 175$
Ceva's theorem 31^2
chain rule $66, 129, 161^1, 201$
chessboard   15, 33, 143
chessnut   33
Chiu, Christie   77
Cipra, Barry 106^C
circle(s)   47, 78, 94, 171, 185, 204
    lattice points inside   102
circulant matrix   207
coin flipping   115
collinear points   140
coloring, integers   90
    cubes and hypercubes   104
    edges of a cube   121
    squares on a checkerboard   33, 61

391

Colwell, Jason 38^1
commissioners 2
commutative ring 135
comparison test 115, 133, 145
complex number(s) 47^C, 69, 82^2, 89^2, 104^2, 131, 178^2, 195^2, 207
composition of functions 40, 42, 66, 85, 144, 160, 181, 194
concavity 150, 156
Condorcet's paradox 2^C
configuration, of points 20, 95
congruences (see modular arithmetic)
constructing examples 175^C
convergent sequences 87, 103, 174, 201, 203
convergent series 10, 42, 133, 145, 157, 186, 193
convex 29, 52, 72, 93, 169, 172^2
corner square 33
cross product 3^2
cube(s) 121
  black and white 104
cube root of unity 195^2
cubic curve 44, 64
cubic polynomial 7, 116, 178, 180, 181
cycle(s) 36, 85, $195^{1,2}$
cyclotomic polynomials 207
cylinder 150, 159

dancers 55
decimal representation 190
deer 22
de Moivre's theorem 69
Descartes' rule of signs 80^1
determinant(s) $3^{2,3}$, 11, 51, 83, 107, 126, 207
die 98, 206
difference set modulo n 198
differential equation(s), linear 111
  separable 122, 197
digital watches 46
digits 6^2, 21, 35, 53, 62, 79, 99, 165, 190, 192
dimension, of vector space 76, 83, 141^2, 199^2
discriminant 7^3, 116, 124, 168, 172^1
distance(s) 8, 23, 56, 57, 132, 171, 191, 202

divisible, by 27 11
  by 3 or 9 53
  by 6 128, 195
  by 2^n 100
  by $x^2 + y^2 - 1$ 185
division algorithm 199^1
dodecahedron 98
dominoes 197
dot product 25^1, 103^2, 105, 152

edges of a cube 121
eigenvalues/eigenvectors 89^2, 96, 111^C, 195^1
election 2
elementary matrices 51
ellipse(s) 124, 172
envelope(s) 189
error, in trapezoidal estimate 119^C
  round-off 37^C, 108
essentially different 98, 121
Euclidean algorithm 180^1, 199^2
Euler ϕ-function 62^2
Euler's formula (for polyhedron) 206
even, odd function(s) 66, 107
even-subscripted Fibonacci sum 208^1
expected, value 115, 197
  net return 189

Fahrenheit 108
fair coin 115
Fast Eddie 205
Fermat, Pierre de 1^C
Fermat's Little Theorem 137
Fibonacci numbers 109, 138^C, 208
finite geometry 20
fire hydrants 148
Firey, William 49^2
fish 22, 86
five of diamonds 63
Fjelstad, Paul 208^1
foreign agent 15
function(s) 107, 157, 193
  continuous 4, 58, 101, 159, 160
  differentiable 17, 34, 119, 129, 141
  one-to-one and onto 85
  rational 199
  tent 160
functional equation 107, 149, 187

INDEX **393**

Fundamental Theorem of Calculus 84, 119, 130^1, 159^C

Galois theory 180^3
gambling game 189
game 68, 121, 189, 205
gamma function 182^C
Gaussian elimination 199^2
generating function 104^2, 184^C, 197, 203^C
geometric series (see series, geometric)
golden ratio 109, 134, 138, 187
graph 4, 49, 110, 139, 204
    of permutations 106
greatest common divisor 28, 85, $180^{1,2}$

greedy strategy 16^2, 20
grid, points 165
    cubes 104
grouse 22

hats 112
Heuer, Gerald A. 3^2
homogeneous
    linear equation(s) 107^C, 199^2
    polynomials 207
hydrants 148
hyperbola 136
hypercubes 104

I-point 167
ice houses 86
icosahedron 105
inclusion-exclusion principle 13^1, 22
indeterminate form (see l'Hôpital's rule)
induction (see mathematical induction)
infinite descent 1^C, 170
infinite product 177, 203
inhomogeneous linear equation 107^C
inscribed polygon 60, 78^C, 94
insect 171
integer(s), as solution(s) of equations 1, 7, 18, 32, 113, 123, 195, 206
    satisfying given conditions 6, 9, 21, 28, 35, 65, 71, 79, 88, 90, 162, 164, 200
    as a sum/difference of cubes 75
    one step apart 24

integral, definite 34, 39, 58, 119, 142, 159, 175^1, 182
    improper 74^2, 127, 163, 171, 173, 179, 186
    indefinite 14
    multiple 171, 182^C
integration by parts 14^3, 58^1, 74^2, 179, 182
interchanges 30, 55
Intermediate Value Theorem 58^1, 80^1, 86, 116, 124^2, 139, 159, $178^{1,3}$, 181, 204
interpolation 50^C, 116
inverse (co)sine 45, 152
inverse tangent 142, 179

Jerison, Meyer 103^2
Jessie and Riley 26

kernel (see null space)
Klamkin, Murray S. 25^2
Kuperberg, Greg 192^C, 193, 195^3, 205^C
knight's tour 143

L-point 167
Lake Wohascum 171
lattice points 102, 132^3, 195
law of cosines 56, 60, $132^{3,4}$, 202
law of sines 38^2, 60, 134
leap year 36
l'Hôpital's rule 58^2, 74^2, 94, 119, 130^1, 133^1, 146, 159^C, 205, 207
licenses 22
limerick 123
limit comparison test 133^1, 186
line(s) 140, 191
    tangent 44, 64, 70, 168
line segment(s) 31, 57^1, 167
linear programming 23^C
linear transformation 72^1, 124^2, 141^2, 172^2
linearly (in)dependent 76, 83, 126, 141, 207
locus, of centroid 47
    (see also points, set of)
loudspeakers 8
Luks, Eugene 14^3, 75^1, 109^2, 131, 180^2

MAA Student Chapter   105, 171
Maclaurin series   (see series, Taylor)
marks   176
Mathcamp $73^2, 77$
Mathematica   163
mathematical induction   40, 48, 53, 61, 65, 67^1, 68, 85, 87, 90, 96, 99, 101, 104, 128, 138, 146, 147, 152, 158, 161^1, 189, 192, 194, 199^1, 203, 208
matrix/matrices   11, 20, 51, 83, 89^2, 91, 96, 195^1, 207
mayor   184
Mean Value Theorem 44^1, 58^2, 80^2, 133^C, 201
median 8^1, 29^3, 31, 134, 175^2
mesh   165
method of infinite descent 1^C, 170
microfilms   15
mirror   25
missile silos   15
modular arithmetic 21^C, 62^2, 91^C, 104^2, 126, 128, 192, 198
multiple root 44^2, 64, 124^1, 180^2
multiplicative magic square   200

n-gon(s)   60, 94, 169
n-space   76, 104, 152
   ($n = 3$)   25, 83, 93, 105, 117
Nelson, Les 3^3
Newton's method   17, 181
noncollinear points   95
normal line(s)   54
Nuchi, Huggai 29^2
null space 141^2, 199^2

obtuse angles   117
octahedron   202, 206
odds   205
one step apart   24
one-to-one correspondence   36, $67^{3,4}$, 72^1, 85, 112^2, 139^2
optimal strategy   121, 205
orthogonal complement 152^2
outdoor concert   8

p-series   10
Pappus' theorem 175^2
parabola(s)   3, 54, 81, 168
paradox   2, 163

parallel pentagon   72
parallelogram   172
parametrization by arc length 130^2
parity (even-odd)   1, 26, 30, 54, 68, 91, 123^1, 143, 170
partial derivatives   129
partial fraction decomposition   74, 147
partial product   177
partial sum(s)   42, 114, 155, 197, 205
Pascal's triangle   61
Pellian equation 18^C
pentagon   29, 72, 98, 105
perfect square (see square)
period(ic) 62^I
   of function   41, 49^1, 157, 193, 204
   of sequence   136, 137^I, 144, 181, 187, 190, 203
permutation(s) 30^C, 36, 73, 76, 106
Perry, John 112^2
picnic   171
plane, complex 47^C
   in 3-space   25, 105
   transformation of $29^1, 72^1$
pogo stick   5
points, approachable   188
   in the plane   56, 64, 95, 132, 136, 140, 188
   rational   132
   set of   23, 49, 57, 78, 110, 172, 191
polar coordinates 103^2, 110, 122, 171
polygon(s)   29, 94, 169
polyhedron   206
polynomial(s) 14^4, 40, 50, 70, 97, 187, 199
   cubic   7, 44, 54, 64, 116, 178, 180, 181
   cyclotomic   207
   degree 5   59
   degree 6 139^2
   homogeneous   207
   in two variables 123^2, 185
   linear   119
   quadratic 44^1, 168, 178
polynomial identity 75^1, 207
Poonen, Bjorn 130^2
prime factors   21
prism   202

probability   15, 68, 108, 112, 126, 165, 184, 189, 205
projection 25^1, 117, 152^2, 202
Pythagorean triples 118^2, 131, 170, 185^1

quadrilateral   52, 95, 158

random(ly)   15, 108, 171, 184, 189
random walk 205^C
rational coefficients   180, 199
rational function   199
rational number(s)   132, 144, 153, 174, 183, 190, 199
rational point   132
rational root 97^1, 180
rational root theorem 97^1
ray(s)   117
   of light   25, 151
recurrence (relation) $67^{1,2}$, 89^2, 109^2, 115, 126^C, 128, 147, 149, 152^1, 182, 184, 197, 203 (see also sequence(s), given recursively)
reflection 8^1, 25, 143, 151
region(s), area of   94
   number of   81
regular
   dodecahedron   98
   icosahedron   105
   n-gon   60, 94, 169, 206
   octagon   169
   octahedron   202, 206
   pentagon   72, 105
   tetrahedron   206
relatively prime   28, 62^2, 88, 162, 199^2
Reznick, Bruce 57^2, 161^2
Richards, Ian 195^2
right turn (shallow)   82
ring   135
Risch algorithm 14^C
Rolle's theorem 80^2, 156, 204
root(s), integer   7, 50
   multiple 44^2, 64
   of unity 195^2, 207
   rational 97^1, 180
   real   70, 80, 97, 178, 187
rotation 124^2
Round Lake   86

round-off error 37^C
Ruth, George Herman ("Babe")   37
scalenity   134
Schwenk, Allen   43
secant method 138^C
second derivative test (see concavity)
seed   181
sequence(s), given recursively   17, 84, 96, 101, 109, 125, 137, 138, 144, 166, 174, 181, 187, 194, 201, 203, 208
   limit of   17, 87, 103, 138, 166, 174
   of integers   137, 155, 192
   of points   82, 136, 167, 188
   of quadrilaterals   158
series, convergence of   10, 42, 115, 133, 145, 157, 186, 193
   geometric   5, 28, 68^2, 82^2, 92, 94, 156, 177, 184^C, 190^1, 197, 207
   sum of   74, 92, 114, 147, 177, 182, 190, 207
   Taylor (Maclaurin) 66^C, 74, 84, 142^1, 161, 177, 197
   telescoping   114, 147
skew lines   191
slugging percentage   37
snowmobile race   30
socks   184
solution(s), in integers   1, 18, 123
   of equation   19, 41, 80, 97, 156, 204
   of system   18, 27, 69, 139^2
sorting   36, 106
sound level   8
space   (see n-space)
splitting field 180^3
square(s), magic   200
   of chessboard   15, 33, 143
   perfect 7^3, 24, 28, 32, 113, 118, 123, 131, 170
squeeze principle   101, 103^1, 128, 146^1, 154, 155, 173, 179, 198
step function(s)   108, 186
strategy   68, 121, 189, 205
strings
   of 0's and 1's   71
   of X's and O's   128
subdivision   148

subset(s), averages of   77
  disjoint union of   16
  minimal size of   164, 198
  number of   13, $67^{3,4}$
  sum of   9, 120
superinteger   192
swimming races   20
switching integration and summation 74^2, 142^3, 173^I
symmetric   86, 121, 143
symmetric functions 7^2
symmetry, of polyhedron   206
system of equations   18, 27, 37^C, 69, 107, 199^2

T-number   194
tangent line(s) 38^1, 44, 64, 70, 124, 168
Taylor polynomial 66^C
Taylor series   74, 84, 142^1, 161, 177
telescoping series 45^2, 114, 118^1, 147
tent function   160
thermometer   108
tiling   29, 33, 104
time/temperature display   108
total fluctuation   73
Town Council   148
transposition 30^C
trapezoid   119
trapezoidal estimate 119^C
triangle(s)   23, 47, 52, 105, 134, 151, 175
  area of   3
  equilateral   8, 202
triangle inequality 132^3

trisector   31
two-player game   68, 121

uniform convergence 142^3
uniformly continuous 58^3
unit circle   26, 185
United States Mathematical Olympiad 7^2

vector(s) 3^2, 25, 45^1, 76, 93, 103^2, 105, 117, 126, 152
vector space   83, 141^2
volume of revolution   159, 175

watches   46
Waterhouse, William C. 141^2
well defined   201
wild turkey   22
win-lose game   205
winning strategy   121
Wohascum, Bridge Club   63
  Center   31, 98, 108, 148, 184
  College   105, 163
  County   15, 46, 86
    Board of Commissioners   2
    Fish and Game Department   22
  Folk Dancers   55
  High   20
  Lake   171
  Municipal Park   8
  National Bank   108
  Puzzle, Game and Computer Den   98, 121
  Snowmobile Race   30
  Times   30, 36

zero divisors   135

About the Authors

Mark Krusemeyer has been teaching at Carleton College since 1984 and, during the summers, at Canada/USA Mathcamp since 1997. He has served two three-year terms as a problem setter for the William Lowell Putnam Mathematical Competition, and he has recently succeeded Loren Larson as an Associate Director of that competition. Over the years, Mark's research interests have varied from commutative algebra to combinatorics; he has also written a textbook on differential equations. He has received an award for distinguished teaching from the North Central Section of the MAA. Mark plays and teaches recorder, and his other activities include mountain scrambling, duplicate bridge, and table tennis.

George Gilbert grew up in Arlington, Virginia, and went on to earn degrees in mathematics from Washington University and Harvard. His interest in problem solving, begun as a student contestant, has led to his posing problems for the Putnam, for national high school contests, and for the Konhauser Problemfest and the Iowa Collegiate Mathematics Competition. He was Problems Editor of *Mathematics Magazine* from 1996–2000. He has been a professor in the Department of Mathematics at TCU for the past 20+ years. His two grown-up sons and he each claim to be the best poker player in the family.

Loren Larson is Professor Emeritus at St. Olaf College in Northfield, Minnesota. His interest in problem-solving dates to his early days of teaching when he engaged students in seminars that featured problems from the Putnam and the *American Mathematical Monthly*. Solutions were submitted under the byline *St. Olaf College Problem Solving Group*. Twenty years of such experience culminated in a book *Problem-Solving Through Problems* which isolated and illustrated the most common problem-solving techniques encountered in undergraduate mathematics. Loren has served the MAA as a governor, a reviewer of books and articles, two five-year terms as Problem Editor for *Mathematics Magazine* and twenty-six years as Associate Director of the William Lowell Putnam Exam. He presently enjoys woodworking and specializes in crafting artistic mathematical puzzles.